# 4th International Conference on Safe Production and Use of Nanomaterials

# (Nanosafe 2014)



**Journal of Physics: Conference Series Volume 617**

## Grenoble, France
## 18 – 20 November 2014

**ISBN: 978-1-5108-0584-2**
**ISSN: 1742-6588**

**Printed from e-media with permission by:**

Curran Associates, Inc.
57 Morehouse Lane
Red Hook, NY 12571

**Some format issues inherent in the e-media version may also appear in this print version.**



Printed by Curran Associates, Inc. (2015)

For permission requests, please contact the Institute of Physics
at the address below.

Institute of Physics
Dirac House, Temple Back
Bristol BS1 6BE UK

Phone: 44 1 17 929 7481
Fax: 44 1 17 920 0979

techtracking@iop.org

**Additional copies of this publication are available from:**

Curran Associates, Inc.
57 Morehouse Lane
Red Hook, NY 12571 USA
Phone: 845-758-0400
Fax: 845-758-2634
Email: curran@proceedings.com
Web: www.proceedings.com

# TABLE OF CONTENTS

## NEW APPLICATIONS OF NANOMATERIALS



## EXPOSURE



## DETECTION AND CHARACTERIZATION

## TOXICOLOGY



## ENVIRONMENTAL INTERACTIONS OF NANOMATERIALS



## NANOMATERIAL RELEASE

Journal of Physics: Conference Series **617** (2015) 011001 doi:10.1088/1742-6596/617/1/011001

This volume contains a collection of contributions presented at the 4th International Conference on Safe Production and Use of Nanomaterials (NANOSAFE 2014) held in Grenoble, France, from 18th to 20th November 2014.

The issues of fast progress in the field of Nanosafety are up to the potential benefits that nanotechnology can bring to mankind. Making more efficient - more sustainable - easier to share mineral resources, increasing the yields of new energy technologies, enabling drugs that act selectively and locally are just few examples of the wide range of nanomaterial applications that currently benefit humanity.
Nevertheless, the dynamic development of nanomaterials requires the adhesion from the general public who rightly demand major progresses in Nanosafety as a prerequisite.

This is our exciting responsibility and challenge!

Following the successful outcome of the three past international conferences on **safe production and use of nanomaterials: Nanosafe 2008, 2010 and 2012,** the organizing committee has the pleasure to welcoming you again to Minatec, Grenoble with some of the most famous specialists in the field.
This year, two new topics have been added dealing with the "New Application of Nanomaterials" and Nano-responsible Development" in addition to the usual issues addressed in previous Nanosafe conferences such as Expology, Detection and Characterization, Toxicology, Environmental Interactions, Nanomaterials Release, Life Cycle Analysis, Regulation and Standardization, Risk Management.

The debates in 2012 proved highly successful so this formula has been kept in 2014 with 3 round tables: **Nano-Responsible Development, Risks and Benefits for the Environment, Toxicology Progress.**

In this 4th edition, there were more than 330 registered participants from 28 different countries including 160 oral presentation covering the whole Nanosafety issues in 12 sessions, satellite workshops and round tables. This high number of participants makes this edition one of the most successful conference of the series.

Finally, 2014 was for the organizing committee, a great year as we ramp up the Nano Safety "PNS" platform at Minatec, with a brand new building spread out over 2000 $m^2$ of laboratories and more than 50 doctors and engineers dedicated to Nanosafety.

F. Tardif, J-F. Damlencourt, F. Schuster and V. Gaultier

# Peer review statement

All papers published in this volume of *Journal of Physics: Conference Series* have been peer reviewed through processes administered by the proceedings Editors. Reviews were conducted by expert referees to the professional and scientific standards expected of a proceedings journal published by IOP Publishing.

# *In vivo* study of novel nanocomposite for prostate cancer treatment

**C P Silveira[1*], A J Paula[2], L M Apolinário[3], W J Fávaro[3,4] and N Durán[1,4,5]**

[1]Chemistry Institute, Universidade Estadual de Campinas (UNICAMP), Campus Universitário Zeferino Vaz, s/n, Cidade Universitária, CEP 13083-870, Campinas, SP, Brazil,
[2]Department of Physics, Universidade Federal do Ceará (UFC), Avenida da Universidade, 2853, Benfica, CEP 60020-181, Fortaleza, CE, Brazil,
[3]Department of Environmental Engineering, Universidade Estadual Paulista (UNESP), Avenida Três de Março, 511, CEP 18087-180, Sorocaba, SP, Brazil,
[4]Farmabrasilis R&D Division, Brazil,
[5]Nanobioss, Chemistry Institute, Universidade Estadual de Campinas (UNICAMP), Campus Universitário Zeferino Vaz, s/n, Cidade Universitária, CEP 13083-870, Campinas, SP, Brazil

*Email: camila.silveira@iqm.unicamp.br,

**Abstract.** Because of high occurrence of side effects of conventional anticancer drugs, new therapies have been widely studied, such as the use of non-usual molecules and the association with nanomaterials. In this context, sildenafil is a molecule that has been lately considered as a co-adjuvant in cancer therapy because it inhibits an enzyme with a potential role in tumor progression, phosphodiesterase-5. Taking into account the combination of nanoparticles and sildenafil, we produced systems based on mesoporous silica nanoparticles, Pluronic F-127 and sildenafil as the main therapy for prostate cancer.

## 1. Introduction

A major problem in cancer treatments is the high incidence of side effects intrinsic to anticancer drugs, which end up affecting the patients' lives and the therapies themselves. In an attempt to overcome this issue, the association of these drugs with non-usual molecules that act through different mechanisms as well as the association with nanomaterials are promising alternatives. In this way, sildenafil (SIL) is a molecule that inhibits an enzyme which expression is increased in many carcinomas, phosphodiesterase-5 (PDE-5) [1]. Moreover, it activates nitric oxide expression and apoptosis [2,3]. Among nanomaterials, mesoporous silica nanoparticles (MSNs) are good candidates to be used as drug carriers once they are safe and biocompatible [4]. They have high superficial area and great pore volume, very appropriate to carry bioactives [4]. They provide a sustained release of the drug, preventing it of premature degradation and increasing the effectiveness of drug delivery [4,5].

Another strategy to ensure a better drug distribution is the association with materials that can provide prolonged release [6]. In this perspective, Pluronic F-127 (PF-127) is a FDA-approved biocompatible block copolymer that has a transition sol-gel temperature near room temperature [7].

That is, it is liquid at low temperatures (e.g. refrigerator temperatures) and it gelates at room temperature, which makes it very suitable to reach *in situ* gelation [6], applying the liquid formulation inside the patient body for it to form a hydrogel in there.

In order to obtain preliminary results about the antitumor potential of the association MSNs + SIL + PF-127, the effect of SIL on prostate tumor cells and the role of the nanoparticles in this process, we developed three systems containing these components and tested them in rats with chemically induced prostate cancer. The MSNs were used as nanocarriers for SIL, to reach a sustained release of the drug and PF-127 was used as the matrix of the hydrogel, to achieve *in situ* gelation to reach a long-term release of the components.

## 2. Methods

*Synthesis and characterization of mesoporous silica nanoparticles (MSNs)*

The MSNs were synthesized through on a sol-gel method based on the methods of Stöber [8] [9] and Bein [9] [10] following the protocol developed by Paula et al. (2012) [10] [8]. This approach allow to synthesize spherical and monodisperse nanoparticles with high colloidal stability in aqueous medium. So, to synthesize the MSNs, 0.75 g of cetyltrimethylammonium bromide (CTAB) were added to a 0.050 mol $L^{-1}$ ammonium hydroxide solution of pH 11. Then, 3.2 mL of absolute ethanol were added and after 15 minutes under magnetic stirring, 2.5 mL of tetraethyl orthosilicate (TEOS) were added. The reaction was maintained under reflux at 60°C for two hours. A centrifugation step was carried out to separate the products (60 minutes at 18,400 rcf). To extract the CTAB, the products were resuspended in 90 mL of absolute ethanol and 10 mL of hydrochloric acid were added (1:9 HCL:Ethanol volume ratio), sonicating the mixture for 10 minutes. Finally, to obtain ethanolic suspensions of the nanoparticles, the mixture was centrifuged (60 minutes at 18,400 rcf) following two washing steps with absolute ethanol and resuspended in absolute ethanol.

Transmission electron microscopy (TEM) in the bright field mode was used to analyze size distribution and morphology of MSNs (TEM-BF, Zeiss Libra 120, operating at 80 kV). Surface area, pore volume and pore diameter were obtained through nitrogen-sorption assays (Accelerated Surface Area and Porosimetry System ASAP 2020 micromeritics) using BET (Brunauer-Emmet-Teller) to calculate surface area and BJH (Barret-Hoyner-Halenda) to calculate pore diameter, both using $N_2$ adsorption branch. The pore volume was calculated from the single-point value adsorbed at P/P0 = ~0.94.

*Synthesis and characterization of hybrid systems*

To evaluate the antitumor potential of the hybrid systems, the effect of SIL in tumor cells and the role of MSNs in the association, we developed three systems. The only difference between them was the presence and concentration of the MSNs (Table 1). The systems were produced by adding SIL to the PF-127 solution in an ice bath under magnetic stirring, then adding the nanoparticles lastly to avoid aggregation. The mixture was kept under magnetic stirring at 4°C overnight. All systems were produced in physiological saline solution (0.9% w/v NaCl) to keep the same osmotic pressure than the animals' cells.

The concentration of SIL was calculated based on Das et al. (2010) [1] [2] to reach 5 mg of drug per kilogram of animal body weight, considering a dose application of 0.3 mL and an animal body weight of 150 g (average). The PF-127 concentration was chosen according to the desired range of gelation temperature ($T_{gel}$): the optimum $T_{gel}$ would allow easy manipulation of the systems, avoiding gelation inside the and quickly gelation of the systems after applied intraperitoneally in the animals, avoiding diffusion to unwanted locations and premature release. So, the systems were characterized regarding their $T_{gel}$ by adding 5 mL of the liquid mixture in a beacker in an ice bath with a thermometer. The mixture was heated in a constant rate and the temperature in which the bar stopped moving was considered the gelation temperature.

To obtain information on the potential for prolonged release of the systems, the release profile was studied *in vitro* through a membraneless dissolution method. To do so, 0.5 mL of ht hybrid systems were added to a weighed vial and incubated at 35°C (which is the body temperature of mice used in the *in vivo* assays), using a dry bath, to reach thermal equilibrium. So, the vials were weighed and 0.5 mL of the release medium (NaCl 0.9% w/v) were put carefully in order to avoid mixture. At pre-determined intervals, the release medium was removed, the vial was weighed and a new release medium was put to avoid saturation. The dissolution rate was calculated in terms of weight loss against time.

*In vivo model*

The *in vivo* study was accomplished using seventeen 7-week-old male Fischer 344 rats, obtained from the Multidisciplinary Center for Biological Investigation (CEMIB) at University of Campinas (UNICAMP). The prostate cancer induction followed a new protocol based on Fávaro et al. (2014) [11 - 11] and was performed in 13 animals. This new approach allows the animals to be ready for treatment in approximately four months. The animals received a daily subcutaneous injection of 100 mg $kg^{-1}$ testosterone cypionate diluted in 0.5 mL of peanut oil for three days. Then, the animals were anesthetized with 5 mg $kg^{-1}$ of 2% xylazine hydrochloride and 60 mg $kg^{-1}$ of 10% ketamyne hydrochloride in order to perform 0.5 cm suprapubic incision and inoculation of 0.2 mL of 15 mg $kg^{-1}$ of *n*-methyl-*n*-nitrosourea (MNU) dissolved in 0.3 mL of 1 M sodium citrate (pH 6.0) and 25% PF-127, which allows *in situ* gelation of the solution. After one week, the animals received subcutaneous injections of 5 mg $kg^{-1}$ testosterone cyprionate diluted in 5 mL of peanut oil on alternate days for 120 days.

Of the total number of animals, 4 animals comprised the healthy control group (Control, n=4, received no cancer induction) and 4 comprised the cancer group (Cancer, n=4, received no treatment), both groups received physiological saline (NaCl 0.9% w/v) for treatment. The rest of the animals were divided in 3 groups (n=3): Group S1 was treated with S1 system, Group S2 was treated with S2 system and Group S3 was treated with S3 system. The treatment consisted of a weekly-dose of intraperitoneal inoculation of 0.3 mL of the respective treatment for 30 days. To prevent gelation inside the needles, the systems were kept in ice bath before each application. The animals received water and the same solid diet *ad libitum* (Nuvilab) and were allocadted in single solid-bottom boxes lined with wood shavings in a room with controlled light and temperature (12 hours light and 12 hours dark, 20-25°C). The experimental protocol followed ethical principles in animal research. After treatment, the animals were euthanized and the occurrence of macroscopic changes were observed.

## 3. Results and Discussion

*Synthesis and characterization of MSNs*

It is possible to see through TEM images (Fig. 1a) that MSNs have spherical morphology and size distribution from 45 to 75 nm. Nitrogen-sorption experiments using BET and BJH methods revealed that the superficial area was 970 $cm^2$ $g^{-1}$ and the pore volume was 1.6 $cm^3$ $g^{-1}$, with an average pore diameter of 4.8 nm. The nitrogen-sorption isotherm (Fig. 1b) shows an adsorption pattern that resembles a type IV isotherm (IUPAC) with a subtle stepwise behavior around 0.4 $P/P_0$, which is characteristic of well-ordered mesoporous materials [12].

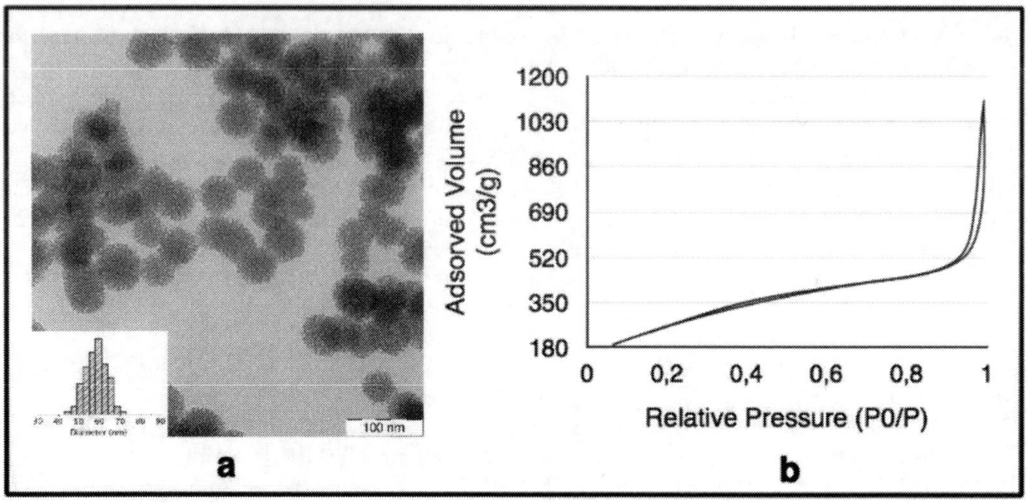


**Figure 1. a)** TEM images of MSNs with size distribution histogram (measuring at least 100 nanoparticles) **b)** Nitrogen sorption isotherm of MSNs

*Synthesis and characterization of hybrid systems*
The hybrid systems were characterized regarding their $T_{gel}$ (Table 1). All systems have a $T_{gel}$ that fits the temperature range preferred, which allows easy manipulation of them during dose applications and a rapid gelation inside the animal body (which temperature is around 35°C).

**Table 1.** Description of the hybrid systems.

| System | Components and final concentration | Gelation temperature ($T_{gel}$, °C) |
|---|---|---|
| S1 | PF-127 (18%)<br>SIL (4.16 mg $mL^{-1}$) | 21.0 ± 0.5 |
| S2 | PF-127 (18%)<br>SIL (4.16 mg $mL^{-1}$)<br>MSNs (1.0 mg $mL^{-1}$) | 20.0 ± 0.5 |
| S3 | PF-127 (18%)<br>SIL (4.16 mg $mL^{-1}$)<br>MSNs (5.0 mg $mL^{-1}$) | 19.0 ± 0.5 |

The potential for prolonged release was confirmed during the *in vitro* dissolution experiment (Fig. 2). Considering the concentration gradient induced by cells permanently on the hydrogel, it is expected that the hybrid system will delay the release of the components after the application.

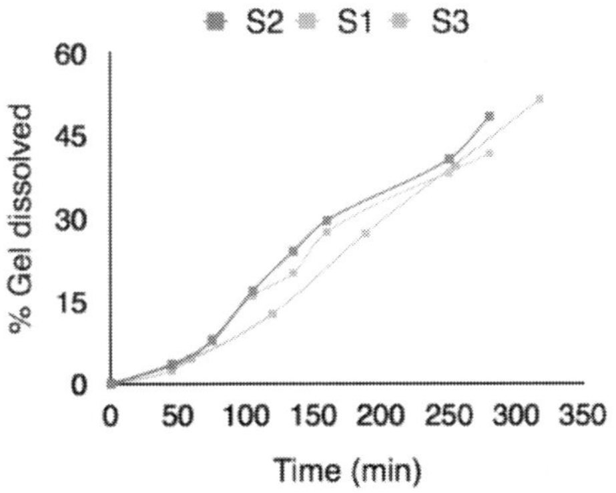


**Figure 2.** *In vitro* gel dissolution profiles of the systems.

The gelation of the system after its inoculation leads to a prolonged release of the components as the gel dissolves inside the animal body, which is a key factor to get a better drug distribution in tumors [13, 14]. It is known that the drug access to tumor cells is achieved via bloodstream [13, 14]. So, if there is no continuous infusion, only drugs that are near the blood vessels can be killed and cells that are more distant are able to survive and regenerate the tumor [13, 14]. Oppositely, a continuous infusion permits a longer residence time of the drug, ensuring a better distribution so a larger number of cells are reached (Fig. 3).

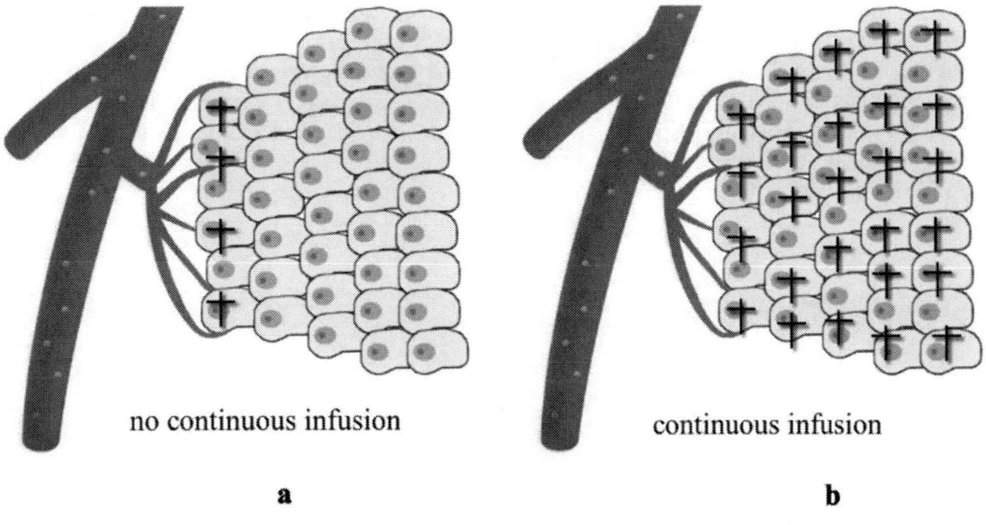


**Figure 3.** Schematic illustration of drug distribution in **a)** no continuous infusion and **b)** continuous infusion (prolonged-release)

*In vivo macroscopic analyses*

After treatment, macroscopic changes regarding organ conditions were observed in order to get preliminary information on the antitumor potential of the systems (Table 2).

**Table 2.** Observed macroscopic changes of five experimental groups.

| Macroscopic observation | Control | Cancer | S1 | S2 | S3 |
|---|---|---|---|---|---|
| No apparent macroscopic changes | 100% | - | - | - | 66,6% |
| Lesion in prostate base | - | 100% | 100% | 66,6% | 33,3% |
| Metastasis | - | 100% | 100% | 33,3% | - |

It is possible to observe in Table 2 a decrease in frequency of metastasis and prostatic lesions that is clearly proportional to MSNs concentration. This calls the attention for the role of the nanoparticles in this process, suggesting that the MSNs might be determinant in a possible improvement of tumor conditions.

It is known that nanoparticles are submitted to the enhanced permeability retention (EPR) effect, that allow macromolecules and particles up to 100 nm to accumulate in interstitial space of tumor cells [15]. This may increase the delivery of SIL, leading to a combined effect from the drug and the nanoparticles on tumor cells.

## 4. Conclusions

To get preliminary results of the antitumor potential of the association of SIL, MSNs and PF-127, we developed three systems that differed only by the MSNs concentration and treated rats with chemically induced prostate cancer. Preliminary macroscopic analyses showed an interesting decrease in frequency of metastasis and prostatic lesions that was proportional to MSNs concentration. The results suggest that the MSNs might be determinant in the treatment. Further studies are necessary to confirm the antitumor potential of the systems, but in a first moment, we can associate the results to a combined effect from SIL and MSNs, caused by EPR effect along with sustained and prolonged release of the drug. This avoids premature degradation of the drug and enables its cell internalization, achieving a better drug distribution and increasing its efficacy. Thus, we open a new perspective of using MSNs as carriers for SIL in a new platform for prostate cancer treatment, using SIL as the main therapy.

## 5. Acknowledgements

Supported from CNPq and FAPESP.

## 6. References

[1] Das A, Durrant D, Mitchell C, Mayton E, Hoke N N, Salloum F N, Parl M A, Qureshi I, Lee R, Dent P and Kukreja R C 2010 *Proc. Natl. Acad. Sci. USA* **107** 1802
[2] Das A, Ockaili R, Salloum F and Kukreia R C 2004 *Am. J. Physiol. Heart Circ. Physiol.* **286** 1455
[3] Das A, Xi L and Kukreja R C 2005 *J. Biol. Chem.* **280** 12944
[4] Mai W X and Meng H 2013 *Integr. Biol.* **5** 19
[5] Chen F, Hao H, Shi S X, Goel S, Valdovinos H F, Hernandez R, Theuer C P, Barnhart T E and Cai W B 2014 *Sci. Reports* **4** 1
[6] Heilmann S, Küchler S, Wischke C, Lendlein A, Stein C and Schäfer-Korting M 2013 *Int. J.*

*Pharm.* **444** 96
[7] Dumortier G, Grossiord J L, Agnely F and Chaumeil J 2006 *J. Pharm. Res.* **23** 2709
[8] Stöber W, Fink A and Bohn E 1968 *J. Colloid Interf. Sci.* **26** 62
[9] Kecht J, Schlossbauer A and Bein T 2008 *Chem. Mater.* **20** 7207
[10] Paula A J, Montoro L A, Souza Filho A G and Alves O L 2012 *Chem. Commun.* **48** 591
[11] Fávaro W J, Apolinário L M, Caballero N E D, Garcia P V, Bueno C P S 2014 PIBr 10 2014 023118 8
[12] Neímark A V, Sina K S W and Thommes M 1997 Handbook of heterogeneous catalysis Volume 1, ed Ertl G, Knözinger H and Wetikamp J (Weinheim: Wiley-VHC) 721
[13] Maeda H, Wu J, Sawa T, Matsumura Y and Hori K 2000 *J. Control. Release* **65** 271
[14] Trédan O, Galmarini C M, Patel K and Tannock I F 2007 *Cancer Inst.* **99** 1441
[15] Maeda H 2003 Biomedical aspects of drug targeting, Volume 1, ed Muzykantov V P et al. (New York: Kluwer Academic Publishers) 211

Journal of Physics: Conference Series **617** (2015) 012002 doi:10.1088/1742-6596/617/1/012002

# Preparation, Characterization and Tests of Incorporation in Stem Cells of Superparamagnetic Iron Oxide

**P S Haddad[1*], T N Britos[1], L M Li[2] and L D S Li[3]**

[1] Exact and Earth Sciences Departament, Universidade Federal de São Paulo, CEP 09913030, Diadema, SP, Brazil,
[2] Neurology Department, Universidade Estadual de Campinas, CEP 13083-970, Campinas-SP, Brazil,
[3] Pediatric Endocrinology Laboratory, Universidade Estadual de Campinas, CEP 13083-970, Campinas-SP, Brazil

*Email: haddadps@gmail.com, haddad.paula@unifesp.br

**Abstract**. Superparamagnetic iron oxide nanoparticles (SPIONs) have been produced and used as contrast-enhancing agents in magnetic resonance imaging (MRI) for diagnostic use in a wide range of maladies including cardiovascular, neurological disorders, and cancer. The reasons why these SPIONs are attractive for medical purposes are based on their important and unique features. The large surface area of the nanoparticles and their manipulation through an external magnetic field are features that allow their use for carrying a large number of molecules such as biomolecules or drugs. In this scenario, the present work reports on the synthesis and characterization of SPIONs and *in vitro* MRI experiments to increase their capacity as probes for MRI applications on stem cells therapy. Initially, the SPIONs were prepared through the co-precipitation method using ferrous and ferric chlorides in acidic solution. The SPIONs were coated with two thiol molecules such as mercaptosuccinic acid (MSA) and cysteine (Cys) (molar ratio SPIONs:ligand = 1:20), leading to the formation of a stable aqueous dispersion of thiolated nanoparticles (SH-SPIONs). The SH-SPIONs were characterized by Fourier transform infrared spectroscopy (FTIR), X-ray diffraction (XRD), transmission electron microscopy (TEM), and vibrating sample magnetometry (VSM). The results showed that the SH-SPIONs have a mean diameter of 14 nm and display superparamagnetic behavior at room temperature. Preliminary tests of incorporation of SH-SPIONs were evaluated stem cells. The results showed that the thiolated nanoparticles have no toxic effects for stem cells and successfully internalized and enhance the contrast in MRI.

## 1. Introduction

Nanostructured materials have been proposed for an increasing number of biomedical applications. Among these materials, superparamagnetic iron oxides nanoparticles (SPIONs), in particular magnetite ($Fe_3O_4$), have attracted the attention of the scientific community [1-2]. SPIONs have been effectively applied in several biomedical applications, such as in diagnosis [3], drug delivery [4-5] and hyperthermia [6]. These nanoparticles are interesting because present two fundamentals characteristics. The first feature is the presence of magnetic monodomains, i.e. these nanoparticles show superparamagnetic behavior at room temperature and act as a single magnetic domain. Individual nanoparticle has a large constant magnetic moment and behave as a giant paramagnetic atom, rapidly responding to applied magnetic fields with negligible residual magnetism and coercivity [7], therefore superparamagnetism is a term that describes an analogy of the small magnetic moment of the paramagnetic atom with large magnetic moment of the nanoparticle. Another peculiarity is the large surface area on which it is possible to functionalize molecules of interest, such as biomolecules. Furthermore these system present low toxicity, they can be guided *in vivo* to specific target by an external magnetic field. SPIONs are of comparable size to important biological systems; display low sedimentation rates and improved thorough tissues [4]. The majority of studies *in vitro* explores physicochemical characteristics of the nanoscale particles, such as shape, size, and surface coating. However, studies have shown that the surface of the nanoscale particles changes after of the interaction with the surrounding environment [8-9].

Coprecipitation of $Fe^{2+}$ and $Fe^{3+}$ salts in alkaline conditions has been used for the synthesis of magnetite nanoparticles ($Fe_3O_4$) [10]. The coating of these systems is of great importance with regard to cell incorporation [11-12]. Usually, coated magnetic nanoparticles have been found to be nontoxic, whereas the simple particles present some toxic effects, such as oxidative stress [13].
The effects of SPIONs coated with biocompatible ligands containing functional groups such as COOH and SH as well as sizes of these nanoparticles are critical determinants in toxicological studies [14].

Cell labeling with SPIO substances is an increasingly common method for *in vivo* cell separation as well as detection by magnetic resonance image (MRI) [15]. Considering interest in biomedical applications of SPIONs, the present paper reports the synthesis and characterization of water soluble magnetic iron oxide nanoparticles and first *in vitro* MRI experiments. The other aim of this work is to optimize SPIONs analyses for MRI applications on stem cells therapy. SPIONs were prepared through the co-precipitation of iron salts, followed by titration with ammonium hydroxide and the suspension formed was separated magnetically. In order to functionalize their surfaces with biocompatible molecules, the SPIONs were coated with two molecules with two hydrophilic molecules containing thiol groups: mercaptosuccinic acid (MSA) and cysteine (Cys) in a molar ratio iron: ligand 1:20 giving stable aqueous dispersion of thiolated nanoparticles (SH-SPIONs). SH-SPIONs were characterized by X-ray diffraction (XRD), quantification of thiols group on the surface of particle by titration with DTNB and transmission electron microscopy (TEM). The magnetization analysis was carried out by isothermal magnetic measurements in applied field. Preliminary studies of incorporation into stem cells and MRI were also accomplished.

## 2. Methods

*2.1.Synthesis of $Fe_3O_4$nanoparticles*
SPIONs were synthesized by using a co-precipitation method, as previously reported [10]. In brief, 4.0 mL of $FeCl_3{\cdot}6H_2O$ and 1.0 mL of $FeCl_2{\cdot}4H_2O$ (molar ratio 2:1), prepared in 1.0 mol/L HCl, were mixed and stirred, while a volume of 50 mL of $NH_4OH$ (0.7 mol/L) was added as precipitator. At this stage, the black suspension formed was decanted magnetically followed by the addition of thiolated (MSA and Cys) ligands in molar ratio $Fe_3O_4$: ligand 1:20. This mixture was then stirred for 1 hour. The dispersion was centrifuged and the new precipitate was washed several times with ethanol leading to a SPIOns covered with MSA and Cys.

*2.2. Cell preparation and SPIONs up-take*
Stem cells were cultured in Dulbecco's modified Eagle's medium (DMEM; Invitrogen) supplemented with 10% fetal bovine serum (FBS) and penicillin/streptomycin (complete medium). Cells were plated in 6 well plates and treatment was performed at 80–90% confluence in serum free medium with 50 µg/ml of SH-SPIONs in water. A negative control without treatment was carried out. After 1, 3, 6 and 20 h, cells were harvested by scrapping them from the plate for optic microscopy staining and magnetic resonance.

*2.3. Determination of the amount of thiol groups on the surface of SPIONs*
The quantity of thiol groups (–SH) on the surface of the nanoparticles was measured by the DTNB reaction, based on the absorbance at 412 nm ($\varepsilon$ = 14,150 $molL^{-1}cm^{-1}$) of the 2-nitro-5-thiobenzoate anion ($TNB^{2-}$) generated in the reaction of –SH groups with DTNB [16]. SH-SPIONs were added to 3.0 mL of DTNB (0.01 $molL^{-1}$) in PBS buffer (pH 7.4) containing 1 $mmolL^{-1}$ of ethylenediaminetetraacetic acid. After 5 min of incubation, the suspensions were filtered by centrifugal ultrafiltration using a Microcon centrifugal filter device containing ultrafiltration membranes (MWCO 10-kDa molar mass cut-off filter, Millipore, Billerica, MA, USA). The supernatant was placed into a quartz cuvette, and the intensity of the absorption band at 412 nm was measured in an Uv-Vis Spectrophotometer (Agilent, model 8553, Palo Alto, CA, USA). The experiments were carried out in triplicates.

*2.4. X-ray diffraction (XRD)*
The diffractograms were obtained with approximately 200 mg of powdered MSA-$Fe_3O_4$ or Cys-$Fe_3O_4$ onto a glass substrate of 2×2 cm. The measurements were performed in reflection set-up; with a conventional X-ray generator (CuK$\alpha$ radiation of 1.5418 Å and a graphite monochromator) coupled to a scintillation detector. The angular scanning performed on all samples ranged from 20 to 80° with 0.05° step-width at 5 s per angle.

*2.5. Transmission Electron Microscopy (TEM)*
Photomicrographies of the nanoparticles were obtained using a Philips CM200 transmission electron microscope (TEM) with an energy dispersive spectrometer, operating at 160 kV. The samples for observation were suspended in water, and then a drop of the supernatant dispersion was placed on an amorphous carbon film supported by a copper grid.

*2.6. Vibrating Sample Magnetometry (VSM)*

The magnetic curves of the SH-SPIONs were calculated at room temperature, and the apparatus was calibrated to a Ni pattern. The magnetization measurements were carried out on a known quantity of powdered sample, which slightly pressed and conditioned in cylindrical Lucite holders.

*2.7. Magnetic Resonance Image (MRI)*
The MRI scan of labeled cells was performed using a 2T scanner (Elscint, Prestige, Haifa, Israel) at FCM, UNICAMP. It was used as a head coil and acquired two different pulse sequences. The parameters of gradient recall-echo T2 was: TR=500 ms, TE=10 ms, flip angle=18, 2 number of excitations, matrix=256×190; and the fast spin-echo T2 was: TR=4800 ms, TE=128 ms, flip angle=170, 1 number of excitation, matrix=256×256. Both sequences had field of view adjusted of the size of plaque and the thickness of slice was 3 mm. MRI experiments were carried out adding nanoparticles dispersed in agarose and also cells before and after nanoparticles incorporation.

## 3. Results and Discussion

The mixture of $Fe^{3+}$ and $Fe^{3+}$ salts in acidified aqueous solution led to the formation of $Fe_3O_4$, according to equation 1

$$Fe^{2+}_{(aq)} + 2\,Fe^{3+}_{(aq)} + 8\,OH^{-}_{(aq)} \rightarrow Fe_3O_{4(s)} + 4H_2O_{(l)} \qquad (1)$$

Fig. 1 shows the schematic representation of the formation of thiolated $Fe_3O_4$ nanoparticles coated with mercaptosuccinic acid (MSA) or cysteine (Cys). It had been used an excess of thiolated ligands to total covering of the $Fe_3O_4$ surface.

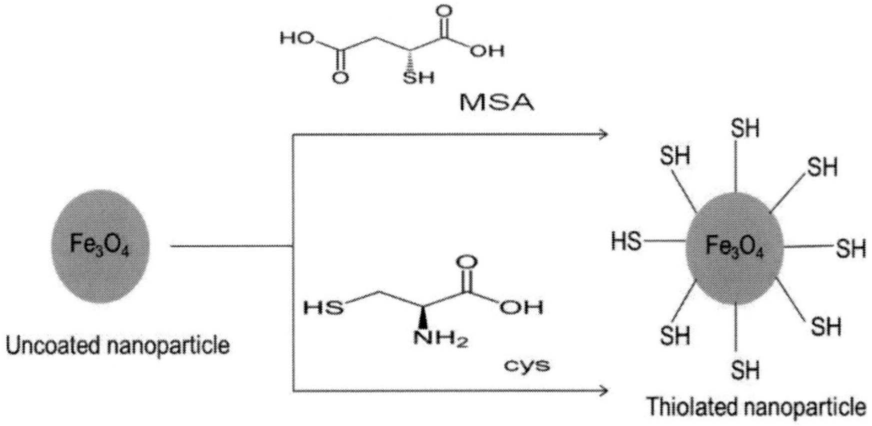


**Figure 1.** Schematic representation of the formation of thiolated $Fe_3O_4$ nanoparticles coated with mercaptosuccinic acid (MSA) or cysteine (Cys).

The amounts of free thiol (-SH) groups on the surface of SH-SPIONs were determined by the reaction of the thiolated nanoparticles with a thiol-specific reagent, DTNB [16]. Values of 28 ± 8 μmol SH/g of MSA-SPIONs and 12 ± 8 μmol SH/g of Cys-SPIONs were found.

Fig. 2 shows the X-ray diffractograms from two synthesized samples: SPIONs coated with MSA and coated with Cys, respectively, along with their characterization as the inverted spinel structure [17]. The diffraction measurements demonstrated that the crystallographic structure of SH-SPIONs corresponds to magnetite ($Fe_3O_4$) (JCPDS 20–596). Crystallite sizes from 12 to 14 nm were obtained from Scherrer's equation [18] by taking into account the (311) Bragg's reflection.

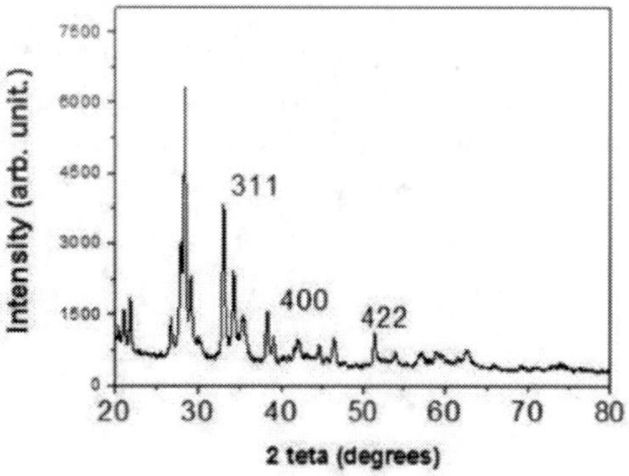


■**XRD**: X-ray powder diffraction from a representative sample of a thiolated $Fe_3O_4$ nanoparticle coated with cysteine, displaying the Bragg peak reflections of magnetite.

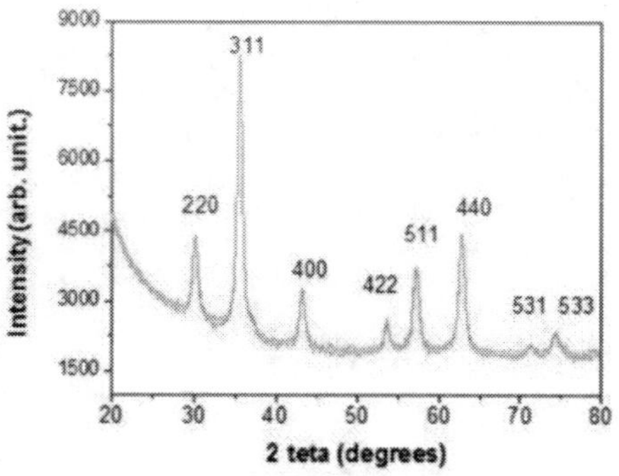


■**XRD**: X-ray powder diffraction from a representative sample of a thiolated $Fe_3O_4$ nanoparticle coated with MSA, displaying the Bragg peak reflections of magnetite.

**Figure 2**. X-ray powder diffractograms from a representative sample of SH-SPIONs coated with MSA and Cys, displaying the Bragg peak reflections of magnetite.

The morphologies and sizes distribution of the SH-SPIONs were analysed by transmission electron microscopy (TEM). Fig. 3(A) shows a representative image from sample Cys-SPIONs. It can be observed that the nanoparticles exhibit spherical continuous, however they present aggregation. It was also to observe in some particles a dark-core light-shell type of contrast. Such a kind of contrast may be associated to differences in the electronic density between iron core and the coating suggesting that the nanoparticles are formed by a Fe-oxide shell with an inner oxygen-poor core. This core–shell morphology is in agreement with the formation of $Fe_3O_4$ [19]. The histogram of the size distribution showed in Fig. 3(B) revealed a narrow distribution with a mean diameter of 14 nm and polidispersity of 8%.

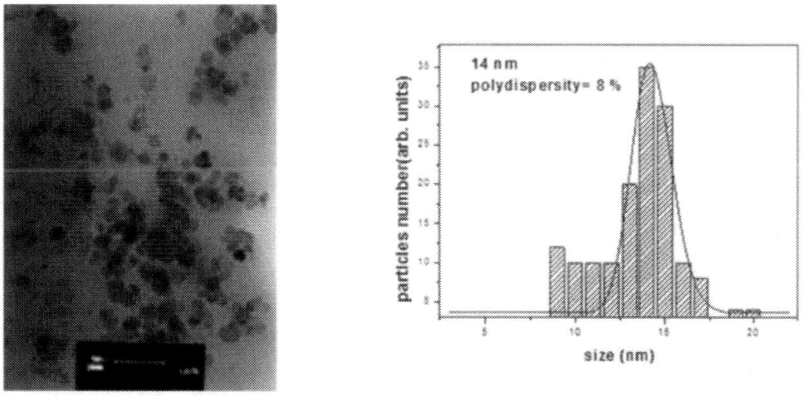


scale 100 nm

A B

**Figure 3.** (A) Transmission Electron Microscopy (TEM) of Cyst-SPIONs; (B) sizes distribution of Cys-SPIONs.

Finally, magnetic characterizations were also carried out to observe the magnetic behavior of samples. All SH-SPIONs exhibited the typical superparamagnetic behaviour (Fig. 4) expected for magnetic nanoparticles in this size range. The coercive field is approximately 0 Oe. Values of saturation magnetization around 60 emu/g were found, compatible with that found for $Fe_3O_4$ magnetite.

*1.2. Phytotoxicity of GSH or GSNO-nanoparti*

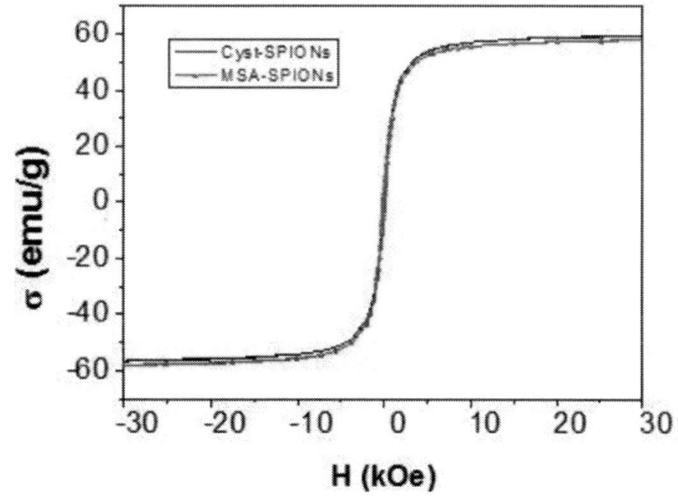


**Figure 4**. Magnetic curves of samples Cys-SPIONs (black) and MSA-SPIONs (red)

Journal of Physics: Conference Series **617** (2015) 012002 doi:10.1088/1742-6596/617/1/012002

Preliminary tests *in vivo* of incorporation were performed. Fig. 5 shows the incorporation of Cyst-SPIONs sample (0.05 mg/mL) for 3 hours in stem cells. Fig 5 (A) represents control cells and the arrows (Fig. 5 B) show the incorporation of aggregated Cys-SPIONs in cells. Results not shown showed that thiolated nanoparticles have no toxic effects for cells [20].

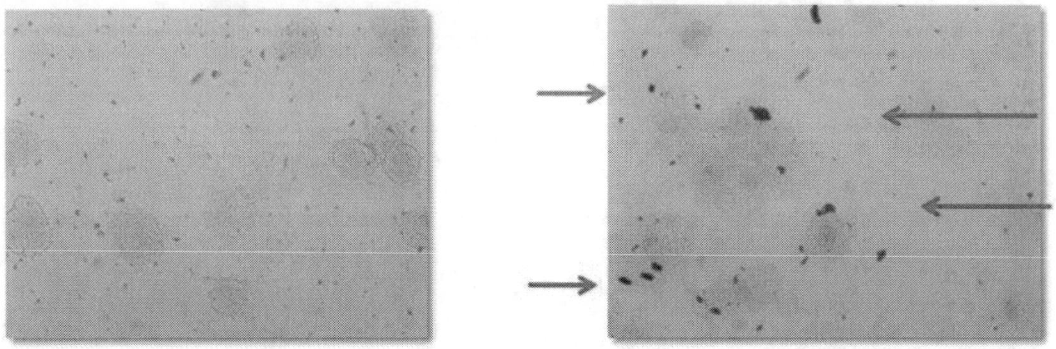

A B

**Figure 5**. Incorporation of Cys-SPIONs in stem cells for 3 hours. (A) control cells; (B) Cyst-SPIONs (0.05 mg/mL) incorporated in stem cells.

Figure 6 shows the incorporation curves of SH-SPIONs (0.05mg/mL) up to 20 hours. The curves were obtained by Prussian blue technique, where the Fe (III) reacts with a composition of 2% HCl acid mixed and 2% $K[Fe(CN)_6]$ producing $Fe4[Fe(CN)_6]_3$, the so called Prussian blue, which is an insoluble blue reaction product deposited inside the cell at the location of the Fe (III) ions. The blue stained area/cell (in pixels) was determined by thresholding as an estimate of the number of nanoparticles phagocytized. It can be seen that the major nanoparticle up-take by the cells occurs in the first 3 hours of incubation, for both samples. It can be seen that the SPIONs containing MSA has a larger incorporation area. This fact suggests an influence of amine group of cysteine molecule in phagocytosis mechanism. Further testing will be performed to elucidate the mechanism of internalization of the particles.

Journal of Physics: Conference Series **617** (2015) 012002 doi:10.1088/1742-6596/617/1/012002

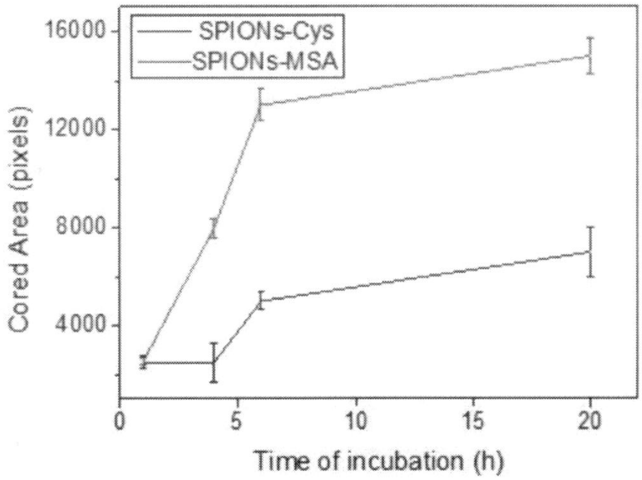


**Figure 6.** Incorporation curves of SH-SPIONs (0.05 mg/mL) in stem cells up to 20 hours: MSA-SPIONs (red curve); Cys-SPIONs (black curve)

Initial essay by magnetic resonance image (MRI) of stem cells containing nanoparticles with MSA incorporated for 3 hours (Fig. 7 a) revealed an increase of negative contrast (Fig. 7 d and e). The control wells in agarose (Fig. 7 b and c) showed no similar pattern of signal loss or reduction. Adjustments in T2-weighted pulse sequence and combination with post-processing imaging techniques (texture analysis) can further improve detection of labelled cells with these particles using clinical MR scanners.

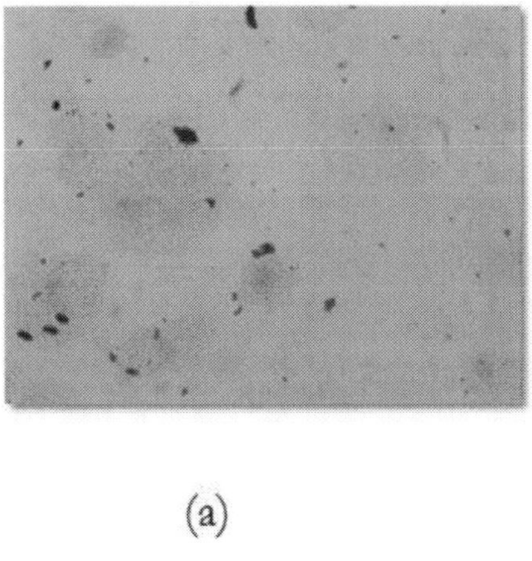


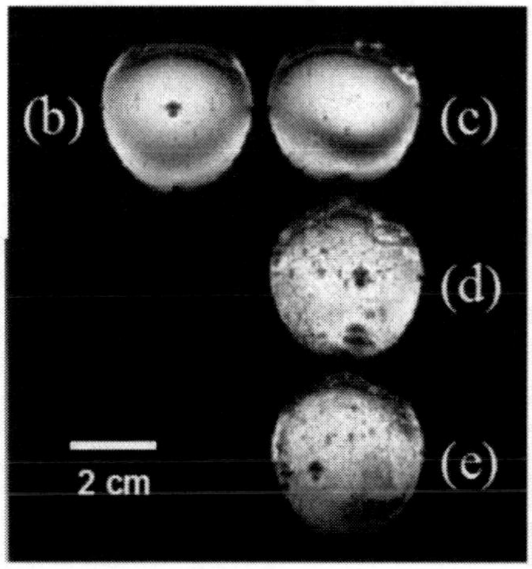


**Figure 7.** (a) Incorporation curves of MSA-SPIONs (0.05 mg/mL) in stem cells for 3 hours; (b) and (c) agarose; MRI tests of MSA-SPIONs (0.05 mg/mL) in stem cells for 3 hours (d) and (e).

Journal of Physics: Conference Series **617** (2015) 012002 doi:10.1088/1742-6596/617/1/012002

## 4. Conclusions

This work describes the preparation and characterization of SH-SPIONs and their impact as contrast agents. 15. These systems investigated were crystalline and presented low polydispersity with superparamagnetic behavior. The advantages of this new approach are the low toxicity of SH-SPIONs in stem cells and the increase of the contrast in MRI. SH-SPIONs act as diagnostic are extremely attractive for diverse biomedical applications.

## 5. Acknowledgements

Supported from FAPESP (Process 2014/13913-07), CNPq and CAPES.

## 6. References


[1] Haddad, P S and Seabra A B 2012 Biomedical Applications of Magnetic Nanoparticles *Iron Oxides: Structure, Properties and Applications* Nova Science Publishers vol. 1, pp 165−188

[2] Chen W, Cao Y, Liu M, Zhao Q, Huang J, Zhang H, Deng Z, Dai J, Williams D F and Zhang Z 2012 *Biomaterials* **33** 7895

[3] Ferro-Flores G, Ocampo-Garcia B E, Santos-Cuevas C L, Morales-Avila, E and Azorin-Veja E 2014 *Curr. Med. Chem.* **21** 124

[4] Gajalakshmi P, Priya M K, Pradeep T, Behera J, Kandasamy M, Madhuwanti S, Saran U and Chatterjee S 2013 *Toxicol. Appl. Pharmacol.* **269** 121

[5] Cai X H, Wang C L, Chen B A, Hua W J, Shen F, Yu L, He Z M, Shi Y Y, Chen Y, Xia G H, Cheng J, Bao W, Zhang Y and Wang X M 2014 *J. Biomed. Nanotechnol.* **10** 251

[6] Patil R M, Sheta P B, Thorat N D, Otari S V, Barick K C, Prasad A, Ningthoujam R S, Tiwale B M and Pawar S H 2014 **802** 4514

[7] Lu A N, Salabas E L and Schut F 2007 *Angew. Chem.* **46** 1222

[8] Lynch I, Cedervall T, Lundqvist M, Cabaleiro-Lago C, Linse S and Dawson K A 2007 *Adv. Colloid Interface Sci.* **134** 167

[9] Nel A E, Madler L, Velegol D, Xia T, Hoek E M, Somasundaran P, Klaessig F, Castranova V and Thompson M 2009 *Nat. Mater.* **8** 543

[10] Haddad P S, Duarte E L, Baptista M S, Goya G F, Leite, C A P and Itri R 2004 *Progr. Colloid Polym.Sci.***128** 232

[11] Wilhelm C, Billotey C, Roger J, Pons J N, Bacri J C and Gazeau F 2003 *Biomaterials* **24** 1001

[12] Jian W, Lai K L, Wu Y and Gu Z W 2014 *Arch. Pharm. Res.* **37** 825 129

[13] Mejias R, Gutierrez L, Salas G, Perez-Yague S, Zotes T M, Lazaro F J, Morales M P and Barber D F 2013 *J. Controlled Release* **171** 225.

[14] Hong S C, Lee J H, Lee J, Kim H Y, Park J Y, Cho J, Lee J, Han D H 2011 *Int. J. Nanomed.* **6** 3219

[15] Gupta A K and Gupta M 2005 *Biomaterials* **26** 3995

[16] Seabra A B, Martins D, Simões M M S G, da Silva R, Brocchi M and de Oliveira M G 2010 *Artif. Organs* **34** E204

[17] Chandra S, Nigam S and Bahadur D 2014 J. *Biomed. Nanotechnol.* 10 32

[18] Klug H P and Alexander L E 1974 X-ray Diffraction Procedures for Polycrystalline and Amorphous Materials, JohnWiley & Sons, vol 1 pp 1- 140

[19] Haddad P S, Rocha T R, Souza E R, Martins T M, Knobel M and Zanchet D *J. Colloid Interface Sci.* 2009 **339** 344

[20] Seabra AB, Pasquoto T, Ferrarini ACF, Santos MD, Haddad PS and de Lima R. *Chem Res Toxicol* 2014 **27** 1207

Journal of Physics: Conference Series **617** (2015) 012003 doi:10.1088/1742-6596/617/1/012003

# Analysis of multivariate stochastic signals sampled by on-line particle analyzers: Application to the quantitative assessment of occupational exposure to NOAA in multisource industrial scenarios (MSIS)

**JM López de Ipiña[1], C Vaquero[1], C. Gutierrez-Cañas[2], DYH. Pui[3]**

[1] TECNALIA - Parque Tecnológico de Alava, 01510 Miñano/Spain
[2] UPV – EHU - Dpt. Chem. & Environ. Eng. University of the Basque Country, Alameda Urquijo s/n Street, 48013 Bilbao/Spain
[3] UMN-PTL - 111 Church St SE, Minneapolis, MN 55455MN 55455/USA

E-mail: jesus.lopezdeipina@tecnalia.com

**Abstract**. In multisource industrial scenarios (MSIS) coexist NOAA generating activities with other productive sources of airborne particles, such as parallel processes of manufacturing or electrical and diesel machinery. A distinctive characteristic of MSIS is the spatially complex distribution of aerosol sources, as well as their potential differences in dynamics, due to the feasibility of multi-task configuration at a given time. Thus, the background signal is expected to challenge the aerosol analyzers at a probably wide range of concentrations and size distributions, depending of the multisource configuration at a given time. Monitoring and prediction by using statistical analysis of time series captured by on-line particle analyzers in industrial scenarios, have been proven to be feasible in predicting PNC evolution provided a given quality of net signals (difference between signal at source and background). However the analysis and modelling of non-consistent time series, influenced by low levels of SNR (Signal-Noise Ratio) could build a misleading basis for decision making. In this context, this work explores the use of stochastic models based on ARIMA methodology to monitor and predict exposure values (PNC). The study was carried out in a MSIS where an case study focused on the manufacture of perforated tablets of nano-$TiO_2$ by cold pressing was performed.

## 1. Introduction

As comprehensively reviewed by Kuhlbusch [12], different approaches have been used to obtain information about the exposure to NOAA: (a) Studies based on real workplaces and (b) process based studies in simulated workplaces and of simulated work processes. However, as they clearly pointed out, data about industrial processes working with nanomaterial products are still scarce [2,11,20] and the overview studies referred to herein give no clear results whether the actual nanomaterial was released or not. The increasing importance and use of NOAA in a diversity of industrial processes requires to build a consensus about both the determination of exposure scenaria and the specific needs in developing a method for the distinction of the background aerosol.

The currently widespread use of samplers and portable analyzers for continuous measurement of Particle Number Concentration (PNC) [5,6], the availability of strategies and measurement procedures [1,3,17,20], reference values (NRV) [18] and rules for decision [4], makes PNC a very suggestive metric for its use in the quantitative risk assessment of occupational exposure to airborne nanoobjects and their agglomerates and aggregates (NOAA). This metric is equally suggestive for monitoring industrial plants, to detect malfunctions and potential emergency situations in industrial processes manufacturing or handling manufactured nanomaterials or nano-enabled products. Furthermore, analysis and management of time series can be used to control exposure and operation in these industrial processes.

However in all cases, the effective discrimination of the signal emitted by the sources with respect to the background is the main challenge of this approach [13,15,20], particularly in multisource industrial scenarios MSIS with high levels of background aerosol. A distinctive characteristic of these specific scenaria is the spatially complex distribution of aerosol sources, as well as their potential differences in dynamics, due to the feasibility of multi-task configuration at a given time. Thus, the background signal is expected to challenge the aerosol analyzers at a probably wide range of concentrations and size distributions, depending of the multisource configuration at a given time.

Measurement of PNC in workplaces differs from traditional sampling hygienic techniques because it requires the use of continuous devices - ELPI+, SMPS, CPC, OPS, Nanoscan, among other - which results in time series data, showing the evolution of the selected metric during the sampling time and whose length is proportional to the sampling frequency [5,6]. The comparability of the data provided by instruments based on different principles has been studied showing that divergence generally arose at the extremes of their working size range and at low number concentration. [16]. These differences were attributed to a response to the different operating procedures used by the instruments and/or the different particle types sampled. ELPI generally showed agreement with the SMPS, FMPS and APS, however this agreement is much poorer at the upper and lower ends of the ELPI working range. For the smallest particle size, the overall uncertainty is 20% (95 % confidence interval), decreasing towards larger sizes and levels at 40 nm particle diameter to 12 %. Regarding the TSI portable CPC 3007 the measurement accuracy specified by the manufacturer is ± 20 %.

The PNC nano reference value (NRV) for nano$TiO_2$, defined as an 8 h time weighted average concentration adjusted for the background concentration, is 40.000 particles/$cm^3$ [18]. In addition, OELs mass based for nano $TiO_2$ (respirable fraction) have been proposed by NIOSH (0,3 mg/$m^3$) [12] and recently by project Scaffold (0,1 mg/$m^3$) [17].

At present, little attention has been paid on the statistical treatment of time series recorded by on-line analyzers of NOAA at workplaces. The analysis of metrics most commonly used, PNC and Particle Size Distribution (PSD), has been based on graphics and quantitative analyzes have been limited to the use of simple statistical parameters [7]. In signal processing and time series analysis there are numerous techniques in both time and frequency domains [8,19]. One of the most popular approaches are the stochastic ARIMA models. These models have been widely used in studies of air pollution, mainly for prediction. More recently ARIMA models have been applied to study time series collected by continuous aerosol analysers of NOAA, in order to identify the effect of a specific task on the concentration levels or to compare the level of concentration in repeated experiments [7]. In classical theory of Box-Jenkins [19], ARIMA models are listed using the standard notation of ARIMA (p,d,q)(P,D,Q), where p is the order of autoregression, d is the order of differencing (or integration), and q is the order of moving-average, and (P,D,Q) are their seasonal counterparts. These models have some limitations; they presume weak stationarity (Time series recorded by on-line analyzers of NOAA are rarely stationary), equal-spaced intervals of observations, no missing values and at least 30 to 50 observations. From stationary series, ARIMA methodology produces comprehensive models that not only explain the underlying generation process of the series themselves, but also can be exploited for further forecasting. In fact, these models may be further used to provide engineering feedback and design automated control systems in complex industrial scenaria.

In this context, this work focuses on exploring the applicability of PNC metric for quantitative assessment of exposure to nano $TiO_2$ in a MSIS. A second objective is aimed to know the background aerosol and its contribution to mask the effective discrimination of the signals at source. And finally, the ultimate goal is focused on the use of stochastic models based on ARIMA methodology to predict exposure values (PNC), based on the analysis of net signals (source minus background) obtained from raw time series recorded by continuous analyzers implemented in the MSIS.

## 2. Exposure scenarios

This study was carried out at the production plant of Bostlan SA, located in Mungia, (Spain). The company manufactures products for casting, among which alloying tablets and mini tablets for the aluminum industry, obtained by mixing and pressing metallic powders (Cr, Cu, Fe, Mn, Ni, Ti). All stages of the process (dosing, mixing, pressing, etc.) are controlled by computer, featuring a high degree of automation. In the framework of project EHS Advance [10], a case study focused on the manufacture of perforated tablets nano-$TiO_2$ was performed. Currently, Bostlan SA does not manufacture these products, so exposure scenarios lasted for a pilot batch of tablets (about two hours).

**Figure 1.** Layout and production lines of the company Bostlan SA. Inside the red circle, the exposure scenario 1 (EE1).

Measurements of $TiO_2$ were conducted in two exposure scenarios (Table 1): 1) Weighing $TiO_2$, 2) Production of $TiO_2$ tablets and 3) Finishing $TiO_2$ tablets. Positions 1 and 2 perform a single task (T1- Weighing or T2- Pressing) and are located in the production hall (Figure 1), where coexist with routine production of another type of tablets (EE1). WP3 is located in a separated room devoted to maintenance work (6 x 4 x 3 m), with two entrances (EE2). The four tasks described above involve the participation of two workers (W1 and W2). In EE1, W1 performs the weighing of $TiO_2$ (T1) and then W2 press the tablets (T2). After completion of T2 in EE1, W2 moves to EE2, to perform T3 and T4.

**Table 1.** Characteristics of exposure scenarios

| Exposure scenario | Location | Workplaces | Tasks | Workers |
|---|---|---|---|---|
| EE1 | Production hall | WP1. Weighing $TiO_2$<br>WP2. Production of $TiO_2$ tablets | T1. Weighing $TiO_2$ powder<br>T2. Pressing $TiO_2$ powder | W1. Worker 1<br>W2. Worker 2 |
| EE2 | Maintenance room (6 x 4 x 3 m) | WP3. Finishing $TiO_2$ tablets | T3. Drilling tablets<br>T4. Packaging tablets | W2. Worker 2 |

### *2.1. Exposure scenario 1 (EE1)*

Task 1 - Weighing $TiO_2$ powder (T1). The amount of $TiO_2$ (powder) required for producing a tablet is manually weighed on an electronic scale. The $TiO_2$ loading on the weighing container, is done manually with a small shovel from the raw material bag (Figure 2). The operator W1 uses an average of about 50 seconds for each weighing. Exposure measurements were performed during 11 weighing of material. A pallet next to the workplace allows intermediate storage of the raw material bags.

Task 2– Pressing $TiO_2$ powder (T2). An automatic vertical axis manufacturing press was enabled in manual mode for the manufacture of tablets by pressing $TiO_2$. The press is located 3 m from WP1. The manual loading of $TiO_2$, press drive, machine manipulations during successive pressing cycles and the final manual removal of the tablet are performed by the same operator (W2) (Figure 2). Exposure measurements were performed during the manufacturing of 12 tablets. Each tablet requires between 2 and 4 minutes and several pressing cycles and intermediate manipulations. The press has a local exhaust ventilation system (suction speed of 0,23 m / s).

**Figure 2.** From left to right and top to bottom: 1) Weighing $TiO_2$ powder (T1); 2,3) Cold pressing of $TiO_2$ powder at vertical axis press (T2); 4) Resulting $TiO_2$ tablet at the end of machine cycle; 5) Drilling tablets (T3); 6) Packaging in lots of 5 tablets (T4).

*2.2. Exposure scenario 2 (EE2)*

Task 3 - Drilling tablets (T3). This task makes a hole in the center of the tablet using a vertical drilling. Each drilling operation requires approximately 10 to 20 seconds. Some tablets broke down during operation and then the worker (W2) cleaned the waste. Exposure measurements were made during drilling of 32 tablets.

Task 4. - Packaging tablets (T4). Once drilled, tablets are carefully wrapped with aluminized paper in packs of five and stored in box. Exposure measurements were carried out during the preparation of five packs of tablets.

*2.3 Nanomaterial handled in exposure scenarios*

The nanomaterial selected for the case study is nanoparticulate titanium dioxide manufactured by EVONIK, with Aerosil ® process (flame hydrolysis), under the trade name AEROXIDE ® $TiO_2$ P 25. According to MSDS provided by the manufacturer, this product is a highly dispersed titanium dioxide, with a crystallographic composition of 80% anatase and 20% rutile and an average primary particle diameter of 21 nm. However primary particles are not present in isolated form, but primarily as aggregates or agglomerates, and the average diameter of the resulting particles, typically is in the range of near-micron scale and well above 100 nm.

## 3. Measurement set-up and methodology

For the characterization both of background aerosols and aerosols at sources, direct measurement instrumentation was deployed (Table 2) by the case study. Instruments 1 to 4 were connected in rack, to ensure the simultaneous capture of aerosol in a fixed sampling point. A portable termohygrometer to monitor relative humidity and temperature during the measurements was placed at the sampling points of background aerosols, in EEI and EE2. Additionally a portable anemometer was used to monitor air currents and measure the capture speed of the local exhaust ventilation installed in the press. The measurement strategy followed the procedures provided by NANOGEM [1] that displays three hierarchical levels of evaluation. Operational constraints in Bostlan SA - it is a unique experimental lot of tablets scheduled for a particular day –, leads to the unfeasibility of repeated measurements. Thus, the tier 3 (expert evaluation) was directly deployed. Measurements were designed to provide tangible evidence of the presence or absence of $TiO_2$ in the atmosphere of the workplace, by combining advanced on-line instrumentation (Table 2) and off-line analysis of aerosol samples (SEM, ICP-MS, EDX, etc). The measurement strategy also included the determination of aerosol background by its simultaneous and continuous measurement in two selected locations.

This paper covers exclusively time series recorded by on-line instruments numbered as #1, #2 and #5 in Table 2. In both exposure scenarios, instruments #1 and #2 were dedicated to the registration of aerosol at source and #5 for registration of background aerosol. The selected metric in all cases was the PNC.

**Table 2.** Instrumentation deployed for on-line monitoring of aerosols at workplaces and instruments selected for the present paper (#1, #2 and #5).

| | Instrument and model | Metrics | Measuring range (nm) | Concentration range (#/cm$^3$) | Remarks |
|---|---|---|---|---|---|
| 1 | CPC (TSI 3775)* | Total particle number concentration | 4 -> 3000 | 0 - $10^7$ | These instruments shared a common sampling point |
| 2 | ELPI+ (Dekati)* | Total particle number concentration and aerosol size distribution (channels) | 6 - 10000 | Depending on the most concentrated stage | |
| 3 | OPS (TSI 3330) | Total particle number concentration and size distribution | 300 - 10000 | 0 - 3000 | |
| 4 | Aerotrak 9000 (TSI) | Surface area of particles deposited in the lungs | 10 -1000 | | |
| 5 | CPC (TSI 3007)* | Total particle number concentration | 10 - 1000 | 0-$10^5$ | Measurement of background aerosol |

(*)Instrumentation selected for this paper

First, measurements in T1 were performed over a period of approximately 98 minutes, placing the sampling inlet next to the scale (40 cm). The following measurements were performed for 50 minutes in T2, positioning the inlet, 40 cm from the axis of the press. While performing these tasks, the instrument that measured the background aerosol was located about 8 meters of the working area. After finishing T1 and T2, the rack of instrumentation moved to EE2. In this room T3 and T4 were performed. In both cases, the sampling inlet was located at a distance of about 30 cm from the sources. While performing these tasks, background concentration was measured on the opposite side of the room, about 6 m from T3 and 3 m from T4. In all cases the sampling inlet was positioned closed to the breathing zone of workers W1 and W2.

Statistical analysis of time series was performed with the statistical package SPSS 22.0. This software includes a time series modeler allowing to build custom non seasonal or seasonal ARIMA models for time series, with or without a fixed set of predictor variables, and producing forecasts. The module includes also an expert modeler that automatically identifies and estimates the best-fitting ARIMA model for one or more dependent variable series. Both conventional trial-error and expert modeler approaches were combined in this study.

## 4. Time series and modelling

### *4.1 Time series recorded in EE1*

At EE1, the measurement log includes a first period without tasks named Period of Inactivity 1 (POI1). This period can be divided into two sections: POI1.1 preceding the start of the industrial activity in plant and POI1.2 in which all production processes are running. During period POI1, the rack of instrumentation was positioned at the location of T1. During the whole measurement period (POI+ T1 + T2), the ELPI+ measured significantly higher levels of particles than the other instruments (Figure 3). PNC raw signals provided by equipment at source evolve approximately in parallel with a continuous offset of 15.000 #/cm$^3$ between both signals. Both paths have several very marked relative maxima, the two most important are symmetric (110/120.000 #/cm$^3$) and recorded during POI2. Minimum values of PNC, around 25.000 #/cm$^3$, were recorded at the beginning of the measurement, when industrial plant processes had not begun to operate yet (POI1). PNC at source before starting T1, recorded during the POI2 under MSIS conditions, can be considered, for all purposes, as background

Journal of Physics: Conference Series **617** (2015) 012003 doi:10.1088/1742-6596/617/1/012003

concentrations during this period. The amplitude of maximum values (up to 120000 #/cm$^3$) provides a clear picture of the strong influence exerted by the background aerosol during tasks T1 and T2.

**Table 3.** Percentage of PNC with an aerodynamic diameter less than 100 nm, 380 nm and 1000 nm in time series (from data provided by ELPI +

| Task | ≤100 nm | ≤380 nm | ≤1000 nm |
|---|---|---|---|
| T1 - Weighting | 92,85 | 99,85 | 99,96 |
| T2 - Pressing | 91,43 | 99,83 | 99,95 |
| T3 - Drilling | 86,81 | 99,85 | 99,96 |
| T4 - Packaging | 84,10 | 99,78 | 99,94 |

If we separate the whole PNC signal recorded by the ELPI+ during T2 into two size ranges, "small particles" (≤ 1µm) and "large particles" (> 1µm) (Figure 3), it has been shown that the first signal replicates perfectly the whole signal, both in morphology and amplitude, including almost all of the aerosol particles. The second signal presents a smoothing path on which are several peaks, with maximum at about 176 #/cm3, coinciding temporally with the completion of tasks T1 (2724 data) and T2 (2727 data) and with no significant reflection in the total signal.

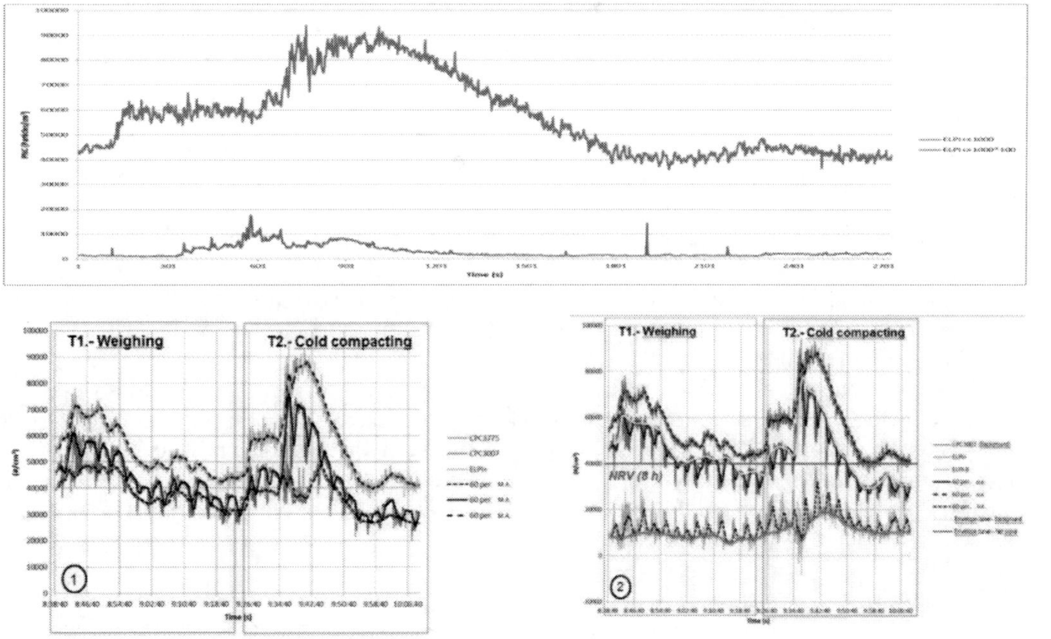


**Figure 3.** Top; PNC time series recorded by ELPI + during T2 showing the contribution of submicron and supermicron particles to the aerosol. 1) Raw time series captured at source and bakground during T1 and T2; 2) Net signal time series (source minus background) and its smoothed envelope (in green).

The measurement of background aerosol at EE1 begins during the completion of T1. If a section of this record is segmented and analyzed in detail, it appears that the general trend is altered by a set of periodic oscillations (Figures 3.1). During T1 and T2, reasonable agreement was noted between the signal measured at source and the smoothed background envelope calculated from CPC 3007 raw data, once the influence of the periodic signal has been suppressed. The offset between the two signals oscillates around 10.000 -15.000 #/cm$^3$. However during pressing (T2), these signals are significantly separated, presenting opposite evolutions which increase the offset until 60.000 #/cm$^3$. With respect to the pulsating component recorded by the background analyzer, it was initially attributed to the impact of manufacturing processes and mobile diesel machinery present in the production hall. However, after analyzing the industrial process in detail, the results point to a spurious component of the aerosol analyzer signal due to non-damped vibrations at the measurement point.

*4.2 Time series recorded in EE2*

Data recording started before T3&T4 (Figure 4), as denoted by POI2, period of inactivity 2. During POI2 the measuring rack was placed at the sampling point T3. Satisfactory agreement was reached between signals from different instruments. PNC raw data provided by equipment at source evolve in parallel, with an offset of about 7000 and 12000 #/cm$^3$. ELPI+ plot shows an increasing steepness from around 32000 #/cm$^3$ up to 54000 #/cm$^3$ and then a gradual decreasing to values of PNC below obtained at the beginning of the measurement (4000 #/cm$^3$). The background PNC shows a general agreement with the above mentioned signals recorded at source, a significant noise component in the ascending segment of the graph and, a maximum of 44000 # /cm$^3$. Data measured before beginning T3 can be considered as values of background aerosol. None of the three recorded PNC signals (source and background), appears influenced by potential releases during tasks T3 (2661 data) and T4 (1491 data).

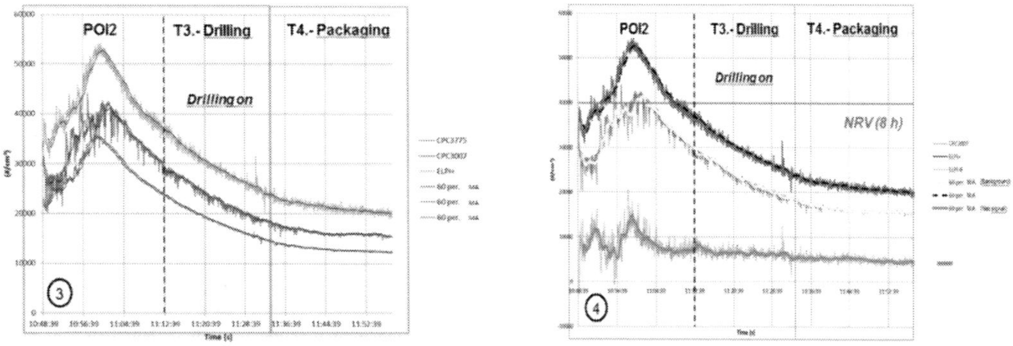


**Figure 4.** From left to right: 3) Raw time series captured at source and background during T3 and T4; 2) Net signal time series (source minus background), (in green).

*4.3 Time series of net signals*

Net signals were calculated by subtracting the background from the signal at source and initially were supposed to represent the contribution of NOAA releases to PNC from tasks (Figure 4a, 4b).

In EE1 the resulting temporal profile of net PNC during T1 and T2 also shows a serrated profile, shaped by the raw background signal (Figure 3.2). The smoothed net signal envelope exhibits during T2, a clear maximum around 38000 #/cm3. Its offset with respect to the corresponding maximum of the background, could suggest a contribution of NOAA released to the PNC. Nevertheless, this cannot conclusively confirmed due to the high level of background aerosol concentration. In this context, the resulting signals obtained for T1 and T2 cannot be directly interpreted as the net contribution of potential releases to the workplace PNC, since these signals may combine in a indistinguishable way the true NOAA releases, the effects of the different instrumental response between the collocated analyzers at source, the increase in the concentration of the background aerosol or a combination of all three.

In EE2, the net PNC signal obtained for T3 and T4 (Figure 4.4) exhibits significant peaks before starting T3 (drilling). The rest of the signal is damped on a level approximately constant at around 7000 # /cm$^3$ - corresponding to the difference of amplitude between the raw signals - without significant contributions of tasks evaluated.

**Table 4.** Summary statistics of time series for PNC

| Time Serie | Sampling period | N | Instrument | Max | Min | P25 | Mdn | P75 | Range | Mean | s | CV |
|---|---|---|---|---|---|---|---|---|---|---|---|---|
| T1 Weighing | 8:38:40 9:24:03 (45' 23' ') | 2724 | CPC 3775 | 49660 | 26800 | 34610,0 | 37970,0 | 45707,5 | 22860,0 | 39376,6 | 5843,8 | 14,8 |
| | | | CPC 3007 | 69095 | 26286 | 37447,0 | 42304,0 | 51453,3 | 42809,0 | 44118,4 | 8652,6 | 19,6 |
| | | | ELPI+ | 78049 | 38051 | 47326,3 | 51049,0 | 62993,8 | 39998,0 | 54533,3 | 9238,2 | 16,9 |
| | | | Net signal | 28676 | 2075 | 7745,3 | 9405,0 | 12393,5 | 26601,0 | 10414,9 | 3932,7 | 37,8 |
| T2 Pressing | 9:24:04 10:09:30 (45'26' ') | 2727 | CPC3775 | 61480 | 25470 | 28370,0 | 35130,0 | 39600,0 | 36010,0 | 34987,5 | 6583,3 | 18,8 |
| | | | CPC 3007 | 84167 | 19854 | 31077,0 | 38031,0 | 48736,0 | 64313,0 | 42499,9 | 13776,3 | 32,4 |
| | | | ELPI+ | 94068 | 35974 | 43184,0 | 56315,0 | 66998,0 | 58094,0 | 57448,6 | 15786,2 | 27,5 |
| | | | Net signal | 39806 | 576 | 10891,0 | 13991,0 | 17816,0 | 39230,0 | 15037,9 | 5848,3 | 38,9 |
| T3 Drilling | 10:48:39 11:32:59 (44' 20'') | 2661 | CPC 3775 | 35750 | 13990 | 18630,0 | 23680,0 | 28295,0 | 21760,0 | 23915,0 | 6019,3 | 25,2 |
| | | | CPC 3007 | 44113 | 17495 | 23090,5 | 28013,0 | 35311,5 | 26618,0 | 29141,1 | 6977,7 | 23,9 |
| | | | ELPI+ | 54214 | 20045 | 29639,0 | 36396,0 | 42353,5 | 34169,0 | 36748,5 | 8245,2 | 22,4 |
| | | | Net signal | 20775 | 344 | 6142,3 | 6937,0 | 8506,3 | 20431,0 | 7631,6 | 2466,6 | 32,3 |
| T4 Packaging | 11:33:00 11:57:50 (22'50'') | 1491 | CPC 3775 | 14770 | 11950 | 12520,0 | 12769,0 | 13340,0 | 2820,0 | 12965,4 | 595,6 | 4,6 |
| | | | CPC 3007 | 18582 | 14864 | 15795,0 | 15988,0 | 16760,0 | 3718,0 | 16309,6 | 750,0 | 4,6 |
| | | | ELPI+ | 25117 | 19006 | 20523,0 | 21193,0 | 22079,0 | 6111,0 | 21365,2 | 1086,8 | 5,1 |
| | | | Net signal | 8325 | 3413 | 4606,0 | 5037,0 | 5445,0 | 4912,0 | 5055,6 | 617,8 | 12,2 |

*4.3 Time series modelling*

In the case of the net PNC signal along T1 & T2, the operational problems described above, hinder their further modelling. Thus, the following discussion about time series modelling will refer to the net PNC along tasks T3 and T4.

Table 5 summarizes the ARIMA models evaluated to model time series T3 (2661 data) and T3&T4 (4152 data). As shown in Figure 4 both series were not stationary, so we proceeded to differentiate once the original series (d), obtaining in both cases resulting stationary series. Then, from the graphs of autocorrelation and partial autocorrelation functions (ACF and PACF), p and q parameters of the regular models (without seasonal component) were estimated. The models initially identified were the ARIMA (1,1,2) for T3 and ARIMA (2,1,1) for T3&T4. In both cases the models were over fitted - increasing by one unit the parameters p and q - and new models were tested to verify if they provided a better fit. The model initially selected for T3 (1,1,2) was also tested for T3&T4. All models included a constant in the initial estimation, which was suppressed in all cases, because it was not significantly different from zero. The models were recalculated again without constant. The ARIMA (2,1,2) model was removed from the final selection for both time series, because several model parameters showed no significant coefficients.

**Table 5.** Summary of models parameters and the goodness of fit of the models predictions (no significant coefficients between brackets; in bold, the two models finally selected).

| Time serie | Model ARIMA | Constant | AR1 | AR2 | D | MA1 | MA2 | $R^2$ | RMSE | MAPE | MaxPE | BIC | Ljung-Box |
|---|---|---|---|---|---|---|---|---|---|---|---|---|---|
| T3 (2661) | **(1,1,2)** | [1,990] | **0,568** | **-** | **1** | **0,552** | **0,251** | **0,890** | **831,655** | **9,196** | **745,177** | **23,459** | **Indepen.** |
| | (2,1,1) | [2,325] | 0,780 | -0,228 | 1 | 0,768 | - | 0,890 | 831,184 | 9,165 | 671,516 | 13,457 | Indepen. |
| | (2,1,2) | [2,079] | 0,612 | [-0,052] | 1 | 0,594 | 0,198 | 0,890 | 831,896 | 9,197 | 732,753 | 13,462 | Indepen. |
| T3&4 (4152) | (1,1,2) | [0,862] | 0,599 | - | 1 | 0,602 | 0,233 | 0,912 | 705,470 | 8,075 | 822,586 | 13,124 | Indepen. |
| | **(2,1,1)** | **[1,104]** | **0,794** | **-0,218** | **1** | **0,798** | **-** | **0,912** | **705,418** | **8,049** | **728,790** | **13,124** | **Indepen.** |
| | (2,1,2) | [1,032] | 0,700 | [-0,124] | 1 | 0,700 | [0,108] | 0,912 | 705,394 | 8,063 | 766,609 | 13,128 | Indepen. |

Normalized BIC criterion (Bayesian Information Criterion) was used among the remaining four models to select the best-fitting model and more parsimonious, with auxiliary criteria for goodness of fit $R^2$, RMSE, MAPE and MaxPE. The Ljung-Box criterion was used to demonstrate the independence of residuals (significant in all cases). At the end of the process the ARIMA (1,1,2) for T3 and (2,1,1) for T3&T4 were selected, although both models provide substantially similar fitting. In figure 5 the goodness of fit of model ARIMA (2,1,1) is shown for a representative section of time series T3&T4.

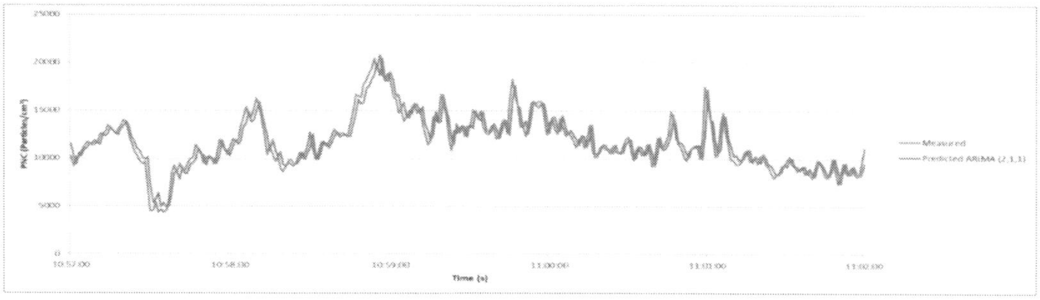


**Figure 5.** Fit between PNC measured in T3&T4 and predicted using a model ARIMA (2,1,1).

## 5 Conclusion

PNC method applicability to MSIS. The PNC is insensitive to emissions in the super-micron range, so depending on the type of NOAA will be roughly performant and/or distinguishable from the background. If the background is high, the typical state of the art aerosol analyzers offset can lead to accuracy problems or even masking between the source and the background signals. In MSIS, dynamic background is a critical element especially when going to average and can induce BIAS on averaging. The net signal can be conditioned by physical artifacts (equipment malfunction, high levels of background) that should be identified and filtered prior to mathematical treatment. In these cases, subtraction of 8 h time weighted average concentrations may result in uncertainty in the assessment of exposure value which is then compared with the NRV. This does not mean that in terms of background stable, the PNC method may be applicable in similar terms to those already demonstrated in other industrial environments (e.g. such as stationary processes, labs).

Background signal distinction in real time and time integrated. When there is a dynamic background as demonstrated above, subtraction of 8 h time weighted averages may lack physical meaning and its comparison with the NRV have no sense. During tasks T1 to T4, net concentration calculated during the sampling time presents values always below the NRV proposed for nano$TiO_2$ (40.000 particles/$cm^3$) (Table 4). However fluctuations of PNC in the aerosol at source are not linked to emissions during the evaluated tasks but are caused by changes in the background, related to manufacturing processes operation, diesel transport machinery, etc. Consequently the comparison between the net value of PNC and corresponding NRV has no physical meaning and therefore their application is not appropriate.

Statistical analysis of time series of PNC has been proven to be feasible in predicting PNC evolution provided a given quality of net signals [7]. However the analysis of non-consistent time series could build a misleading basis for decision making.

There is a need for more experimental studies in real MSIS, which could help to refine the procedures for the assessment of occupational exposure to NOAA, according to the well-established tiered approaches. The challenges are closely related to the distinction of the dynamic background. In this sense a promising approach currently under exploration might be the use of SNR parameter (Signal-to-Noise Ratio) for discriminating whether the MSIS pattern allows the applicability of the evaluation approach based on PNC.

The exposure assessment in company Bostlan SA was solved using mass as metric and applying the NIOSH method. The values 8h averaged, were far from OEL established by NIOSH (0,3 mg/$m^3$). In addition, SEM microscopy confirmed the evidence of nano$TiO_2$ agglomerates in the breathing zone of workers during T1 to T4, whose size ranged from approximately 5-10 μm [10].

## References

[1] Asbach C et al. 2012 *NANOGEM: Standard Operation Procedures*. Federal Ministry of Education and Research.

[2] Brouwer D K 2010 Exposure to manufactured nanoparticles in different workplaces. *Toxicology*

269(2):120–127.

[3] Brouwer D, Berges M, Virji MA, Fransman W, Bello D, Hodson L, Gabriel S, Tiele-Mans E. 2012 Harmonization of Measurement Strategies for Exposure to Manufactured Nano-Objects; Report of a Workshop, *Ann. Occup. Hyg.* 2012: 1–9.

[4] Brouwer D, Boessen R, Klein-Entink R, Moehimann C, Bard D, Bekker C, Van Duuren-Stuurman B 2014 Underpinning of decision criteria in a tiered approach strategy. *European Conf. on Nanotechnologies and Nanomaterials: opportunities and challenges*, Brussels.

[5] ISO 28439:2011 *Workplace atmospheres. Characterization of ultrafine aerosols/nanoaerosols. Determination of the size distribution and number concentration using differential electrical mobility analysing systems.*

[6] ISO/TS 12901-1:2012 *Nanotechnologies -- Occupational risk management applied to*

[7] Klein R H, Fransman W and Brouwer, D H 2011 How to statistically analyze nano exposure measurement results: using an ARIMA time series approach. *J Nanopart Res* (2011) 13: 6991-7004.

[8] Klein R H, Bekker C, Fransman WF and Brouwer, DH Analysis of time series of particle size distributions in nano exposure assessment, *Journal of Aerosol Science* (2015) 81: 62-69.

[9] Kuhlbusch T, Asbach C, Fissan H, Goehler D, and Stintz M, Nanoparticle exposure at nanotechnology workplaces: A review. *Particle and Fibre Toxicology*, 2011, 8:22.

[10] Lopez de Ipiña, JM, Vaquero C, Galarza N, 2014 . *EHS Advance – Occupational Guide*, Project deliverable, 168 pp

[11] Möhlmann C, Welter J, Klenke M, Sander J Workplace exposure at nanomaterial production processes. *J Phys Conf Ser* 2009, 170:012004.

[12] NIOSH 2011 *Occupational Exposure to Titanium Dioxide*. Current intelligence bulletin 63, DHHS (NIOSH) Publication No. 201–160.

[13] Ono- Ogasawara M., Serita F. and Takaya M., Distinguishing nanomaterial particles from background particulate matter for quantitative exposure assessment, *J. Nanopar. Res* 2009, 11, 1651-1659

[14] Pamachandran G, Ostraat M, Evans D, Methner M, O'shaughnessy P, D'arcy J, Geraci C, Stevenson, Maynard A, Rickabough K 2011 . A strategy for assessing workplace exposures to nanomaterials. *J Occup Environ Hyg* 8:673–685.

[15] Peters TM, Elzey S, Johnson R, Park H, Grassian VH, Maher T, O'shaughnessy P 2009. Air-borne monitoring to distinguishing engineered nanomaterials from incidental particles for environmental health and safety. *J Occup Environ Hyg* 6:73–81.

[16] Price HD, Stahlmecke B, Arthur R, Kaminski H, Lindermann J, Däuber E, Asbach C, Kuhlbusch TAJ, BéruBé KA, Jones T P 2014 Comparison of instruments for particle number size distribution measurements in air quality monitoring, *Journal of Aerosol Science* 76 (2014) 48-55.

[17] Scaffold project 2014 *Strategies and methods for exposure measurement (inhalation and dermal)*. Project deliverable D3.5, 22 pp.

[18] Social and Economic Council 2012 *Provisional nano reference values for engineered nanonomaterials*, SER - Advisory report 12/01 | mMarch 2012, 18 pp.

[19] Shumway RH, Stoffer DS 2011 *Time series analysis and its applications*, 3rd Edition, Springer, 596 pp.

[20] Vogel, U., Savolainen, K., Wu, O., Van Tongeren, M., Brouwer, D. and Berges, M. 2014 *Handbook of nanosafety. Measurement, Exposure and Toxicology*. Elsevier, 357 pp.

**Acknowledgment**

Research carried out by projects SCAFFOLD and EHS Advance were made possible thanks to funding from European Commission through FP7 (GA 319092) and Basque Country Government through ETORTEK Programme.

Journal of Physics: Conference Series **617** (2015) 012004 doi:10.1088/1742-6596/617/1/012004

# EXPOSURE SCENARIO LIBRARIES AS A TOOL FOR EXPOSURE ASSESSMENT

**Araceli Sánchez Jiménez[1], Shahzad Rashid[1], Derk Brouwer[2], Wouter Fransman[2], Carlos Fito[3], George Boulougouris[3], Martie van Tongeren[1]**

(1) Centre for Human Exposure Science (CHES), IOM, UK. Araceli.sanchez@iom-world.org, (2) TNO, The Netherlands; (3) ITENE, Spain

**Abstract.** The development of nanotechnology has reached a point where it is being widely applied, and numerous nanomaterials and nano-enabled products are handled across a broad range of industrial sectors. Exposure extends beyond occupational settings as products containing nanomaterials are used by different consumer groups.

Despite the knowledge on their toxic effects is growing there is still not OEL for most NMS and therefore the precautionary approach is still used where levels are kept as low as possible Therefore there is a need to assess workers and consumers exposure.

## 1 Occupational exposure assessment to nanomaterials

There are several measurement strategies for nanomaterials.
NanoGEM approach, developed by a group of German institutes, structured in three tiers, where tier one is information gathering, tier two involves a basic campaing and tier tree a detailed in-depth measurement survey

The French approach is structured in five tiers, and also included off-line analysis as dustiness test.
NIOSH has prepared specific guidance on the assessment of CNT and CNF And also for $TiO_2$.
There is also a general approach the Nanoparticle Emission Assessment Technique,
There are differences in the details of the approaches, the instrument used and assessment of the background but all follow a tiered approach where:

- Tier one consist of information gathering and assessment of potential release scenarios
- Tier two release and potential exposure is confirmed for which a basic campaign is required
- Tier three, detailed exposure concentrations are measured

The difference between tier 2 and 3 can be just the use of a couple more instruments. And longer measurement periods.

The move from tier 1 to tier 2 is more substantial, tier one is a desk-base exercise whereas tier two involves a visit to the industrial site in the case of an occupational scenario and the use of instruments, personnel who knows how to use the equipment, and all of that has financial implications. So, the decision has to be well informed. And whatever method you use it has to be a systematic.

## 2 Tier one tools

Control banding tools; like nanosafer, sttoffemanager nano and the nanocontrol banding tool for occupational exposure and nanoriskcat for consumers. These tools provide a hazard and a exposure band and an estimation of the risk. However, the hazard information is not always available, they do not cover all the domains, and what is more important they have not been validated with measurement data.

Hristozov developed a model base on weight of evidence. The model covers occupational exposure and it overestimates the exposure.

In short this consists of defining lines of evidence and assigning a weight to estimate an overall Or one can also use a read-across approach.

Read-across is a technique for data-gap filling where endpoint information from one chemical is used to predict the same endpoint for another chemical which is considered to be similar in some important aspect relating to that endpoint. Read-across may be for a qualitative or quantitative result.

*Read-across exposure*
Use of measurement data from analogous scenario in a conclusion on release/exposure or risk assessment.

*What are exposure scenario libraries?*
Collection of exposure scenarios with summaries of measurement data where scenarios can be searched using different descriptors and the user can find a match for the scenario of interest and read-across from the scenario library. I have to say all data is anonymised so companies cannot be identified. There are several projects developing or using data libraries.

## 3 MARINA Library

In the project MARINA, which stands for Managing the Risks of Nanomaterials we have developed the exposure library further. The template for data collection is more detailed and information is coded to facilitate the search.

This is an example: it includes information on the substance, the activity, the room conditions, other factors affecting the release and the exposure. The measurement data consists of summary statistics for the different instruments used and also descriptive information of SEM, TEM analysis if that is available.The library will be available online. It allows to search using the type of scenario, occupational, consumers or service life, CES name, name of the material, the life cycle and operating procedure. At the moment it as data on 11 NMs and most of the scenarios are for occupational exposure.

**Limitations**

- Quality of the exposure scenario information
- Quality check: completeness & transparency,
    - representativeness of the measurements
- Measurements are not collected in a harmonised way
- Might lead to different outcomes if reading from two different ES
- Uncertainty can be described but not quantified
- Criteria to read-across

This method of reading approach exposure information is being developed further in other projects.
In the Guidenano

1. Further population of the library with more ES
2. Algorithm to measure the similarity of the studied ES and the library ES
3. Library will be used as a tier one method in the Exposure Assessment module of the GuideNano tool.

As part of SUN (Sustainable nanotechnologies) work is being done on dermal exposure and this will also be put into a Library which will include dermal transfer efficiencies from surface to hand and hand to mouth.

## 4 Conclusion

- Exposure assessment is an essential part in risk assessment/management
- Need to be done in an intelligent, tiered approach
- Systematic approach to move to tier two to carry out a monitoring survey
- ES libraries are a useful tool for this process

Journal of Physics: Conference Series **617** (2015) 012005 doi:10.1088/1742-6596/617/1/012005

# Nanoparticle exposure biomonitoring: exposure/effect indicator development approaches

**C Desvergne[1], M Dubosson[1], M Lacombe[2], V Brun[2], V Mossuz[1]**

[1] Univ. Grenoble Alpes, CEA, Nanosafety Platform, Medical Biology Laboratory (LBM), 17 rue des Martyrs, F-38054 Grenoble, France
[2] Univ. Grenoble Alpes, CEA, INSERM, IRTSV, U1038 Exploring the Dynamics of Proteomes Laboratory (EDyP), 17 rue des Martyrs, F-38054 Grenoble, France

Email: veronique.mossuz@cea.fr

**Abstract.** The use of engineered nanoparticles (NP) is more and more widespread in various industrial sectors. The inhalation route of exposure is a matter of concern (adverse effects of air pollution by ultrafine particles and asbestos). No NP biomonitoring recommendations or standards are available so far. The LBM laboratory is currently studying several approaches to develop bioindicators for occupational health applications. As regards exposure indicators, new tools are being implemented to assess potentially inhaled NP in non-invasive respiratory sampling (nasal sampling and exhaled breath condensates (EBC)). Diverse NP analytical characterization methods are used (ICP-MS, dynamic light scattering and electron microscopy coupled to energy-dispersive X-ray analysis). As regards effect indicators, a methodology has been developed to assess a range of 29 cytokines in EBCs (potential respiratory inflammation due to NP exposure). Secondly, collaboration between the LBM laboratory and the EDyp team has allowed the EBC proteome to be characterized by means of an LC-MS/MS process. These projects are expected to facilitate the development of individual NP exposure biomonitoring tools and the analysis of early potential impacts on health. Innovative techniques such as field-flow fractionation combined with ICP-MS and single particle-ICPMS are currently being explored. These tools are directly intended to assist occupational physicians in the identification of exposure situations.

## 1. Introduction

The use of engineered nanoparticles (NP) is in constant growth in international research and industry. However their toxicity has not yet clearly been identified and the precautionary principle is currently recommended in France by the Public Health Council (HCSP, "Haut Conseil de la Santé Publique") owing to alert signals, such as a similar toxic action between the asbestos fibres and certain carbon nanotubes (CNT) as evidenced in animals [1, 2] or the known health effects of ultrafine particle pollution. With respect to these data, the inhalation exposure route now seems the most preoccupying for NPs.

Besides, carbon nanotubes and, by way of example, titanium dioxide (TiO2) widely used as a pigment and opacifier in the composition of many everyday consumer products (sunscreens, tooth paste, confectionery, paints, drugs…) is assumed to feature certain toxic effects similar to those of asbestos. The tests conducted on mice and on human cells show that in nanometric form, titanium dioxide produces a pro-inflammatory activity on the lungs and the peritoneum, like asbestos and silica [3]. In addition, titanium dioxide was classified as a possible carcinogenic agent in humans (2B) by the International Agency for Research on Cancer (IARC) in 2006.

Although uncertainties remain in the field of occupational risk assessment, strict measures for protection of workers potentially exposed to NPs must be implemented. For this purpose, an exposure monitoring procedure is of the essence. In addition to atmospheric monitoring, biological monitoring is a suitable approach to optimise the determination of individual exposure to occupational toxic substances. In fact, work atmosphere monitoring (atmospheric metrology) versus limit values to be observed is an indispensable process for risk assessment and prevention. It also allows the external dose to be measured. However, it is affected by certain limitations notably due to the integration of the sole respiratory tract, as well as the application of limit values to pure substances and not to the preparations, or to multiple exposures, and the failure to integrate individual absorption, metabolism and susceptibility factors.

Biological monitoring includes the use of exposure bioindicators and effect biomarkers. Exposure biomarkers are informative of the subjects' exposure level. They allow the assessment of the toxic substance quantities which actually penetrated into the body (internal dose) through the respiratory tract, but also through dermal or even digestive routes. Therefore, it is possible to determine the substance itself or its metabolites in the biological fluids, and this is constitutive of exposure biomarkers. Effect biomarkers are intended to highlight potential biological alterations linked with exposure.

Figure 1 illustrates the exposure and effect biomarkers in the continuum between exposure and pathology.

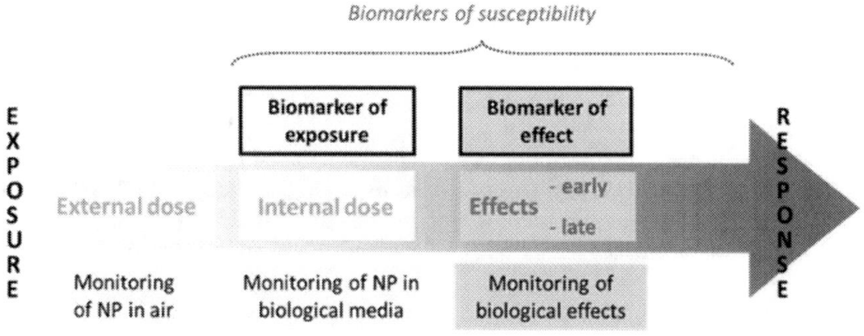


**Figure 1.** Continuum between exposure and pathology and exposure monitoring.

Generally, biomonitoring (which regroups biometrology and biotoxicology) offers the interest of integrating the various absorption routes, as well as multi-exposures (either occupational or not), with their possible synergy. Besides, it integrates the individual susceptibility factors and the actual exposure conditions such as physical activity, amongst others, which causes a hyperventilation and the resulting increase in lung absorption of toxic substances, as well as the use of individual protective gear (breathing mask, gloves, protection garments).

In the case of nanoparticles, no biological monitoring standards or recommendations are currently in effect, although this monitoring process is indispensable in the general approach of nanoparticle health impact prevention [4]. The remaining question is the choice of biological matrices to be used. Occupational monitoring refers to non- or minimally invasive matrices (conventionally urine and blood) but in the case of nanoparticles, the toxicokinetic data acquired in animals tend to prove that only very few nanoparticles could be found in these matrices after inhalation, owing to a translocation process which is demonstrated but however remains quantitatively low, as well as a quick internalisation in different organs after passage into the blood stream [5]. Therefore, we are studying other biological matrices, and the development of non-invasive exposure monitoring tools specific to the inhalation exposure route would be particularly useful.

A first non-invasive approach consists of sampling nasal secretions on a tissue flag for the analysis of potentially inhaled nanoparticles, in order to provide a "yes/no" type inhalation indicator at the workstations. This method is already in use for this type of indication in the case of radioactive particles [6]. Besides, concentrations of nickel dose-dependents with atmospheric exposure to nickel particles were found in workers' nasal mucous membrane [7], indicative of a significant retention of inhaled particles at nose level. In the case of nanoparticles, the deposit curves for particles inhaled in the respiratory tract proposed by the ICRP (International Commission on Radiological Protection) confirm that nanoparticles are especially retained at nose level, all the more so as they are smaller.

A second non-invasive approach consists of sampling exhaled breath condensates (EBC) which are a biological environment of interest for respiratory tract in connection with exposure by inhalation. In fact, the exhaled breath is in equilibrium with the water vapour at body temperature. Owing to the large lung exchange surface, 400 mL water are daily lost by evaporation. The exhaled aerosol is formed of microscopic droplets which appear at the respiratory mucus surface which covers the respiratory tract, due to turbulence phenomena between inhaled and exhaled air in the bronchi and alveoli. The exhaled air contains more than 99% water and a small fraction of volatile and non-volatile compounds from the respiratory tract. The exhaled air can be condensed in the form of liquid (the EBC) in which various molecules may be searched. The exhaled molecules are rather small, not exceeding 100 kDa, amongst which certain oxidative stress markers or inflammation markers may be found [8, 9]. Exposure biomarkers such as metals were also studied in EBCs in connection with smoking habits [10] or with occupational exposure to metal particles [11, 12]. The remaining question is to determine whether NPs could be detected in EBCs further to exposure by inhalation.

Whereas the development of exposure biomarkers is intended to determine the nanoparticles themselves (or their derivatives) in the biological matrices, the development of precautious effect biomarkers firstly requires the study of the toxicity mechanisms and the associated markers. Inflammation notably seems to result from toxicity linked with exposure to nanoparticles [13]. By way of example, certain bronchoalveolar liquid cytokines are induced by CNT inhalation, notably TNF-$\alpha$, IL-6 and TGF-$\beta$ in an inhalation study in mice concerning single-walled CNT, [14]. In another inhalation study on mice the $TiO_2$ causes the secretion of interleukines IL-1$\alpha$ and IL-1$\beta$ [3]. More generally, cytokines IL-3, Il-4, IL-5, IL-6, IL-10, IL-12, TNF-$\alpha$ and IFN-$\gamma$ are pointed out in the immunotoxicity of nanoparticles [15].

For its part, exposure to nanoparticles emitted by photocopiers causes an increase in proteins IL-6, IL-8, TNF-$\alpha$, IL-1$\beta$, G-CSF, EGF, IL-10, MCP1, fractalkine, and VEGF in nasal lavage [16]. The analysis of the induced expectoration from patients diagnosed with silicosis or asbestosis by comparison with des control subjects indicates that interleukines IL-1$\beta$, IL-6, the TNF-$\alpha$, and the metalloprotease inhibitor TIMP-1 are associated with the impairment of respiratory function in the sick patients [17]. Finally, a significant increase in interleukines IL-1$\beta$, IL-6, IL-8, IL-10 and IL-12p70 in EBCs was underlined in smokers versus non-smokers. The same observation was made in patients suffering from a chronic acute obstructive respiratory disease versus a stable pathology [18].

The search for biomarkers targeted on a toxicity mechanism such as inflammation is a possible approach. However, it would be interesting to combine it with non-targeted approaches, allowing the discovery of new bioindicators. Proteomics, for example, consists of describing all proteins present in specific cells or tissues. Contrary to the genome, the proteome is very variable in the course of time and incurs evolutions dictated by pathological and environmental changes [19]. Over the past 15 years, the quick progress of proteomics based on quantitative mass spectrometry allowed the discovery of many potential biomarkers for different cancers and other diseases [20, 21]. In the field of environmental and occupational health, although the data are sparser, a change in protein expression due to inhaled toxic substances was noted by proteomic approaches. Concerning NPs, the change in protein expression was found in the bronchoalveolar liquid (BAL) after carbon black NP intratracheal instillation in mice, and ZnO NP inhalation in rats [22, 23]. Concerning EBCs: the proteomic study showed that the proteins found in the EBCs featured 83% similarity with the proteins found in the BAL [24], thereby confirming the interest of this matrix to represent the deep lung. Different studies were consequently focused on the search for a protein signature in EBCs for determination of biomarkers according to different conditions (smoking, chronic obstructive bronchopneumopathy, pulmonary emphysema) [24, 25].

In summary, the combined search for exposure and effect bioindicators, bearing on biological matrices such as nasal secretions and the exhaled breath condensates, seems a promising approach for the development of biological NP exposure monitoring tools representative of the respiratory tract, like biomarkers measured in blood and urine which are more informative on a systemic rather than a local illness.

## 2. Equipment and methods

*2.1. Nasal secretion sampling.*
The device is comprised of a small absorbent tissue strip secured to a wooden stem (shaped as a flag) to collect nasal secretions by smear test in each nostril. The tissue is slightly moistened with ultrapure water before sampling. One flag is used for each nostril and each individual.

*2.2. Exhaled breath condensate sampling.*
The LBM laboratory uses the RTube device (Respiratory Research) according to the recommendations proposed by the "American Thoracic Society" and the "European Respiratory Society" [9]. Before using these devices, the LBM laboratory implements an intensive flushing protocol with ultrapure water and a check for absence of background noise by dynamic light scattering.

In brief, the sampling with nose blocked lasts for 15 minutes during which the subject breathes normally through the device. Water rinsing of the mouth was carried out and the subject was asked to abstain from drinking or eating for an hour before the sampling. A survey is also associated and allows these elements to be clarified. After collection, the exhaled breath condensates are immediately stored at -80°C in low absorption tubes. The EBC quantity collected generally varies between 1 and 2 mL in healthy adults.

*2.3. Analysis techniques*

*2.3.1 Exposure biomarkers.* The various techniques used by the LBM laboratory belong to the arsenal of the processes of the LBM and the Nanosafety Platform of CEA Grenoble (PNS, "Plateforme NanoSécurité"). They provide data concerning size, shape and chemical composition. In fact, there is no universal technique for nanoparticle analysis, allowing the determination in one single analysis of the different physico-chemical parameters of the particles required for their characterization. By priority, we combined three analyses, namely: inductively coupled plasma mass spectrometry (ICP-MS) for elemental composition determination, dynamic light scattering (DLS) for granulometric distribution determination, and scanning electron microscopy (SEM) coupled to X-ray fluorescence (EDX) for the determination of particle size/morphology and elemental composition.

The ICP- MS is the reference technique for a multi-element quantitative determination, both specific and sensitive. The LBM laboratory is equipped with a Nexion 300X device from the Perkin Elmer company, with methane reaction cell for the interfered elements such as iron and selenium. The LBM laboratory developed and validated a multi-element analysis technique allowing the simultaneous determination of 17 elements (Zn, Al, Ti, Co, Cu, Zr, Ni, Cr, Ga, In, Pl, Mn, Fe, Se, Cd, Ge, Be). The LBM laboratory determined the analytical LD and LQ as well as the LD and LQ methods (LDm and LQm), taking into account the background noise associated with sampling. Considering the very low concentrations found in EBCs, the analysis by ICP-MS is performed at this stage without prior mineralization. However, for the nasal smear test flag, a mineralization step is required in order to dissolve the flag before analysis by ICP-MS. This soft mineralization is performed in HNO3 in a microwave oven (MW3000 Anton Paar).

The DLS allows the granulometry measurement of particles in suspension in liquids. The LBM laboratory uses a Zetasizer Nano ZS device from the Malvern company (results not provided).
Electron microscopy coupled to X-ray elemental analysis is the reference technique for the qualitative confirmation of the size, morphology and chemical composition of the particles. The SEM produces high resolution images of a sample surface, up to a few nanometres. For its part, the EDX X-photon energy scattering analysis system, associated with the SEM, allows the chemical characterization of constitutive elements in the observed materials.

A SEM model 5500 unit (Scanning Electron Microscope) from the Hitachi company associated with a Noran model EDX system (X-ray analysis system) from the Thermo company was used.

Analyses with the TEM-EDX (Hitachi 5500.) were also performed on certain samplings.

*2.3.2. Effect biomarkers.* For analysis of cytokines in EBCs we used a multiplex technique (Magpix technology Biorad) allowing the analysis of a range of 29 inflammatory markers. The EBC samples were concentrated by freeze-drying (concentration by a factor 12) and the results are expressed in U/ml EBC, where U is a technical unit of the laboratory.

The inflammatory markers sought were classified by family: interleukines (Il 1b, Il 1ra, Il2, Il 4, Il5, Il 6, Il 7, Il 9, Il 10, Il 12, Il-13 , Il-15, Il-17a), chemokines (Il 8, eotaxin, RANTES, IP-10, MCP-1, MIP-1a, MIP-1b) and others (TNF alpha, IFN-g, G-CSF, GM-CSF, PDGF, FGF, VEGF, ICAM-1, VCAM-1).

The proteomic analysis required a previous biochemical treatment of the sample in order to extract the proteins while eliminating the phospholipids. In fact, the droplets of exhaled biological material in the EBCs are comprised of 90% lipids from the surfactant [24], which may impair the proteomic analysis quality. The EBC proteome was characterized by mass spectrometry (LC-MS/MS analyses on a mass spectrometer of OrbiTrap Velos type (Thermo). The analyses were firstly performed on EBC pools (triplicate of 10 different EBCs in each triplicate) owing to the very low concentrations.

### *2.4. Studied populations*

For EBCs, for analysis of metals and particle content, the LBM laboratory firstly took part in a large-scale study in an airport environment whose purpose was to assess the respiratory health of this type of population. This study offered the interest of testing the feasibility of EBC sampling in occupational health care, as more than 200 volunteer employees benefited from this type of RTube sampling. This sampling was associated with a wide range of clinical examinations and a detailed survey. The analysis by ICP-MS and DLS (results not provided) was performed on all samplings.

For the analysis on cytokines, we worked on a population de 16 healthy volunteers representative of the working population (aged between 28 and 56, 15 men, 1 woman, 12 non-smokers, and 1 smoker), and on a professional population of 32 employees in the scope of a confidential study. For the proteomic analysis, a population de 30 healthy volunteers representative of the working population was studied.

For each one of these studies, prior informed consent from the subjects had been obtained either in the scope of a method development process on anonymised samples, or in that of a study approved by the concerned company's Management and occupational physicians.

## 3. Results

*3.1 Nostril smear test flags*

For the flags, the results are essentially of analytical order. The use of the ICP-MS allowed an analytical sensitivity improvement from1 µg to 25 ng by flag.

The LBM laboratory proceeded with analytical developments on other types of nanoparticles.

A first confidential feasibility study was performed on workers in contact with metal oxide nanoparticles. This study produced first significant results through smear test flags on subjects after a workshift on a potentially exposed workstation and after a holiday period.

*3.2 Metals in the exhaled breath condensate*

The elemental concentrations obtained in the EBCs from 211 airport office employees for 17 elements are shown in Table 1.

**Table 1.** Elemental concentrations in EBCs (µg/L) shown as mean, median and range for concentrations above LDm or LQm. Aberrant values characterized by several orders of magnitude above other data were discarded.

| N = 211 | > LDm | | | | > LQm | | | |
|---|---|---|---|---|---|---|---|---|
| µg/L EBC | Mean | Median | Range[a] | n | Mean | Median | Range[a] | n |
| $Zn^{66}$ | 63 | 41 | 25-303 | 110 | 180 | 168 | 85-303 | 18 |
| $Al^{27}$ | 6.9 | 5.2 | 3.2-46 | 121 | 21 | 17 | 11-46 | 13 |
| $Ti^{48}$ | 0.8 | 0.6 | 0.3-3.6 | 88 | 1.8 | 1.5 | 1.1-3.6 | 12 |
| $Co^{59}$ | 0.3 | 0.2 | 0.1-1.1 | 58 | 0.7 | 0.5 | 0.4-1.1 | 16 |
| $Cu^{63}$ | 1.6 | 1.0 | 0.6-16 | 101 | 3.4 | 2.6 | 1.9-16 | 24 |
| $Zr^{90}$ | 0.7 | 0.6 | 0.2-9.7 | 211 | 0.7 | 0.6 | 0.1-4.3 | 211 |
| $Ni^{60}$ | 1.0 | 0.5 | 0.9-2.1 | 135 | 2.4 | 1.1 | 0.8-10 | 37 |
| $Cr^{52}$ | 0.8 | 0.6 | 0.1-8.7 | 193 | 0.9 | 0.7 | 0.4-8.7 | 157 |
| $Ga^{69}$ | 0.2 | 0.1 | 0.04-0.9 | 70 | 0.3 | 0.2 | 0.1-0.9 | 22 |
| $In^{115}$ | 0.1 | 0.1 | 0.01-0.3 | 60 | 0.1 | 0.1 | 0.02-0.3 | 49 |
| $Pt^{195}$ | 0.4 | 0.4 | 0.2-1.2 | 170 | 0.8 | 0.7 | 0.5-1.2 | 35 |
| $Mn^{55}$ | 0.8 | 0.4 | 0.3-7.2 | 75 | 3.2 | 2.4 | 1.1-7.2 | 10 |
| $Fe^{56}$ | 12 | 8 | 3.8-36 | 20 | 22 | 20 | 13-36 | 7 |
| $Se^{80}$ | 0.7 | 0.5 | 0.4-1.2 | 4 | — | — | — | 0 |
| $Cd^{111}$ | 0.2 | 0.1 | 0.1-0.7 | 56 | 0.4 | 0.3 | 0.3-0.7 | 17 |
| $Ge^{72}$ | 0.1 | 0.1 | 0.01-0.5 | 78 | 0.1 | 0.1 | 0.03-0.5 | 65 |
| $Be^{9}$ | 0.1 | 0.1 | 0.01-0.6 | 82 | 0.1 | 0.1 | 0.02-0.6 | 63 |

[a] Range (min-max)

Elemental concentration distribution is shown by way of example for 2 elements in Figure 2.

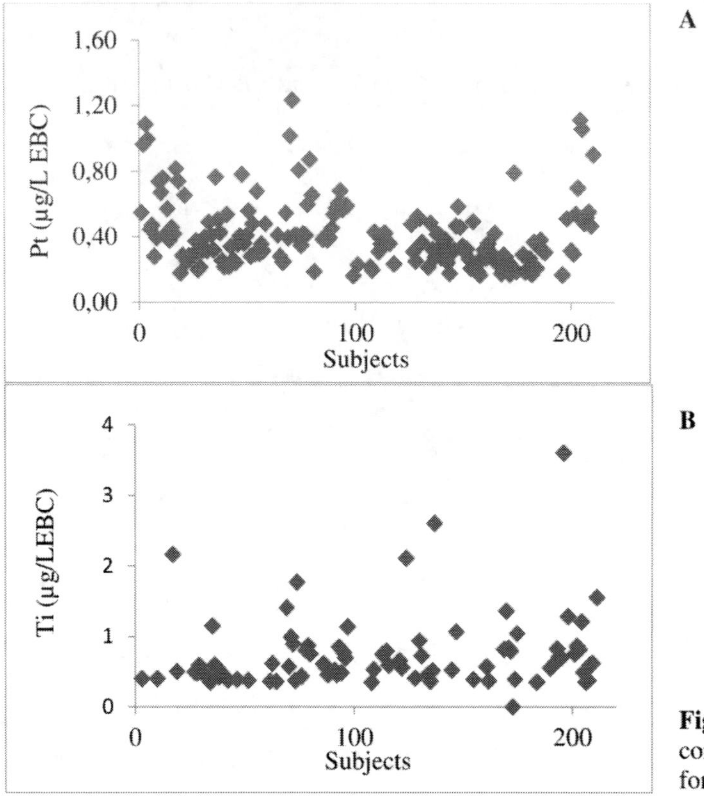


**Figure 2.** Elemental concentration dispersion in EBC for Pt (A) and Ti (B)

TEM-EDX images were shot on certain EBC samples to highlight a carbon nanoparticle within the EBC shown by its Na content (Figure 3).

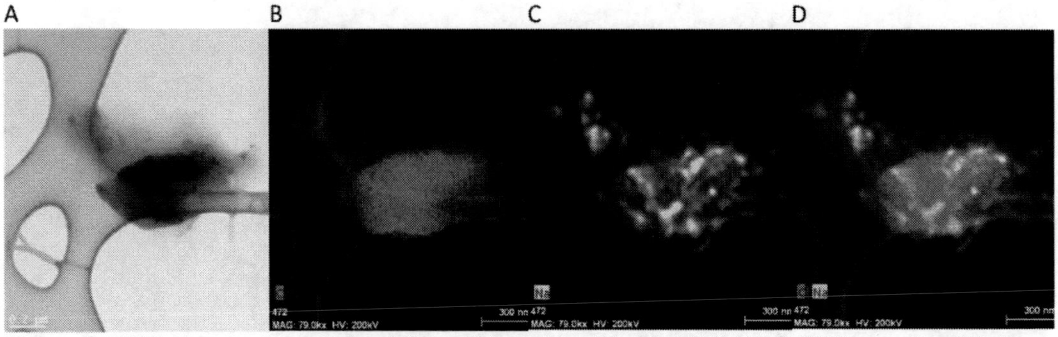


**Figure 3.** TEM picture of a particle of about 500 nm identified in EBC (A).C EDX mapping (B), Na EDX mapping (C), combination of C and Na EDX mappings.

### *3.3 Cytokines in EBCs*

Our first results on volunteers indicate that cytokine levels are very low despite the concentration step by freeze-drying. We expressed our results in technical units instead of mass units before confirmation. An example of the results obtained on the interleukines is shown in Table 2.

The cytokines analyses were also carried out on workers' EBCs, but in strict confidence.

**Table 2: result examples for cytokines on a volunteer population**

| Volunteer | Smoking habit | Interleukines (U/mL EBC) | | | | | | | | | | | | |
|---|---|---|---|---|---|---|---|---|---|---|---|---|---|---|
| | | Il-1b | Il-1ra | Il-2 | Il-4 | Il-5 | Il-6 | Il-7 | Il-9 | Il-10 | Il-12 | Il-13 | Il-15 | Il-17A |
| 1 | NS | 0.0024 | 0.52 | 0.050 | | | 0.012 | | 0.021 | 0.074 | 0.020 | | | 0.0072 |
| 3 | NS | 0.045 | 0.95 | 0.096 | | 0.026 | 0.035 | 0.041 | 0.014 | 0.1152 | 0.050 | 0.033 | | 0.26 |
| 4 | NS | 0.0024 | 0.95 | 0.018 | | | 0.009 | | | | 0.050 | | | |
| 6 | NS | | | | | | | 0.015 | | | | | | |
| 7 | NS | | 0.31 | 0.037 | | | | 0.028 | | | | | | |
| 8 | NS | | 2.1 | 0.050 | | | 0.012 | | 0.038 | 0.032 | 0.017 | | | |
| 9 | NS | | | | | | | | | | | | | |
| 10 | NS | | | | | | | | | | | | | |
| 11 | NS | | | | | | | | | | | | | |
| 12 | NS | 0.11 | | | | | | 0.023 | | | | | | |
| 13 | NS | 0.012 | 0.81 | | | | 0.014 | | 0.067 | | 0.015 | | | 0.21 |
| 16 | NS | 0.0040 | 0.66 | 0.10 | | | 0.006 | | 0.039 | 0.49 | 0.034 | 0.0016 | 0.39 | |
| 14 | S | 0.0016 | 0.13 | | | | | | 0.020 | | 0.023 | | | |
| 15 | S | | 0.73 | | | | | 0.019 | 0.012 | | | | | |
| 2 | S | 0.0024 | 0.38 | 0.050 | | | | | | | 0.0096 | | | 0.036 |
| 5 | S | 0.0016 | 0.52 | 0.12 | | | 0.009 | 0.023 | 0.033 | | | | | |

### *3.4 Proteomics in EBCs*

The last study consists in the proteomic analysis of EBCs that has been conducted in collaboration with the EDYP team. The EBC proteome from healthy volunteers has been determined based on a pooled EBC analysis. Three replicates have been performed on a pool of about 10 EBCs and more than 100 proteins have been described. The data analysis is in progress but some proteins of interest are emerging. In fact, some proteins originate from the deep lung, such as surfactant proteins, and others from macrophages. The next step will consist of improving the method sensitivity for the purpose of individual analysis.

## 4. Discussion

Through this paper, the LBM laboratory proposes various approaches to efficiently develop NP exposure and effect bioindicators. The interest of the search for NPs in biological mediums is the availability of an exposure-specific bioindicator, provided that the NPs can be characterized specifically. This is an analytical challenge, strongly linked with the nature of the NPs. For example, carbonaceous NPs are not easy to characterize specifically as, in fact, the elemental analyses of ICP-MS type fail to allow their detection and as these NPs may be ubiquitously present in the environment. Consequently, it is often necessary to combine different techniques such as ICP-MS, DLS and electron microscopy. The object is then to ensure that the results provided by the ICP-MS and the DLS comply with SEM-EDX or TEM-EDX observations for the purpose of possibly proposing, in the long term, a simplified analysis protocol in which the electron microscopy observation could be avoided, as this technique is heavy to implement owing to a long analysis time for each sample.
For the effect bioindicators, the search for inflammatory markers in the tested range offers the interest of highlighting, not only one marker (which, when isolated, would largely be non-specific) but several markers which, when simultaneously analysed, would provide potentially specific signatures of an exposure. Finally, proteomic analysis allows the exploration of the possible existence of new effect bioindicators as, in principle, it is not focused on suspected targeted mechanisms in the toxicity of NPs.

However, difficulties arose in the development of such engineered nanoparticle exposure and effect biomarkers. At analytical level, one of the main issues that the laboratory is currently addressing is that the EBC is strongly diluted biological medium. Consequently, it requires perfectly optimised and highly sensitive techniques. For example, the levels obtained for cytokines in EBCs are very close to the quantification or detection limits of the technique, despite the freeze-drying step, and this causes severe uncertainty factors. Therefore, for the time being, we are not in a position to conduct relevant comparisons between different types of populations. These comparisons will be planned in the longer term, when this barrier is lifted. This is why at this stage, the LBM laboratory prefers to provide the obtained results in technical unit rather than in concentration. This sensitivity issue is faced with, both for cytokine measurement and for the proteomic analysis which could only be conducted so far on EBC pools, as was already described in the literature [24, 25]. Once the technique is optimised, our objective is to proceed with individual analyses, failing which the use of the identified bioindicators would be compromised. For the elemental analysis by ICP-MS, the very low concentrations dictated the choice not to mineralize the sample firstly, in order to avoid the introduction of an additional sample contamination and dilution source.
As regards the nasal smear test flag, the LBM laboratory had initially performed early analytical developments on ZnO nanoparticles through the TXRF Total X-Ray Fluorescence technique [26]. The technique sensitivity was improved by a factor 40 for ZnO NPs, thanks to the use of the ICP-MS. This result is encouraging and the process is currently being developed for other NPs.
In addition to analytical improvements, an also critical issue is the small number of studies on volunteers conducted so far. Before proceeding with a wider scale study, the LBM laboratory wishes to optimise all these techniques, in order to further obtain reliable reference value on a general population, for comparison with populations of workers exposed to NPs.

The EBC standardisation issue also forms part of the subject to be looked at. On all EBC samples in airport environment, the laboratory performed analyses in µg/L EBC for all metals measured, but also in µg/100 µg sodium in EBCs (results not provided), as sodium is proposed as a standardisation tool by several teams [9]. However, the statistic result analysis is not different, whether it uses the results in µg by litre or in µg/100 µg sodium. The determination of total proteins by µBCA is currently being assessed at the laboratory with a view to standardisation by the total proteins present in the EBC samples. Proteomic analysis may also allow the identification of a protein which could be seen as the relevant candidate for a standardisation of the analyses in EBCs.

In parallel, it is also important to develop techniques for nanoparticle characterization improvement in the biological matrices, which remain complex even when they are diluted, as is the case of EBCs. The behaviour of nanoparticles in the biological fluids will depend both on the nanoparticle properties (solubility, zeta potential, etc.) and on the biological fluid nature itself, notably the presence of more or less large quantities of proteins and/or phospholipids likely to form coronas, thereby making their characterization more complex.
In conclusion, the studies presented in this paper are of interest insofar as they are directly conducted on humans: consequently, they reflect reality, whereas extrapolations from *in vitro* and *in vivo* studies are not always relevant as the behaviour of the nanoparticle is closely linked with its medium or environment. Consequently, the combination of several characterization techniques is of the essence. The LBM laboratory is currently working on the examination of biological mediums by innovative techniques such as ICP-MS Field-Flow Fractionation (FFF-ICP MS) or single-particle ICP MS (SP-ICP MS).
All these tools are intended to be used by the occupational physicians in order to ensure biological monitoring of employees working on engineered nanoparticles, as a complement to exposure traceability.

**References:**

[1] Poland, C.A., et al., Carbon nanotubes introduced into the abdominal cavity of mice show asbestos-like pathogenicity in a pilot study. Nat Nanotechnol, 2008. 3(7): p. 423-8.
[2] Takagi, A., et al., Induction of mesothelioma in p53+/- mouse by intraperitoneal application of multi-wall carbon nanotube. J Toxicol Sci, 2008. 33(1): p. 105-16.
[3] Yazdi, A.S., et al., Nanoparticles activate the NLR pyrin domain containing 3 (Nlrp3) inflammasome and cause pulmonary inflammation through release of IL-1alpha and IL-1beta. Proc Natl Acad Sci U S A, 2010. 107(45): p. 19449-54.
[4] Trout, D.B. and P.A. Schulte, Medical surveillance, exposure registries, and epidemiologic research for workers exposed to nanomaterials. Toxicology, 2010. 269(2-3): p. 128-35.
[5] Mills, N.L., et al., Do inhaled carbon nanoparticles translocate directly into the circulation in humans? Am J Respir Crit Care Med, 2006. **173**(4): p. 426-31.
[6] Castagnet X, Cazoulat A, Briot F, Bohand S, Bohand X and Laroche P 2008 [Alpha emitters inhalation: detection by a process with flags].*Ann Pharm Fr* **66** 140-5
[7] Torjussen W and Andersen I 1979 Nickel concentrations in nasal mucosa, plasma, and urine in active and retired nickel workers.*Ann Clin Lab Sci* **9** 289-98
[8] Barregard, L., et al., Experimental exposure to wood smoke: effects on airway inflammation and oxidative stress. Occup Environ Med, 2008. **65**(5): p. 319-24.
[9] Horvath, I., et al., Exhaled breath condensate: methodological recommendations and unresolved questions. Eur Respir J, 2005. **26**(3): p. 523-48.
[10] Mutti, A., et al., Exhaled metallic elements and serum pneumoproteins in asymptomatic smokers and patients with COPD or asthma. Chest, 2006. **129**(5): p. 1288-97.
[11] Goldoni, M., et al., Chromium in exhaled breath condensate (EBC), erythrocytes, plasam and urine in the biomonitoring of chrome-plating workers exposed to soluble Cr(VI). J Envion Monit, 2010. **12**: p. 442-447.
[12] Goldoni, M., et al., Exhaled breath condensate as a suitable matrix to assess lung dose and effects in workers exposed to cobalt and tungsten. Environ Health Perspect, 2004. **112**(13): p. 1293-8.
[13] Andujar, P., et al., *Respiratory effects of manufactured nanoparticles.* Rev Mal Respir, 2011. 28(8): p. e66-75.
[14] Shvedova, A.A., et al., Inhalation vs. aspiration of single-walled carbon nanotubes in C57BL/6 mice: inflammation, fibrosis, oxidative stress, and mutagenesis. Am J Physiol Lung Cell Mol Physiol, 2008. 295(4): p. L552-65.

[15] Elsabahy, M. and K.L. Wooley, Cytokines as biomarkers of nanoparticle immunotoxicity. Chem Soc Rev, 2013.

[16] Khatri M, D. B, Gaines P, Martin J, Pal A, Gore R and Woskie S 2012 Nanoparticles from photocopiers induce oxidative stress and upper respiratory tract inflammation in healthy volunteers.*Nanotoxicology*

[17] IRSST, Analyse de l'expectoration induite dans la silicose et l'amiantose : corrélation avec la fonction pulmonaire. Etudes et Recherches, 2006. **R-450**.

[18] Gessner C, Scheibe R, Wotzel M, Hammerschmidt S, Kuhn H, Engelmann L, Hoheisel G, Gillissen A, Sack U and Wirtz H 2005 Exhaled breath condensate cytokine patterns in chronic obstructive pulmonary disease.*Respir Med* **99** 1229-40

[19] Vlandanderen

[20] Gillette, M.A. and S.A. Carr, Quantitative analysis of peptides and proteins in biomedicine by targeted mass spectrometry. Nat Methods, 2013. 10(1): p. 28-34.

[21] Lange, V., et al., Selected reaction monitoring for quantitative proteomics: a tutorial. Mol Syst Biol, 2008. 4: p. 222.

[22] Chang, C.C., et al., Proteomic analysis of proteins from bronchoalveolar lavage fluid reveals the action mechanism of ultrafine carbon black-induced lung injury in mice. Proteomics, 2007. 7(23): p. 4388-97.

[23] Juang, Y.M., et al., Changes in protein expression in rat bronchoalveolar lavage fluid after exposure to zinc oxide nanoparticles: an iTRAQ proteomic approach. Rapid Commun Mass Spectrom, 2014. 28(8): p. 974-80.

[24] Bredberg, A., et al., *Exhaled endogenous particles contain lung proteins.* Clin Chem, 2012. 58(2): p. 431-40.

[25] Fumagalli, M., et al., Profiling the Proteome of Exhaled Breath Condensate in Healthy Smokers and COPD Patients by LC-MS/MS. Int J Mol Sci, 2012. **13**(11): p. 13894-910. 2008

[26] Marie-Desvergne, C., et al., *Development of a novel method to detect a potential inhalation of nanoaerosols.* Journal of Physics: Conference Series, 2010. **304**(1).

# Occupational exposure to nano-$TiO_2$ in the life cycle steps of new depollutant mortars used in construction

**C.Vaquero[1], N. Gelarza[1], J.L. López de Ipiña[1], C. Gutierrez-Cañas[2], I. Múgica[2], G. Aragón[2], M. Jaen[3], R. Pina[4], I. Larraza[4], A. Esteban-Cubillo[5], D. Thompson[6], D.Y.H. Pui[6]**

(1)Tecnalia, Leonardo da Vinci 11, 01500, Miñano, Spain; (2) University of the Basque Country, Alameda de Urquijo s/n, 48013, Bilbao, Spain;(3) Tecnan, Los Arcos, 31120;(4) Acciona. Valportillo II, 280108 Alcobendas, Spain; (5) Tolsa Group, Carretera de Madrid a Rivas Jarama, 35, Madrid, Spain, (6) University of Minnesota, 111 Church Street, SE, Minneapolis, MN 55455

E-mail: celina.vaquero@tecnalia.com

**Abstract.** The present work is focused on the measurement of workers exposure to nano-$TiO_2$ in the life cycle steps of depollutant mortars. It has been done in the framework of the SCAFFOLD project, which aims at the management of potential risks arising from the use of manufactured nanomaterials in construction. Main findings can be summarized as follows: (1) The occupational exposure to nano- $TiO_2$ is below 0.3 $mg/m^3$ for all measured scenarios. The highest concentrations were measured during the cleaning task (in the nano- $TiO_2$ manufacturing process) and during the application (spraying) of depollutant coatings on a wall. (2) It was found a high release of particles above the background in several tasks as expected due to the nature of the activities performed. The maximum concentration was measured during drilling and during adding powder materials (mean total particle concentration up to 5.591E+04 particles/$cm^3$ and 5.69E+04 particles/$cm^3$). However, considering data on total particle concentration released, no striking differences have been observed when tasks have been performed using conventional materials in the sector (control) and when using materials doped with nano-objects.

## 1. Introduction

In the recent years MNMs are being considered for various applications in the construction sector and it is expected that its use will grow in the future years. However there are concerns about their potential health effects and how to properly manage them to protect the workers.

A survey developed by FIEC and EFBWW (2009) showed that the majority of construction workers and their employers (~75%) were not aware that they work with nano-products. Additionally, the state of the art shows that there are currently few data on worker exposure to NOAAs (Clark, 2012) and more specifically in this sector. In this sense, it can be mentioned the papers from Van Broekhuizen et al (2011) that monitored several processes in the sector as spraying a liquid window coating, applying a cement repair mortar and nano-concrete drilling.

The present work is focused on the measurement of workers exposure to nano- $TiO_2$ in the life cycle steps of depollutant mortars. It has been done in the framework of the SCAFFOLD project (Grant agreement N° 280535, 2011-2015), which aims at the management of potential risks arising from the use of manufactured nanomaterials in construction. The measurements include the following processes: depollutant mortar fabrication, its application in walls, machining of materials during use and finally, demolition. Additionally, the manufacturing process of nano $TiO_2$ has been considered. Measurements have been performed at pilot and industrial scale in similar conditions to those of real practice.

## 2. Method

The strategy to measure occupational exposure followed the method proposed by NIOSH (Bulletin 63) (NIOSH, 2011). According to it two samples at the personal breathing zone were collected for off-line ICP-MS and SEM/EDX analysis (respirable fraction). Simultaneously, the aerosols released in the activities have been characterized using on-line devices following the tiered approach established by Asbach et al (2012). Measurement equipment included CPC3007 (TSI, 10 nm->1 μm), CPC3775 (TSI, 4 nm->3 μm) W-CPC (TSI, 25 nm->3 μm) ELPI+ (Dekati, 6 nm->10 μm), OPS (TSI, 0,3- 10 μm) and Aerotrak (TSI, 10 nm-1 μm). The background (BG) aerosol was characterized following three different approaches including time series analysis, the spatial approach and comparative study with/without nanomaterials (Kuhlbusch et al, 2011).

Occupational exposure limits considered for nano- $TiO_2$ are 0.3 mg/cm3 (NIOSH, 2011) and 0.1 mg/cm3 proposed in Scaffold project (Stockmann-Juvala, H. 2014). Additionally, total particle concentration (particles/$cm^3$) was compared with nano reference values, NRV, that for nano- $TiO_2$ is 40.000 particles/$cm^3$ (SER, 2012; IFA, 2014).

## 3. Results

The description of the scenarios and the main results of the measurements performed are summarized in the next paragraphs; the table 2 collects the main data measured.

*Scenario1: nano- $TiO_2$ manufacturing process.*

This scenario relates to the manufacturing process of commercial nano- $TiO_2$, Tecnapow- $TiO_2$ (TECNAN). The industrial production process proceeds by batches consisting on five tasks: (T1) reaction (flame process, closed system), (T2) beating, (T3) final product collection, (T4) product transferring to small containers, and (T5) facility cleaning. It was performed in an industrial site with natural and general ventilation. During T1 and T2 the system was closed; the other three activities were performed by an operator using PPEs including respiratory protection (mask FPP3), Tyvek suit, gloves and protection gasses.

The results showed that the $TiO_2$ concentration during the (T5) cleaning task was 0.388 mg/$m^3$ (62 min sampling). However this task is an infrequent process for the company so the occupational exposure calculated to 8 h-TWA was much lower; in this case, considering that the task is performed 5 h/day, 1 day per week the occupational exposure to nano-$TiO_2$ was 0.048 mg/m3. For this task, the SEM analysis showed the evidence of nano-$TiO_2$ agglomerates at the PBZ. On the other side, the measurements of particle concentration showed that mean total particle concentration was between 3.85 E+3-1.33E+4 particles/$cm^3$ during the hole process, although high peaks of concentration up to 1.20 E+5 particles/$cm^3$ were observed during the cleaning task (data from CPC3007, BG corrected).

*Scenario 2: Mortar manufacturing.*

In this scenario it was manufactured three different types of depollutant mortar: material A (control) and material B and C filled with nano-$TiO_2$ (see table 1).

**Table 1.** List of materials manufactured in scenario 2

| Product | Composition | Quantity |
|---|---|---|
| **Material A** | Conventional mortar: 20% cement, 80% sand (approx..) +3,7 kg standard additives | 1 Ton |
| **Material B** | Depollutant mortar: conventional mortar + 8,2 kg nano-$TiO_2$ supported on sepiolite (1) | 1 Ton |
| **Material C** | Depollutant mortar: conventional mortar + 4,1 kg nano- $TiO_2$ (Aeroxide P25, Evonik) | 1 Ton |

(1) The additive for Material B has been developed by TOLSA to achieve better $TiO_2$ dispersions and following the "safe by design" concept (nano$TiO_2$ fixed on sepiolite).

Three batches of products were processed, 1 ton each one. The process includes three tasks: (T1) additive weighing, (T2) additive adding to the hopper and (T3) mortar bagging. It was performed in an industrial site with natural ventilation by an operator using respiratory protection and gloves.
The main data measured are showed in figure 1. As can be observed, the $TiO_2$ concentration during the manufacturing of each material was between 0.008-0.073 mg/m3 (the sampling time was nearly 60 min for each product). In this case the occupational exposure calculated to 8 h-TWA had the same value than the measured concentration because it was assumed the worst case, so the operator was performing that task during the 8-day, 40 h-week. From the data it should be also highlighted that the $TiO_2$ concentration for material C was higher than for material B (nano-$TiO_2$ supported on sepiolite).The analysis of the samples at the PBZ showed the evidence of nano-$TiO_2$ agglomerates during the manufacturing of product C (see figure 2). Finally, regarding the release of particles, the highest particle concentration observed was during adding additives to the hopper (T2) were mean particle concentration was 5.69E+4 particles/cm$^3$ for the material C (filled with nano-$TiO_2$).

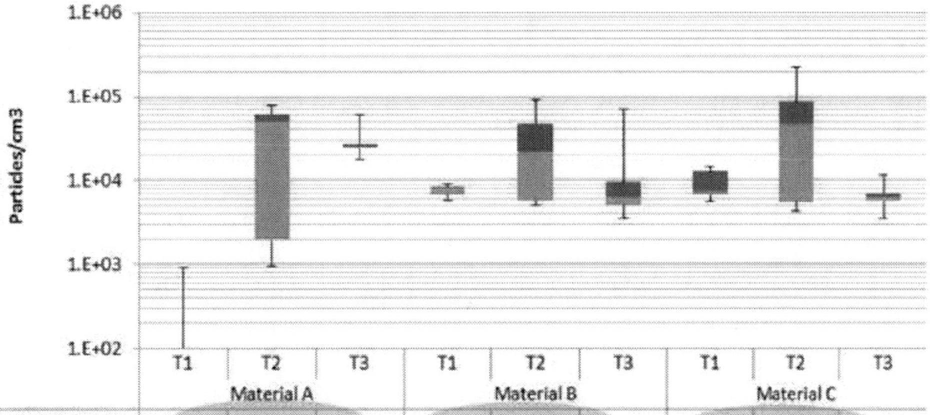


| | T1 | T2 | T3 | T1 | T2 | T3 | T1 | T2 | T3 |
|---|---|---|---|---|---|---|---|---|---|
| | | Material A | | | Material B | | | Material C | |
| 8-hr TWA (mg/m3) | | 0.009 | | | 0.008 | | | 0.073 | |
| TIO2-Conctr. (mg/m3) | | 0.009 | | | 0.008 | | | 0.073 | |
| Sampling time (min) | | 57 | | | 54 | | | 59 | |
| Mean (particles/cm3) | 8 | 35010 | 25447 | 7660 | 30719 | 8493 | 9510 | 56864 | 6334 |

**Figure 1.** Main data measured in scenario 2. Boxplot, background-corrected average number of particles/cm$^3$ during the tasks (data from CPC3007).

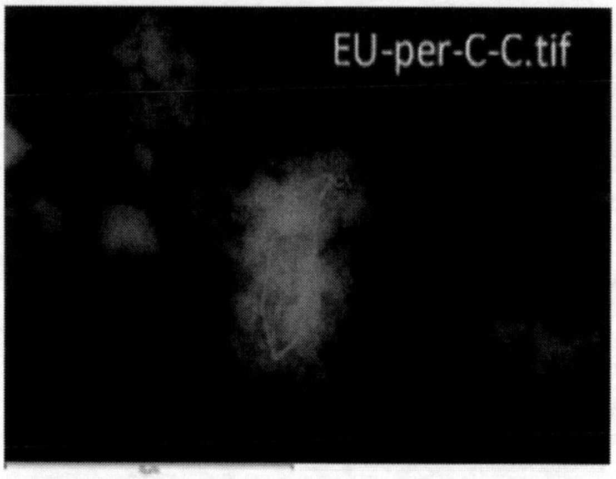


**Figure 2.** SEM image of collected PBZ samples during manufacturing material C, mortar filled with nano-$TiO_2$.

*Scenario 3: Mortar application.*
The three materials previously developed in Scenario 2 were applied in a wall. The process includes three tasks: (T1) adding mortar and mixing with water, (T2) mortar application and (T3) scrapping. The work was performed outdoors by two operators who wear respiratory masks, gloves and work ropes.

Figure 3 shows the main results for this scenario. As can be observed, the $TiO_2$ concentration was between 0.003-0.055 mg/m$^3$ during the time sampled (7-44 minutes), and the occupational exposure 8 h-TWA was calculated assuming the worst case. Similar to previous scenario, the $TiO_2$ concentration for material C was higher than for material B. Regarding the release of particles, the highest particle concentration observed was during adding mortar and mixing with water (T1); mean particle concentration between 1.41E4-1.56E4 particles/cm$^3$ with peaks of maximum values up to 1.34E+5 particles/cm$^3$ for Mat. C.

*Scenario 4. Sol-Gel spraying.*
Two sol-gels coatings were sprayed in a wall: material A was a sol-gel with 1.7% of nano-$TiO_2$ (Aeroxide P25 Evonix); material B was a sol-gel with 1.3% of nano-$TiO_2$ supported on sepiolite. The process was performed outdoor by an operator using a manual spray gun. The operator used PPEs including respiratory protection (mask FPP3), Tyvek suit, gloves and protection glasses.

The results showed that $TiO_2$ concentration at the PBZ during the task was 0.069-0.195 mg/m$^3$ for materials A and B respectively (sampling time 13-15 minutes). Samples taken at the PBZ showed the evidence of nano-$TiO_2$ agglomerates during the two tasks. Mean particle concentration measured was 2.18E+4-2.28E+4 particles/cm$^3$ for the two products and maximum peaks up to 2.57E+5 particles/cm$^3$.

*Scenario 5. Machining tasks (drilling)*
An operator drilled different samples of mortar: mortar A, conventional (control); mortar B filled with 1% nano-$TiO_2$ (Aeroxide); mortar C filled with 1% nano-$TiO_2$ supported on sepiolite; sol-gel A (mortar with a coating containing 1% of nano-$TiO_2$, Aeroxide) and sol-gel B (mortar with a coating containing 1% of nano-$TiO_2$ supported on sepiolite). The task was performed outdoor by an operator using conventional PPEs (mask, gloves and glasses).

In this scenario the $TiO_2$ concentration was not measured because it was expected to find very low concentration of $TiO_2$ due to the short duration of the tasks and low percentage of nano-objects on the materials. The process was monitored using on-line devices to characterize the release of particles.

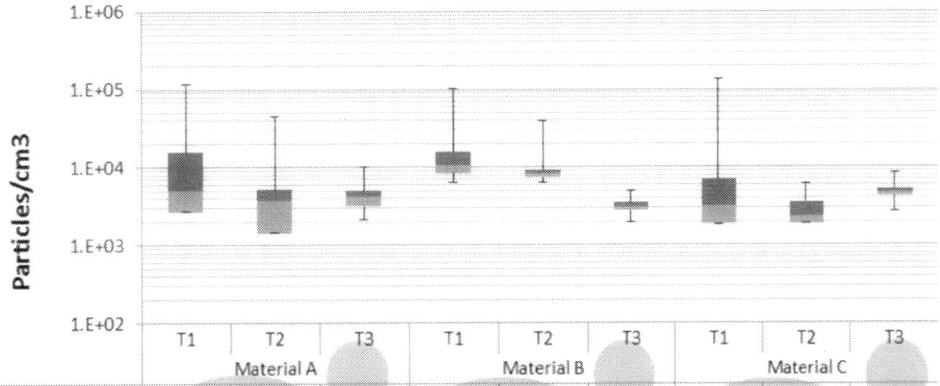


| | Material A T1 | Material A T2 | Material A T3 | Material B T1 | Material B T2 | Material B T3 | Material C T1 | Material C T2 | Material C T3 |
|---|---|---|---|---|---|---|---|---|---|
| 8-hr TWA (mg/m3) | | 0.003 | 0.009 | | 0.016 | 0.017 | | 0.043 | 0.055 |
| TiO2-Conctr. (mg/m3) | | 0.003 | 0.009 | | 0.016 | 0.017 | | 0.043 | 0.055 |
| Sampling Time (min) | | 41 | 10 | | 44 | 6 | | 24 | 7 |
| Mean (particles/cm3) | 14098 | 3725 | 4388 | 15598 | 8467 | 3183 | 14098 | 3725 | 4834 |

**Figure 3.** Main data measured in scenario 3. Boxplot, background-corrected average number of particles/cm$^3$ during the tasks (data from CPC3007).

Data showed that the mean particle concentration was between 1.68E+4 - 5.59E+4 particles/cm$^3$ for all the different materials with maximum values up to 2.38E+5 particles/cm$^3$. It is important to highlight that the SEM analysis of the samples taking during the machining of materials filled with nano-$TiO_2$ did not showed particles of nano-$TiO_2$ free at the PBZ.

*Scenario 6: Demolition*

Two cabins were demolished; cab A was covered with mortar containing 2% of nano-SiO2 (Meyco 685) and cab B was covered with same quantity of mortar (304 kg) containing 2% of additive of nano-$TiO_2$ supported on sepiolite. The process was performed outdoors and two operators were involved: the one in the excavator and a second worker situated near the demolition area (4.5 m from the cab), both wearing conventional PPEs (see figure 4).

The results showed that the $TiO_2$ concentration during the demolition of cabin B was 0.008 and 0.009 mg/m3 for both operators. The measurement of particle concentration showed values of mean particle concentration between 1.02E+4-2.32E+4 for both cabins with maximum value up to 3.70E+5 particles/cm$^3$.

**Figure 4.** Pictures taken during the demolition of the cabs in scenario 6.

**Table 2.** Summary of measurements by scenario/tasks.

| | **Particle concentration (particles/cm$^3$) Data from CPC3007 (BG corrected)** | | | **Occupational Exposure** | | |
|---|---|---|---|---|---|---|
| **Scenario/Tasks** | *Mean* | *Max* | *Min* | *Time (min)* | *$TiO_2$ conc. (mg/m$^3$)* | *8 h-TWA (mg/m$^3$)* |
| **ES1. $TiO_2$ manufacturing** | | | | | | |
| T1. Reaction | 1.06E+04 | 2.48E+04 | 3.37E+03 | 161 | 0.007 | |
| T2. Beating | 4.86E+03 | 8.27E+03 | 2.87E+03 | | | |
| T3. Collection | 3.85E+03 | 5.48E+03 | 1.55E+03 | | | |
| T4. Transferring | 1.33E+04 | 1.97E+04 | 6.54E+03 | | | |
| T5 Facility cleaning | 9.95E+03 | 1.20E+05* | 5.67E+02 | 62 | 0.388 | 0.048 |
| **ES2. Mortar manufact. A, B & C (1)** | | | | | | |
| T1. Weighing | 9.51E+03 | 1.86E+03 | 5.72E+03 | | 0.286 | 0.593 |
| T2. Adding to the hopper | 5.69E+04 | 1.38E+05* | 5.00E+03 | | 0.067 | 0.796 |
| T3. Mortar bagging | 2.54E+04 | 6.11E+04 | 1.77E+04 | | | |
| All tasks | | | | 57-59 | 0.008-0.073 | 0.008-0.073 |
| **ES3:Mortar application A, B & C (1)** | | | | | | |
| T1. Adding and mixing | 1.56E+04 | 1.34E+05* | 6.28E+03 | | | |
| T2. Mortar application | 8.47E+03 | 4.58E+04 | 6.44E+03 | | | |
| T3. Scrapping | 4.83E+03 | 9.99E+03 | 2.68E+03 | | | |
| All tasks | | | | 7-44 | 0.003-0.055 | 0.003-0.055 |
| **ES4: Sol-Gel spraying** | | | | | | |
| Sol-gel A | 2.28E+04 | 2.57E+05* | 0.00E+00 | 13 | 0.069 | 0.069 |
| Sol-gel B | 2.18E+04 | 2.45E+05* | 0.00E+00 | 15 | 0.195 | 0.195 |
| **ES5: Machining (drilling)** | | | | | | |
| Mortar A | 3.01E+04 | 1.76E+05* | 0.00E+00 | | | |
| Mortar B | 2.66E+04 | 2.17E+05* | 0.00E+00 | | | |
| Mortar C | 1.68E+04 | 1.79E+05* | 6.46E+03 | | | |
| Sol-gel A | 3.25E+04 | 1.79E+05* | 3.72E+03 | | | |
| Sol-gel B | 5.59E+04 | 2.38E+05* | 9.72E+03 | | | |
| **ES6: Demolition** | | | | | | |
| Cabin A | 2.32E+04 | 3.70E+05* | 1.08E+04 | | | |
| Cabin B | 1.02E+04 | 3.01E+05* | 0.00E+00 | 37 | 0.009 | 0.009 |

(*) Data above the upper limit of CPC3007; (1) Maximum values measured for mat.A, B and C

## 4. Conclusions

The following main conclusions can be highlighted from this work:

- The occupational exposure to nano-$TiO_2$ is below the OELs for all scenario measured, considering both limits, 0.1 mg/m$^3$ (SCAFFOLD, Stockmann-Juvala, 2014) and 0.3 mg/m$^3$ (NIOSH, 2011).
- The highest mass concentrations of nano-$TiO_2$ (mg/m3) were found during the cleaning task in the nano-$TiO_2$ manufacturing (scenario 1), and during the sol-gel spraying of depollutant on a wall (scenario 4).
- Mean total particle concentration released (particles/cm$^3$) was above NRV during two tasks: machining of depollutant mortar (scenario 5) and when adding additives to the hopper (while manufacturing depollutant mortar, scenario 2). However, these tasks are quite short (in time)

and concentration should be normalized to 8 h day- 40 h week to be compared with exposure limits (e.g. NRV).

- Data on total particle concentration released (particles/cm$^3$) showed no sticking differences using conventional materials (control) and materials filled with nano-objects.
- Results on mass concentration and on number concentration are inconsistent. More research on this issue is needed to achieve a consensus.

These findings contribute to clarify the discussion about whether the incorporation of nano-objects to the construction industry may originate an increase on the risks for workers in this sector.

## 5. References

[1] FIEC, IVAM, EFBWW (2009) Nano-products in the Europe construction Industry. http://www.efbww.org/pdfs/Nano%20-%20GB%20Summary.pdf

[2] Clark K. et al 2012 Limitations and information needs for engineered nanomaterial-specific exposure estimation and scenarios: recommendations for improved reporting practices *J Nanopart Res (2012) 14:970*

[3] Van Broekhuizen et al 2011, Use of nanomaterials in the European construction industry and some occupational health aspects thereof *J Nanopart Res DOI 10.1007/s11051-010-0195-9.*

[4] NIOSH bulletin 63 (2011)- Occupational Exposure to Titanium Dioxide

[5] Asbach et al (2012). NanoGEM Tiered approach for the assessment of exposure to airborne nanoojects in work-places.

[6] Kuhlbusch et al. 2011 Nanoparticle exposure at nanotechnology workplaces: A review *Particle and Fibre Toxicology 2011, 8:22*

[7] Stockmann-Juvala H. et al, 2014 Risk management in the construction industry – derivation of occupational exposure limit values for nanomaterials *NanoTox 2014 Conference, April 23-26 Turkey.*

[8] IFA (2014a) Criteria for assessment of the effectiveness of protective measures. Available at: http://www.dguv.de/ifa/Fachinfos/Nanopartikel-am-Arbeitsplatz/Beurteilung-von-Schutzmaßnahmen/index-2.jsp (sited 17.2.2014).

[9] SER (2012) Provisional nano reference values for engineered nanomaterials. Advisory report 12/01. The Netherlands. Socail and Economic Council. http://www.ser.nl/~/media/Files/Internet/Talen/Engels/2012/2012_01/2012_01.ashx.

Journal of Physics: Conference Series **617** (2015) 012007 doi:10.1088/1742-6596/617/1/012007

# Strategy for the lowering and the assessment of exposure to nanoparticles at workspace – Case of study concerning the potential emission of nanoparticles of Lead in an epitaxy laboratory

**Sébastien Artous[1], Eric Zimmermann[1], Paul-Antoine Douissard[2], Dominique Locatelli[1], Sylvie Motellier[1], Samir Derrough[1]**

[1] Univ. Grenoble Alpes, F-38000 Grenoble, France
CEA-Grenoble, NanoSafety Platform, F-38054 Grenoble, France

[2] ESRF, F-38043, Grenoble, France

E-mail: sebastien.artous@cea.fr

**Abstract.** The implementation in many products of manufactured nanoparticles is growing fast and raises new questions. For this purpose, the CEA - NanoSafety Platform is developing various research topics for health and safety, environment and nanoparticles exposure in professional activities. The containment optimisation for the exposition lowering, then the exposure assessment to nanoparticles is a strategy for safety improvement at workplace and workspace. The lowering step consists in an optimisation of dynamic and static containment at workplace and/or workspace. Generally, the exposure risk due to the presence of nanoparticles substances does not allow modifying the parameters of containment at workplace and/or workspace. Therefore, gaseous or nanoparticulate tracers are used to evaluate performances of containment. Using a tracer allows to modify safely the parameters of the dynamic containment (ventilation, flow, speed) and to study several configurations of static containment. Moreover, a tracer allows simulating accidental or incidental situation. As a result, a safety procedure can be written more easily in order to manage this type of situation. The step of measurement and characterization of aerosols can therefore be used to assess the exposition at workplace and workspace. The case of study, aim of this paper, concerns the potential emission of Lead nanoparticles at the exhaust of a furnace in an epitaxy laboratory. The use of Helium tracer to evaluate the performance of containment is firstly studied. Secondly, the exposure assessment is characterised in accordance with the French guide "recommendations for characterizing potential emissions and exposure to aerosols released from nanomaterials in workplace operations". Thirdly the aerosols are sampled, on several places, using collection membranes to try to detect traces of Lead in air.

## 1. Introduction

Nanoparticles are present in various and numerous fields and activities (medicine, cosmetology, building materials, …). As a result, the number of workers in contact with nanoparticles increases. As collective protections are the first one to be implemented, before individual protections, it becomes necessary to be able to test their efficiency for nanoparticles.

The CEA - NanoSafety Platform (PNS) located in Grenoble works on all matters of safety and security related to the handling of nanomaterials. It conducts in parallel actions of R & D and services missions of measurement and characterisation, expertise, training and intervention in accidental situations. It employs nearly 150 professionals, researchers and experts, whose jobs are to provide a strong support to researchers and partners using nanotechnology, while maintaining the highest level of knowledge and skills in the field of nanosafety.

For this purpose the NanoSafety Platform is developing various research topics for health and safety, environment and nanoparticles exposure in professional activities. The optimisation of containment for lowering the exposition, then the assessment of exposure to nanoparticles is a strategy for safety improvement at workplace and workspace. The lowering step consists in an optimisation of dynamic and static containment at workplace and/or workspace. Generally, the exposure risk due to

Journal of Physics: Conference Series **617** (2015) 012007 doi:10.1088/1742-6596/617/1/012007

the presence of substances containing nanoparticles does not allow modifying the parameters of containment at workplace and/or workspace. Therefore, gaseous or nanoparticulate tracers are used to evaluate performances of containment. Using a tracer allows to modify safely the parameters of the dynamic containment (ventilation, flow, speed) and to study several configurations of static containment. Moreover, a tracer allows simulating accidental or incidental situation. As a result, a security procedure can be written more easily in order to manage this type of situation. The step of measurement and characterization of aerosols can therefore be used to assess the exposition at workplace and workspace.

The case of study, aim of this paper, concerns the potential emission of Lead nanoparticles at the exhaust of a furnace in an epitaxy laboratory: The laboratory makes Lead crystal growth in furnaces. The use of Helium tracer to evaluate the performance of containment is firstly studied. The efficiency of two extraction systems, located at the exhaust of an epitaxy furnace, is evaluated. In addition, the influence of a refractory component, placed at the exhaust of the furnace, to canalise the Lead vapour is studied.

Secondly, the exposure assessment is characterised in accordance with the French guide "recommendations for characterizing potential emissions and exposure to aerosols released from nanomaterials in workplace operations". These recommendations should be proposed to international regulation. The characterization of aerosols includes measurements of the concentration using condensation particle counters and measurements of the size distribution based on electrical mobility of particles.

Thirdly the aerosols are sampled, on several places, using collection membranes to try to detect traces of Lead in air by atomic absorption and X-Ray Fluorescence (XRF).

## 2. 2.1. Characterization of containment

*2.1. Characterization of containment by a gaseous tracer – leak detection*

The implementation of an air blowing associated to an extraction of air with an HEPA filter is the classical configuration of a lab where nanoparticles are handled. This configuration enables to maintaining the pressure in the laboratory lower than the pressure outside to contain inside the lab a potential contamination of nanoparticles. In addition, close to the source of nanoparticles, an extracting system is usually used.

The evaluation principle of the performance of containment is based on the diffusion of a tracer injected at a constant flow rate instead of the source of nanoparticles (Figure 1). An analyser of leak allows detecting the tracer in several areas to determine if a diffusion of nanoparticles in the laboratory might occur in a real configuration. Representative results were obtained with a gaseous tracer when the aeraulic parameters (air blowing, extracting air of room and process, temperature of furnace…) are the same that when the source of nanoparticles is handled. This characterisation is based on the hypothesis that the transport of nanoparticles is principally due to the turbulent diffusion and that the concentration of gaseous tracer in the room is low. Helium tracer is used in this study (see paragraph 3.1.1). Using a gaseous tracer instead of a nanoparticles source enables to optimising aeraulic parameters in complete safety.

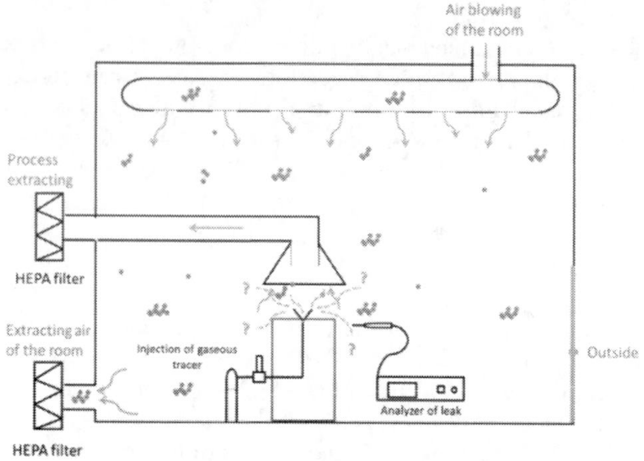


**Figure 1.** Characterisation principle of the containment in a laboratory where nanoparticles are handled

*2.2. Containment characterisation by real-time measurement of potential aerosol emission and by aerosol sampling*

The methodology, first validated by the NanoSafety Platform (PNS), has been deployed at a large scale since the end of 2009 (up 270 workstations have been measured and analysed up to date).

The goals of an aerosols measurement are the identification, the characterization and the assessment of nanosize and microsize aerosols potentially generated, intentionally or unintentionally, during the normal working phases (studied size range: 5nm – 32µm).

At this stage, there is no real consensus on the most representative unit of measure to assess the exposure of workers. The evaluation principle of the exposure in a located area is to use a sampling tube connected to the instruments measuring both concentration and distribution (Figure 2). Thus, the equipments used in this study are detailed in paragraphs 3.1.2 [2]. Nevertheless these equipment cannot detect specifically the Lead.

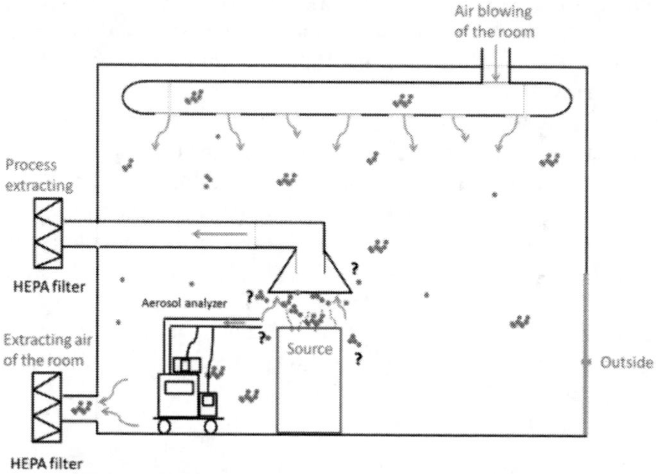


**Figure 2.** Characterisation principle of the exposure

Besides the aerosol measurement, the aerosols are sampled at several places, using collection membranes to detect potential traces of Lead in air by atomic absorption and X-Ray Fluorescence (XRF), see paragraph 3.1.3.

## 3. Materials and method

### *3.1. Method*

#### *3.1.1. The Helium tracer*

The injection of Helium is controlled by a mass flow meter ANALYT MTC-358 (Figure 3-a). The injection flow is 1 L/min at 20°C and is located in the centre of the furnace for simulating an emission of Lead. The concentration of tracer gas is measured by a mass spectrometer Adixen - type ASM 102S. The sampling point is moved into several areas according to the method presented at the paragraph 2.1.

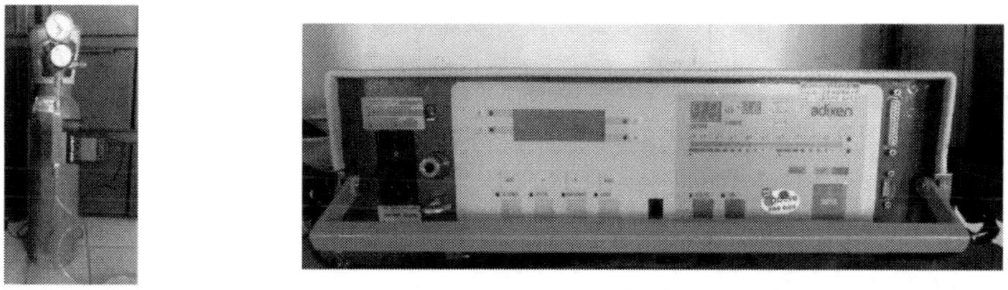


a b

**Figure 3.** Mass flow meter and mass spectrometer

#### *3.1.2. The equipment for the aerosols measurement*

The particles concentration is determined by a Condensation Particle Counters (CPC) 3785 of TSI. The size range of particles detected is from 5nm to 3μm. If an emission of particles is detected when analyzing the CPC results, the next step is the determination of the size distribution of the aerosol emission. A FMPS (model 3091 - TSI) is used in this study. The size range of particles detected is from 5.6nm to 560nm. In addition an optical counter (Dustmonitor model 1.109 – GRIMM) is used to obtain the distribution of submicronic and micrometric particles (range from 250nm to 32μm).

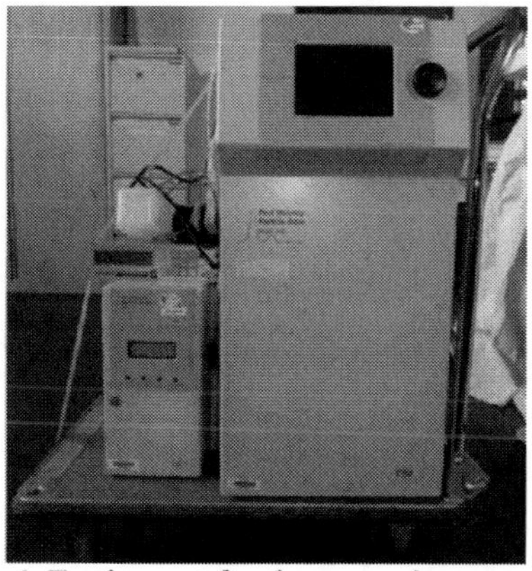

**Figure 4.** Equipment for the aerosols measurement

*3.1.3. The equipment for the determination of Lead*

i. XRF Characterization

X-Ray Fluorescence (XRF) is used for the characterisation of the samples made at workplace and/or workspace. This analytical method is a non-destructive method. The XRF analyses are realised on a Rigaku spectrometer of NanoHunter bench-top apparatus type. The XRF analysis is a surface characterisation technique where a beam of primary X-rays hits a sample with a very small incidence angle (typically 1mrad). The depth of penetration of primary X-rays being very small, the secondary X-ray photons are transmitted through a depth of about 3nm to 5nm. This enables a chemical characterisation of the constitutive elements of the particles which are at the surface of the sample.

ii. ICP-MS

Inductively coupled plasma mass spectrometry (ICP-MS) is used for the characterization of the samples which have been realised at workplace and/or workspace. This analytical method is a destructive method. ICP-MS analysis are realised on an Agilent Technologies 7700 Series. Before analysis the membranes of collection are mineralised by concentrated $HNO_3$.

*3.2. Epitaxy laboratory – Asterix furnace*

The Figure 5 describes the Epitaxy laboratory. Three furnaces are located in the laboratory, Asterix, Solarix and Obelix. For this study only Asterix and Solarix furnaces work. In this paper, only the results concerning Asterix furnace are given. The aeraulic system of the laboratory is made up by three air intakes associated to an air extraction wall with a flow of 1,600 $m^3/h$. Moreover, a dynamic containment is created at the exhaust of each furnace by an extracting system. Two types of extraction systems, an extracting box (Figure 6-a) and an extracting slot (Figure 6-b) located at the exhaust of Asterix furnace, are evaluated with the Helium tracer. The Figure 7-a presents the injection of Helium located in the centre of the furnace to simulate a potential emission of lead. Afterwards, the measurement of a potential aerosol emission is conducted. The Figure 7-b presents an example of localisation of the sampling tube for the aerosol measurement. Finally three sampling are performed on filters exposed for 7 days to the laboratory atmosphere (Figure 5).

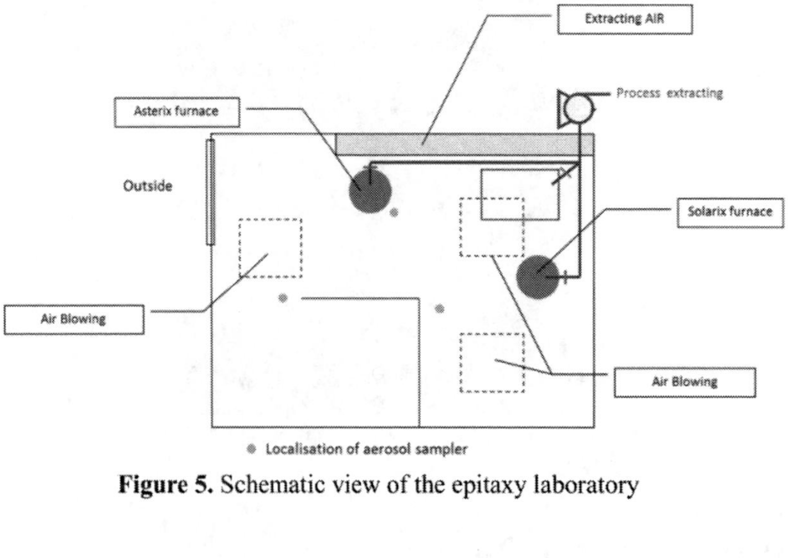


**Figure 5.** Schematic view of the epitaxy laboratory

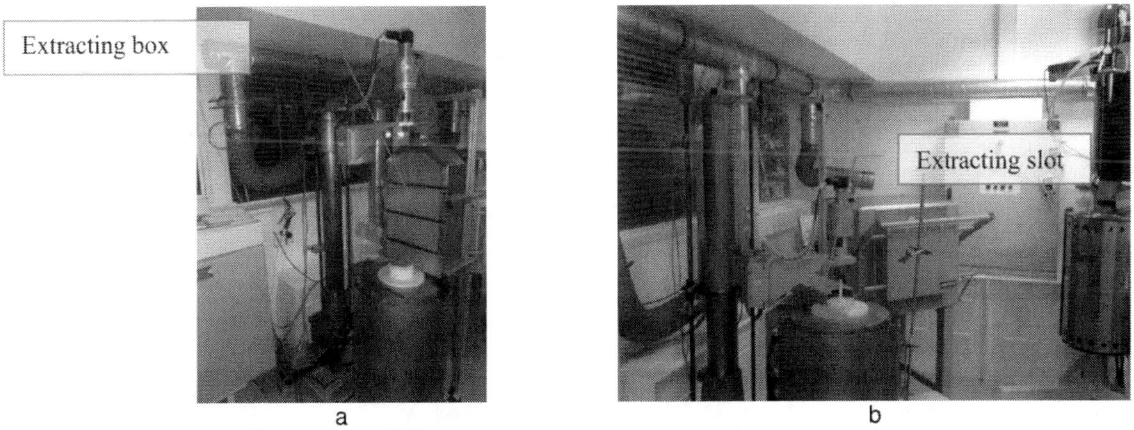


**Figure 6.** Extraction systems of Asterix furnace

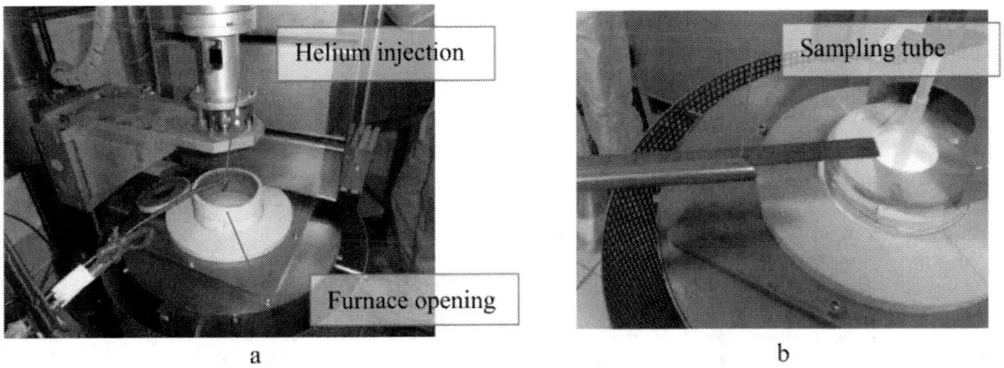


**Figure 7.** Localisation of Helium injection and one example of localisation of sampling tube

## 4. Results

### *4.1. Containment characterisation by a gaseous tracer – leak detection*

Figure 8 shows the evolution of the helium concentration in the area of the Asterix furnace with the extracting box and for a fan extraction speed of 32Hz. A significant concentration of Helium is recorded when the probe is located: at the opposite of the extraction air; between the extraction air and the opening of the furnace; at the back of the furnace; 10cm above the furnace and also at the upper part of the agitator of the furnace. Using the extracting box creates a large area of extraction where Helium is significantly measured. In a normal operation of the furnace some Lead nanoparticles could deposit at the surface of this area. The fan extraction speed increase to 45Hz to increase the extraction flow does not reflect into a significant decrease of the helium concentration for the same measurement locations. Then the fan extraction speed is gradually decreased. The fan extraction speed where significant helium concentration is recorded in the laboratory corresponds to a loss of dynamic containment. This loss of dynamic containment is recorded for a variation of the fan below 15Hz

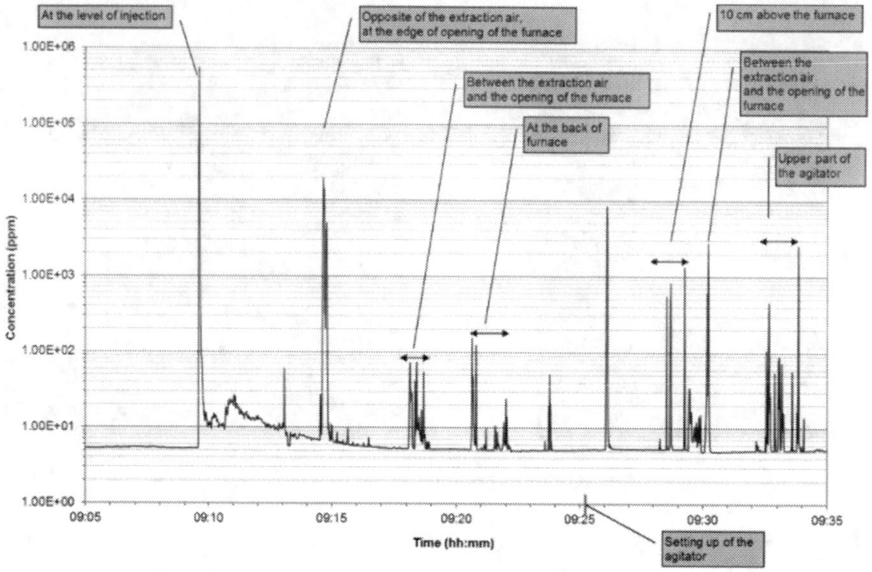


**Figure 8.** Helium concentration with the extracting box for a 32Hz fan extraction speed

Afterward, the evolution of the Helium concentration in the area of the Asterix furnace with the extracting slot for several fan extraction speed is studied (Figure 9). No significant Helium concentration is recorded apart from the measurement realized above the opening of the furnace and at the level of the extracting slot. The concentrations recorded at the opening of the furnace are 10 times lower than the concentrations recorded in the configuration with the extracting box.

A test of a gradual decreasing of the fan extraction speed is conducted to evaluate the los of dynamic containment. This limit is reached when the Helium is extracted by the air laboratory extracting (extraction wall) and not by the air process extracting. With the extracting slot, this limit is around a fan extraction speed of 5Hz.

As a result, the characterisation of a potential aerosol emission will be realised only in the configuration with the extracting slot.

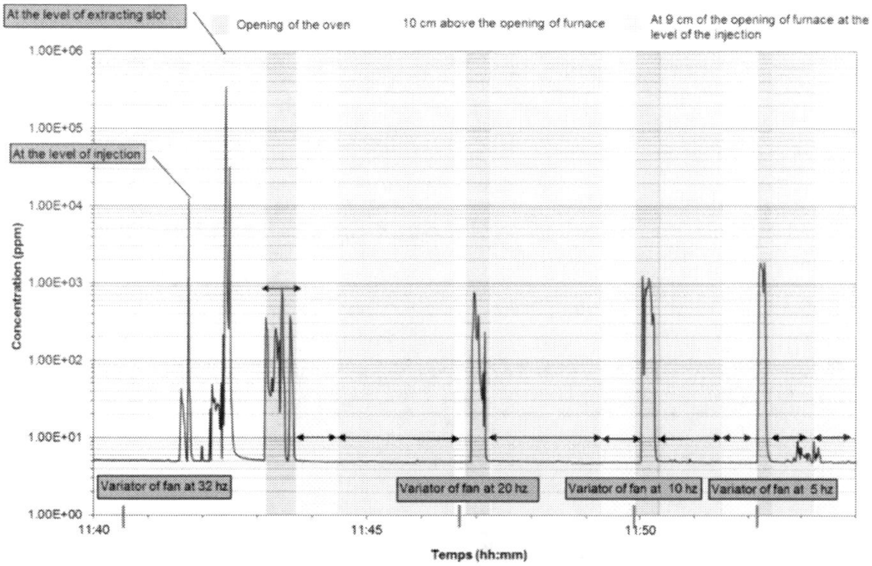


**Figure 9.** Concentration of Helium with the extracting slot for several speed variations of the fan

### *4.1. Characterisation by the measurement of a potential aerosol emission*

Figure 10 shows the evolution of the particles concentration in the area of the Asterix furnace with the extracting slot and for several fan extraction speed. The blue curve shows the evolution of the particles concentration from 5nm to 3µm and the red curve shows the evolution of the particles concentration from 250nm to 32µm. For a fan extraction speed of 32Hz, no significant nanoparticles emission is recorded apart from the measurement realised between the opening of the furnace and a lid. An increase of 1,500 p/cm$^3$ is recorded. Figure 11 shows the distribution at the emission point and in the aerosol background. The emission is characterised by a significant increase of the nanoparticles number around 10nm.

This increase is probably due to the turbulent movement of air caused by the high speed of extracting air associated with a lid. Indeed, this lid was not present at the step of characterisation with a gaseous tracer (see the difference between Figure 7-a and Figure 7-b). This result underlines that the characterisation with the gaseous tracer has to be realised strictly in the same configuration as the configuration where nanoparticles are handled. However, this increase is not recorded for a fan extraction speed at 25Hz or 20Hz. For these two fan extraction speeds no significant nanoparticles emission is recorded.

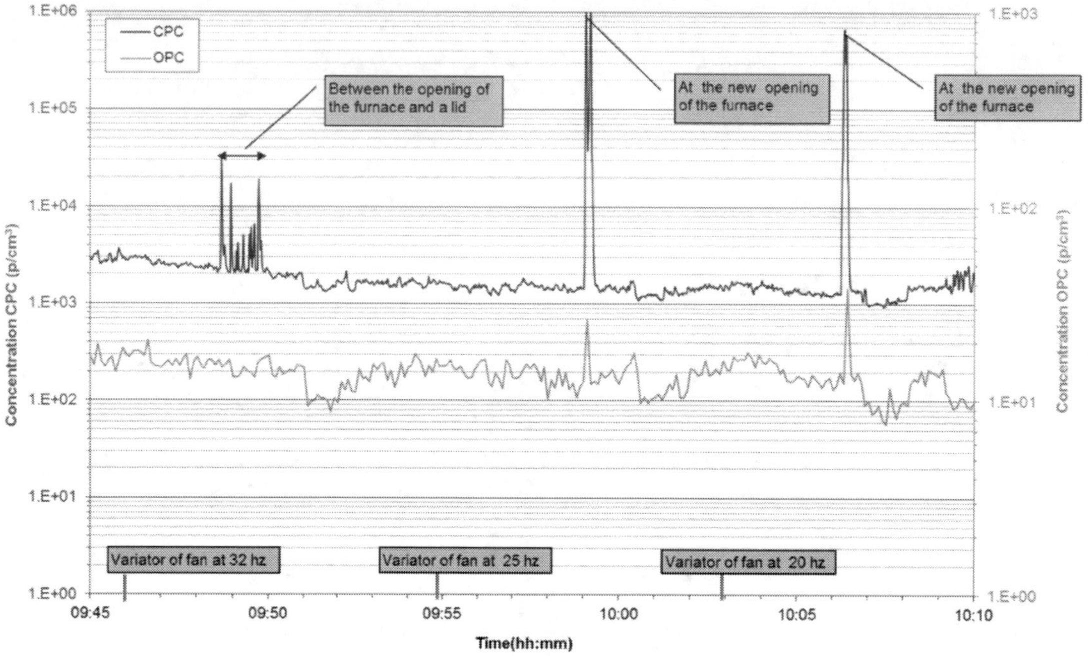


**Figure 10.** Concentration in particles number with the extracting slot for several fan speeds

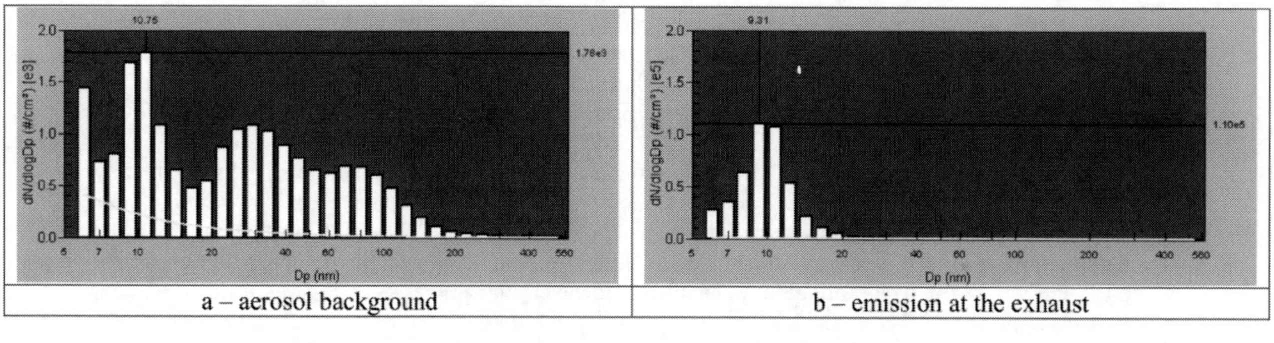


**Figure 11.** Distribution in aerosol background (a) and during an emission at the level of the exhaust of the epitaxy furnace (b)

### *4.2. Aerosol sampling*

The last step of characterisation is the sampling step. The sampling step is conducted for the furnaces with the extraction slot and for a fan speed at 20Hz.

Three collection membranes are used to seek traces of Lead in air. The sampling lasts 7 days with a sampling flow between 3 L/min and 7 L/min. The analysis by X-Ray Fluorescence (XRF) does not highlight the presence of Lead on membrane. Then Analysis by ICP-MS are also conducted. Like the XRF, the ICP-MS does not highlight the presence of Lead on collection membranes. The detection limit of analysis by atomic absorption is 1 ppb. The detection limit resulting for the concentration of Lead in the laboratory is evaluated at 15ng/m$^3$.

## 5. Conclusion

This study shows the strategy for the lowering and the assessment of exposure to nanoparticles at workspace and workplace. The step of containment optimisation with a gaseous tracer determines the success of the lowering of the exposition. In this case of study, using a gaseous tracer enable to optimise the aeraulic parameters with the selection between two types of extraction system (box or slot) for the extracting process and with the optimisation of the fan extraction speed to select the lowest speed as possible. This last parameter is important for the growth process of single crystal films, in maintaining a safer environment.

.

The study underlines that the characterisation with the gaseous tracer to represent the behaviour of aerosol and nanoparticles has to be realised strictly in the same configuration (aeraulic and geometric) as the configuration where nanoparticles are handled. In this case of study, regardless of the method used, no lead was found in the air of epitaxy laboratory.

The three steps of characterisation with using a gaseous tracer with a measurement of aerosol and with a sampling on membrane, ensures a safer environment at the workplace and workspace.

## References

[1] Durand, C., Witschger, O., Le Bihan, O., Reynier, M., Marchetto, A,. Zimmermann, E., Charpentier, D. "French approach for characterizing potential emissions and exposure to aerosols released from nanomaterials in workplace operations", Nanosafe 2012

[2] Zimmermann, E., S. Derrough, D. Locatelli, C. Durand, J. L. Fromaget, E. Lefranc, X. Ravanel, and J. Garrione. "Results of Potential Exposure Assessments during the Maintenance and Cleanout of Deposition Equipment." Journal of Nanoparticle Research 14, no. 10 (October 1, 2012): 1–17. doi:10.1007/s11051-012-1209-6.

[3] Août 2004, French standard, Sorbonnes - Partie 3 : méthodes d'essai de type, NF EN 14175-3, AFNOR

[4] Mai 1972, French standard, Méthode de mesure de l'efficacité des filtres au moyen d'un aérosol d'uranine (Fluorescéine), NF X44-011, AFNOR

Journal of Physics: Conference Series **617** (2015) 012008 doi:10.1088/1742-6596/617/1/012008

# Development of exposure assessment method with the chamber

**N Kato, Y Koyama, H Yokoyama, Y Matsui and M Yoneda**
Kyoto University, Katsura, Nishikyo-ku, Kyoto, 615-8140, Japan

E-mail: Katoh@risk.env.kyoto-u.ac.jp

**Abstract.** This study aims at developing the measurement method of nanoparticle concentration and at getting a representative value of nanoparticle uniform concentration due to chamber ventilation. We conducted a chamber equipped with HEPA filter and control the background nanoparticles concentration by using an adequate ventilation. Then, we used generator to evaluate concentration in the chamber uniformity. We measured background value and source counts at the particle size distribution by SMPS. In addition, we performed numerical analysis with CFD model OpenFoam. As results, we found that there is no aggregate in experimental conditions in this study. Though we confirmed that it is difficult to uniformalise nanoparticle concentration , However we also found simulation results showed higher reproducibility. Therefore, we could assess nanoparticle size distribution and concentration in our chamber at this stage.

## 1. Introduction

Nowadays, the nanomaterials are used in various consumer products [1]. A definition of nanoparticles is different from each country. However, those diameters are approximately less than 100 nm. It was reported that impacts on the health and safety depend on the use of a surface treatment and on the diameter of nanoparticles (NPs) [2-3]. One of the exposure routes through human is to breathe airborne NPs. Products containing NPs release various types of NPs (shape, particle diameter and so on).

Therefore, exposure assessment is difficult to define for products emitting NPs [4]. It is very critical issue to estimate NPs exposure when using products containing NPs in general indoor environments as well as in workplaces [5], and it is necessary to standardize an effective and easy to use exposure assessment method of products containing NPs.

This study aims at developing the exposure assessment method for products containing nanomaterial. Therefore we get a representative value with nanoparticle uniform concentration due to chamber ventilation. We conducted a chamber equipped with HEPA filter and control the background nanoparticles concentration by using an adequate ventilation in the chamber. Then, they become steady state. Thereby, we can obtain the representative value of NPs concentration in the chamber and make an exposure assessment. There are measurement methods which evaluates the quantity of volatile organic chemical substances (VOC) in daily necessities, such as formaldehyde [6-8]. We applied these methods to measure a concentration of NPs. If a ventilation speed is controlled adequately, it is expected that we can measure airborne NPs exposure due to a homogeneous dispersion without aggregate and anybody can measure nanoparticle concentration with ease.

Journal of Physics: Conference Series **617** (2015) 012008 doi:10.1088/1742-6596/617/1/012008

## 2. Methodology

### *2.1. Construction of the chamber*

At first, we show outline map of exposure assessment method (Figure 1).

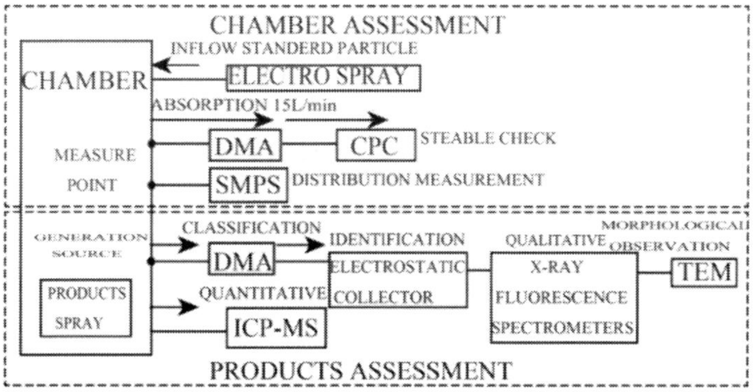


**Figure 1.** Chamber Method Outline Map.

We constructed the chamber equipped with the ventilation (Figure 2). The flames of chamber are stainless steel and inner surfaces are covered with insulating films. There are working glove to work in the sealed state with the front door, a power outlet in chamber and upper window with HEPA filter. Whole inner volume of chamber is ventilated from exhaust pipes on both sides to the upper duct by the airflow for every 12min. In addition, 16 ports for NPs concentration measurement on the right side (Figure 3).

.

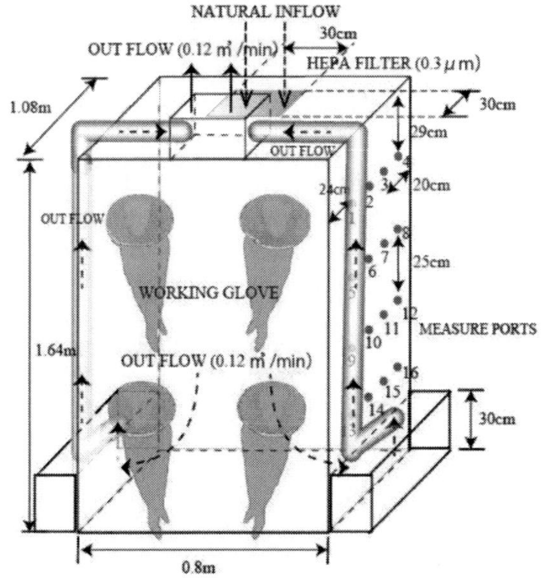


**Figure 2.** Chamber with ventilation and working glove.

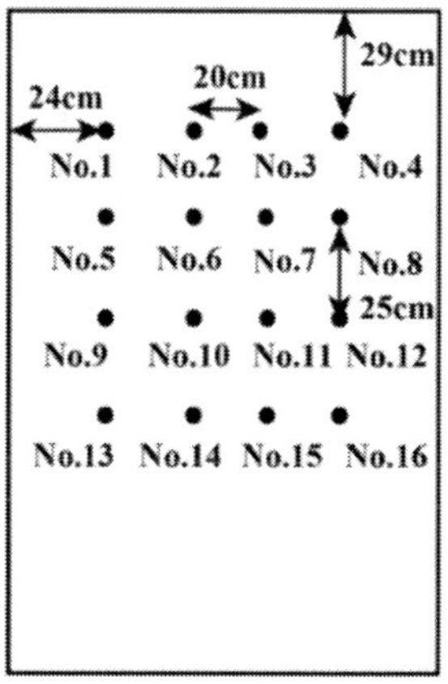


**Figure 3.** Chamber on the right side.

Secondly, we evaluated the chamber constructed, then, we practiced exposure assessment of products containing NPs (Figure 1). This study specialized in part of chamber assessment with measurement by devices and numerical analysis.

*2.2. Assessment of NPs concentration in the chamber by measurement*

We ventilated in the chamber sufficiently until background become low. Next, we measured background value of airborne particles at the NPs size distribution with Scanning Mobility Particle Sizer (SMPS, Model 3910, TSI Inc.) from ports on the right side for a few minutes. Then, we generated standard particles (size 70nm) made from sucrose for assessment with the generator (Model 3480, TSI Inc.) continuously at 12cm distance from port No.7 of the right wall (Figure 3), and checked stable condition at particle size distribution with Condensation Particle Counter (CPC, Model 3775, TSI Inc.). After we confirmed that the value is stable, we measured particle size distribution at 24cm distance from each port in rotation with SMPS for a few minutes every minute.

*2.3. Assessment NPs distribution in the chamber by numerical analysis*

We performed numerical analysis by CFD model (OpenFoam [9]). Governing equation shows (1-2). Calculation conditions in this study were incompressible, and a constant temperature is 298[K] at the measurement condition. We show a calculation condition at Table 1.
On both sides, we made 16 cell outlet which evacuated 0.45[m/s] continuously and inlet on the top of chamber is pressure gradient is zero. We calculated airflow in conditions.

$$\nabla \cdot \vec{U} = 0 \tag{1}$$

$$\frac{D\vec{U}}{Dt} = -\frac{\nabla p}{\rho} + \nu[\nabla^2 \vec{U}] \tag{2}$$

**Table 1.** Calculation conditions.

| head | Conditions |
|---|---|
| Calculation area | 108×102×164 [grid] |
| Discrete method | Finite volume method |
| Initial condition | U=0 [m/s](air velocity) p/ρ=0 |
| Boundary condition | U=0.45 [m/s] Pressure gradient = 0 |
| Calculation time | 20 [minute] |
| Temperature | Constant 298 [k] |
| Kinematic viscosity | 0.000015 [m2/s] |

After calculating for 20 minutes (flow in the chamber was stable about 10 minutes), particle tracers were generated at the same position with generator. We obtain a trajectory of particle tracer in the stable flow then we confirm whether we get the representative value.

## 3. Results

Background concentration showed much low level by the chamber ventilation (Figure 5). The peak of generating particle size distribution is mainly 48.7-64.9nm (Figure 6), and approximately maintained at each port measure data (Figure 7-9). Based on those results, there was no aggregate on this concentration.
However, the concentration was higher in the bottom of Chamber (Figure 9-10). Results of simulation also showed the same tendency which measure by SMPS (Figure 12). The airflow was much weakest in the chamber (Figure 11-12).We obtained results that the distribution of particles generated from

Port No.7 was not uniform, that is rotating in the chamber (Figure 11).

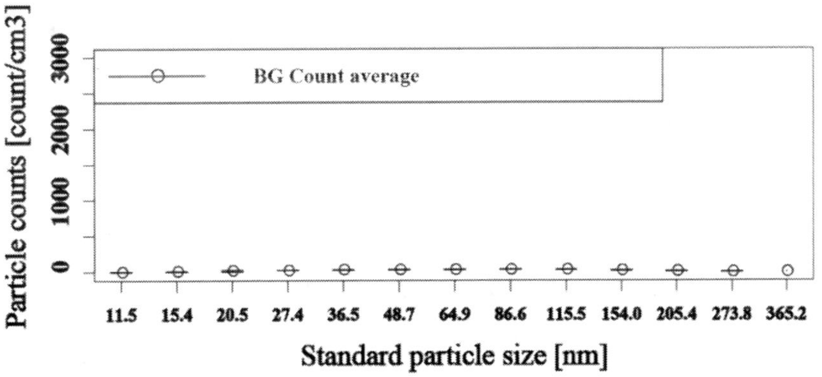


**Figure 5.** Mean and 95% confidence intervals for particle counts at background (n=5).

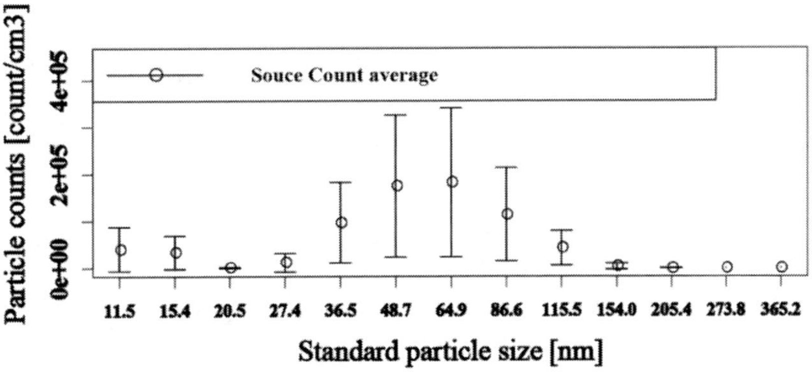


**Figure 6.** Mean and 95% confidence intervals for standard particle counts into port No.7 (n=3).

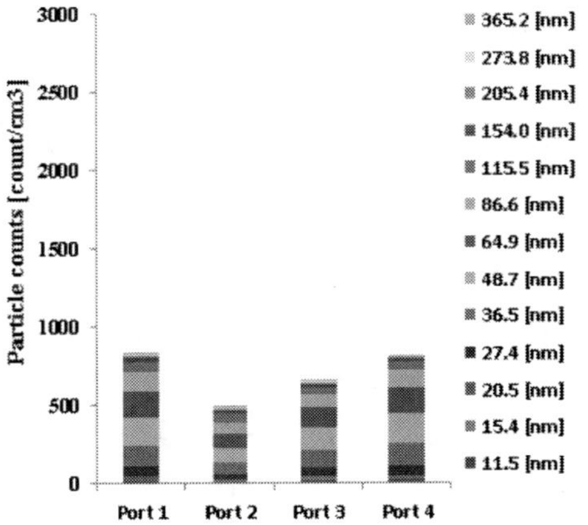


**Figure 7.** Particle size distribution at Port No.1-4 by SMPS.

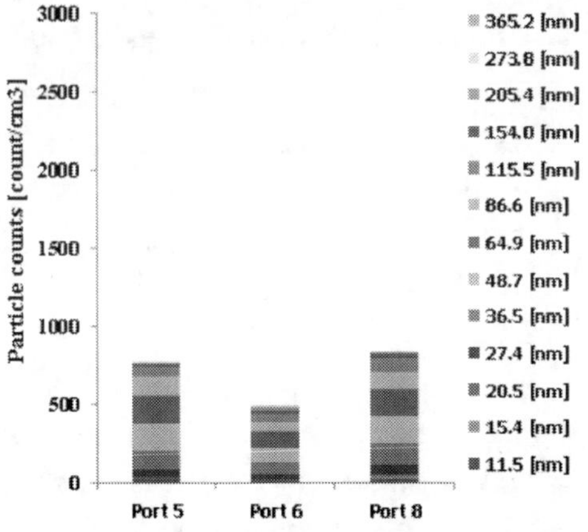


**Figure 8.** Particle size distribution at Port No.5-8 by SMPS.

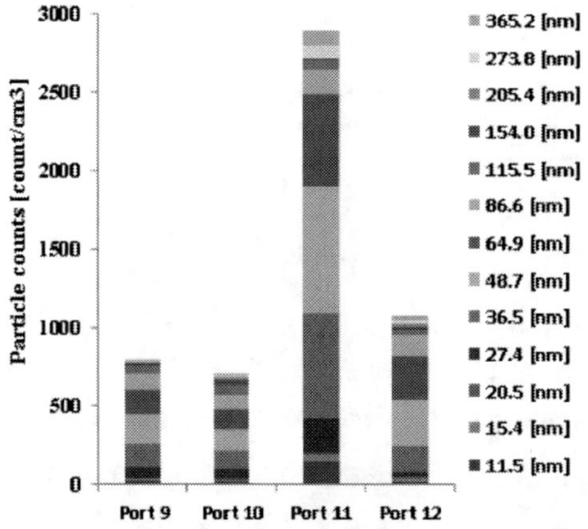


**Figure 9.** Particle size distribution at Port No.9-12 by SMPS

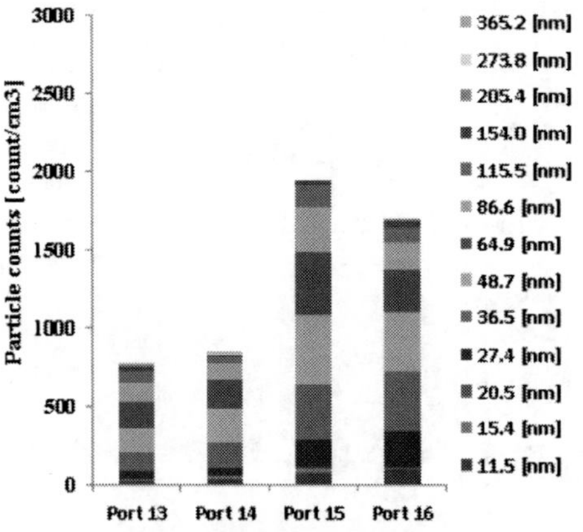


**Figure 10.** Particle size distribution at Port No.13-16 by SMPS.

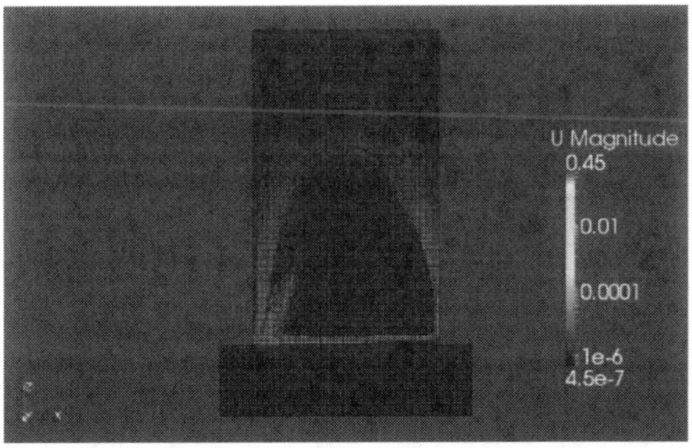


**Figure 11.** Particles trajectory along air advection in front of the chamber.

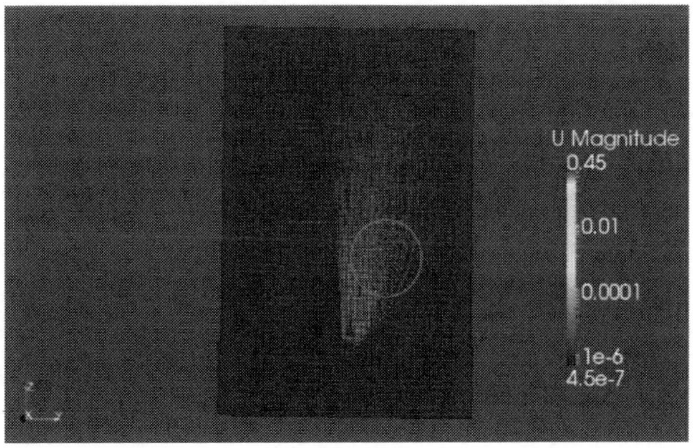


**Figure 12.** Particles trajectory along air advection in right side of the chamber.
The red circle is around Port 11,12,15,16.

## 4. Conclusion

Results did not show uniform concentration of nanoparticles. However, we found that there was agglutination in these conditions (source of sucrose, velocity, concentration of particle generated, temperature). Moreover, the results of simulation corresponded with measurement data by devices. Figure 12 showed many trajectories around Port No.11, 12, 15 and 16 as well as Figure 9-10 estimated high concentration around same points. To be conclusion, we cannot obtain a representative value in this chamber at this stage. However, we obtained that optimum ventilation conditions and numerical analysis is efficient in this study.

Now, we are improving the chamber. We plan to ventilate in the whole chamber with closing upper wall without aggregate and agglomerate (from bottom part to top part).

Therefore, it is expected that nanoparticle concentration is stable uniformity. In addition, we can obtain a representative value easily for anyone.

## References

[1] Hendren C, Mesnard X, Droge J and Wiesner M 2011 *Estimating production data for five engineered nanomaterials as a bais dor exposure assessment. Environ Sci Technol* **45** 2562-

2569

[2] Schmid O, Moller W, Semmler-Behnke M, Ferron GA, Karg E, Lipka J, Schulz H, Kreyling WG, and Stoeger T 2009 *Dosimetry and toxicology of Inhaled ultrafine particles Biomarlers.* **14** 67-73

[3] Stoeger T, Reinhard C, Takenaka S, Schroeppel A, Karg E, Ritter B, Heyder J, and Schulz H 2006 *Instillation of six different ultrafine carbon particles indicates a surface area threshold dose for acute lung inflammation in mice. Environ Haelth Perspect* **114** 328-333

[4] Organisation for Economic Cooperation and Development 2009 Series on the safety of manufactured nanomaterials. Number 10 Identification. compilation,and analysis of guidance information for exposure measurement and exposure mitigation. Manufactured nanomaterials

[5] Klepeis, N.E., Nelson, W.C., Ott, W.R., Robinson, J.P., Tsang, et al 2001 *The National Human Activity Pattern Survey (NHAPS): a resource for assessing exposure to environmental pollutants. Journal of Exposure Analysis and Environmental Epidemiology MayeJune* **11** (3) 231e252

[6] JIS A 1911 2006 Large chamber method – Determination of the emission of formaldehyde for building materials and building related products

[7] JIS A 1912 2008 Large chamber method – Determination of the emission of volatile organic compounds and carbonyl compounds except for formaldehyde for building products

[8] ISO 16000-9 2006 indoor air – Part 9 Determination of the emission of volatile organic compounds from building products and furnishing –Emission test chamber method

[9] OpenFoam http://www.openfoam.com/

# EXPOSURE ASSESSMENT OF MWCNTS IN THEIR LIFE CYCLE

**M Ono-Ogasawara, M Takaya and M Yamada**
Japan National Institute of Occupational Safety and Health, 6-21-1 Nagao, Tama-ku, Kawasaki 214-8585, Japan

E-mail: ono@h.jniosh.go.jp

**Abstract.** Multi-walled carbon nanotubes (MWCNTs) are used as a filler in composites to obtain electrical conductivity, and improve mechanical strength and other properties. However, exposure to MWCNTs may pose health risks because of their size, shape, and insolubility. A quantitative exposure assessment method for CNTs is therefore needed. We have developed a promising carbon analysis method that considers the size distribution of elemental carbon. We conducted exposure assessment according to the lifecycle of CNTs. At the first stage, large quantity of CNTs are handled and exposure to neat CNTs is likely to occur. When large quantity of CNTs are handled, enclosure and automated process are strongly recommended. By applying appropriate measures, CNT concentration can be well controlled. Local exhaust ventilation and less-restrictive enclosures were found to work well during the second stage, which involves handling smaller CNT quantities. At measured sites, MWCNT concentrations were below an occupational exposure level proposed by Nakanishi (i.e., 0.030 mg/m$^3$). This analysis method can also be applied to particles containing MWCNTs. At downstream stages of the lifecycle, neat MWCNTs were not observed and concentrations of embedded MWCNTs were lower than 0.015 mg/m$^3$.

## 1. Introduction

Carbon nanotubes (CNTs) are useful for improving the properties of composite materials, such as electrical conductivity, thermal resistance, and mechanical strength. However, CNTs pose health risks because of their shape, size, crystallinity and low solubility. To date, several occupational exposure limits (OELs) have been proposed by research institutes, for example: 0.03 mg/m$^3$ [1] and 0.001 mg-EC/m$^3$ [2]. These low exposure limits pose several challenges. First, it is difficult to determine such low-level mass concentrations gravimetrically. Second, workplace environments contain particles of nanometer to micrometer size, which are generated from combustion, photochemical reactions, and other means. For this reason, it is not feasible to measure the mass or number concentration of CNTs in isolation from background particles. Third, the recommended exposure limit, 0.001 mg-EC/m$^3$, proposed by the U.S. National Institute for Occupational Safety and Health (NIOSH), is expressed in terms of EC concentration determined by NIOSH Manual of Analytical Methods 5040 [2,3]. This method is sometimes problematic because background carbonaceous particles can show positive effects on the concentration of airborne CNTs. An appropriate analytical method for airborne CNTs requires determining CNTs at the level of 0.001 mg/m$^3$ and distinguishing them from background particles.

EC, the main component of diesel exhaust particles (DEPs), is a surrogate indicator for DEPs. Thermo-optical carbon analysis was developed for assessing the DEP contribution in occupational [4,5] and ambient [6] environments. This method can also monitor airborne CNTs because it measures CNTs as EC, primarily as graphitic carbon. In inhalation exposure studies [7], laboratories [8,9] and workplaces [10–12], EC was used for monitoring airborne CNTs. The key issue in CNT exposure assessments is distinguishing between MWCNTs and ambient particulate matter. Toward this end, we have applied size-segregated sampling and carbon analysis using a modified IMPROVE protocol [13,14] to quantify airborne multi-walled carbon nanotubes (MWCNTs), as shown in **Figure 1.**

Journal of Physics: Conference Series **617** (2015) 012009 doi:10.1088/1742-6596/617/1/012009

MWCNTs easily aggregate/agglomerate into micron-size particles, so micron-size MWCNT aerosols can be determined separately from submicron-size-ECs originating from fuel combustion by taking into account the size distribution of graphitic ECs. This method was applied for exposure assessment in a CNT production facility [15,16].

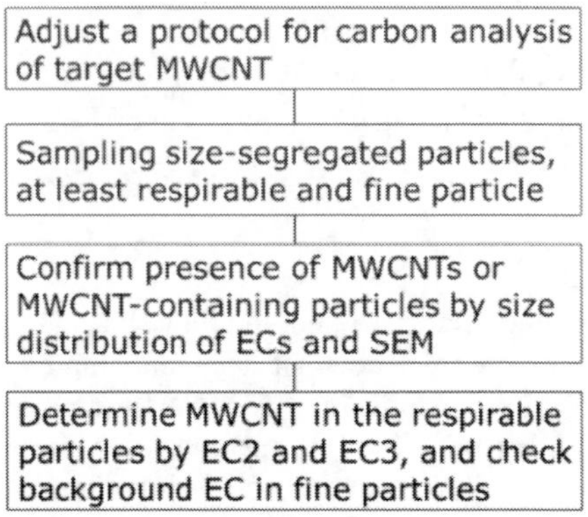


**Figure 1.** MWCNT exposure assessment procedure of using carbon analysis.

The CNT lifecycle can be categorized into six stages according to the handling quantity and dispersal state of CNTs (**Figure 2**). (1) CNT manufacturing is the first stage. It is conducted in an enclosed furnace without oxygen intrusion. Exposure is considered unlikely during normal furnace operation. After production, however, CNT exposure may occur during maintenance of furnaces and manual handling of CNTs. (2) The next lifecycle stage is manufacturing of interim products, such as masterbatches and CNT-dispersed solutions. The equipment scale and handling quantity are smaller than during production. However, CNT powder is processed manually, and vigorous agitation may release CNTs into the environment. (3) The third stage is manufacturing of products. This stage avoids direct handling of CNTs by using interim CNT-containing products manufactured during the second stage. However, some processes, such as solution drying and paint curing, may release CNTs into the air. (4) The fourth lifecycle stage is the processing of products. Here, physical and/or thermal stress is applied to the composite products, in which CNTs are bound to the base polymer. Release of free CNTs from such composites is expected to be low [17,18]. The fifth and sixth stages are (5) use of products and (6) disposal/recycling.

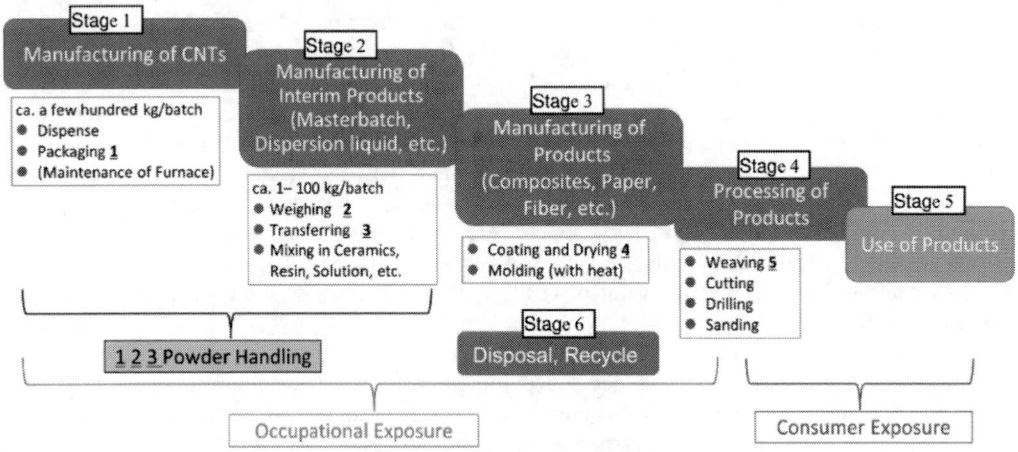


**Figure 2.** CNT lifecycle.

In this study, we conducted exposure assessments over the course of the first four MWCNT lifecycle stages. This was done by carbon analysis of size-segregated particles to test free CNTs and composite fragments containing MWCNTs. All the facilities were well controlled, and exposure concentrations were below the OEL proposed by Nakanishi [1].

## 2. Method

### *2.1. Carbon analysis*

In carbon analysis, carbonaceous substances are analyzed by heating samples in an atmosphere-controlled oven. In our research, we applied the method to quantify MWCNTs separately from background particles. First, a filter sample is heated in a helium atmosphere and the evolved substances are detected with a flame ionization detector (FID) are assigned as organic carbon (OC). Then, the atmosphere is changed to oxygen/helium and temperature is raised in steps to three levels (550°C, 700°C, and 920°C). The resulting oxidized carbon, which is reduced and detected as methane by the FID, is assigned as elemental carbon (EC).

There are various types of MWCNTs, which differ in number of graphene layers, impurity content, functionalization, and so on. Temperature and MWCNT oxidation rate reflect the characteristics of individual MWCNTs. Even though MWCNTs have various characteristics, they can be detected as ECs oxidized at 700°C and/or 920°C, designated as EC2 and EC3, respectively. Therefore, the sum of EC2 and EC3 corresponds to the amount of MWCNTs in the sample. EC2 and EC3 are graphitic carbon [18]. MWCNT fibers with greater diameters are oxidized at temperatures higher than 900°C.

In this study, we analyzed MWCNTs in samples collected on a quartz fiber filter with a carbon aerosol monitor (Sunset Laboratory Inc.). We employed a modified IMPROVE protocol (**Table 1**) to characterize and quantify the MWCNTs.

**Table 1.** Modified IMPROVE protocol.

| | Time (sec) | Oven Temperature (°C) | Oven gas |
|---|---|---|---|
| OC1 | 180 | 120 | He |
| OC2 | 180 | 250 | |
| OC3 | 300 | 450 | |
| OC4 | 300 | 550 | |
| EC1 | 360 | 550 | 2% $O_2$/He |
| EC2 | 600 | 700 | |
| EC3 | 360 | 920* | |

*The temperature is 800°C in the original IMPROVE protocol.

### *2.2. Real-time monitoring*

Monitoring was conducted simultaneously at the same points for area sampling. For monitoring number concentrations at workplaces, we used optical particle counter (OPC; KA-80B, RION Co. Ltd.) or optical particle sizer (OPS; TSI 3330, TSI Inc.) for aggregated/agglomerated particles. We monitored number size distribution of 10 to 400 nm with a scanning mobility particle sizer (SMPS; TSI 3910, TSI Inc.).

### *2.3. Sampling*

Area sampling and personal sampling were performed. Area sampling points were set near the source and a few meters away from the source. Since our carbon analysis method requires size distributions for the EC concentration of the sample, a Sioutas cascade impactor (SKC Inc.) was used to collect size-segregated particles with a quartz fiber filter (2500–QAT, Pall). The impactor can classify particles into five fractions by using a flow rate of 9 L/min. The nominal size of each fraction in terms of their aerodynamic particle diameters is as follows: >2500 nm, 2500–1000 nm, 1000–500 nm, 500–250 nm, and backup stage < 250 nm. The experimentally determined $D_{50}$ of each impaction stage when using a polytetrafluoroethylene substrate was 2,600, 950, 520, and 230 nm [20]. Respirable particles were collected with a personal sampler (NWPS-2500, Sibata Scientific, Co. Ltd.) operated at a flow rate of 2.5 L/min with the same filter as in sampling with the Sioutas cascade impactor.

*2.4 Exposure measurements.*

Measurements were conducted at five workplaces over the course of the first to fourth stages of the MWCNT lifecycle (**Figure 2**). At sites 1–3, neat MWCNT powder is handled. At sites 4 and 5, MWCNT-containing materials and solutions are handled. Site 1: MWCNTs are automatically fed from a reservoir into a plastic bag in an enclosed system installed in the facility. Since the enclosed system has a small opening to check operation, small quantities of MWCNTs may leak into the facility. The sampling point is approximately 2 m from the opening and 1-m high. The sampling duration is the same as the packaging process; about 1 h. The background sample was collected at the same point when no operation was underway. Ambient air can enter the facility. The enclosed system is kept at negative pressure and workers wear Tyvek protective clothing with a hood, goggle, gloves, and respiratory protection equipment. Site 2: MWCNT is manually weighed and carefully transferred into a covered mixing machine. The mixture containing MWCNT is then transferred into a plastic bag. The worker and the machine are in a plastic chamber (**Figure 3**). Area sampling was carried out at 2 m from the worker. Local exhaust ventilation (LEV) equipment is installed near the balance and near the opening of the mixing machine. The LEV creates negative pressure inside the chamber. Site 3: A mixture containing MWCNT in the plastic bag are carefully transferred to another equipment for the next process. The opening is set at an upper stage in the facility. An LEV is near the opening of the equipment. Outside air can intrude inside the facility. At sites 2 and 3, workers wear Tyvek protective clothing, gloves, and respiratory protection equipment. Site 4: Polyester fiber soaked with CNT-containing solution is continuously dried and rolled. The location where mechanical stress is applied to the coated fiber is equipped with an LEV. Workers wear Tyvek protective clothing, gloves, and respiratory protection equipment. Site 5: MWCNT-coated yarn is woven. **Figure 4** shows the arrangement of the operator and sampling point. Air sampling "near" is sampling point 2. Personal sampling was conducted by a researcher walking along with the operator.

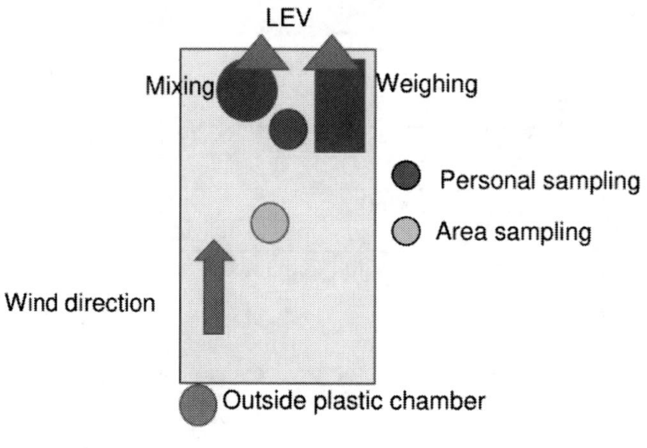


**Figure 3**. Schematic of sampling site 2 in Figure 2.

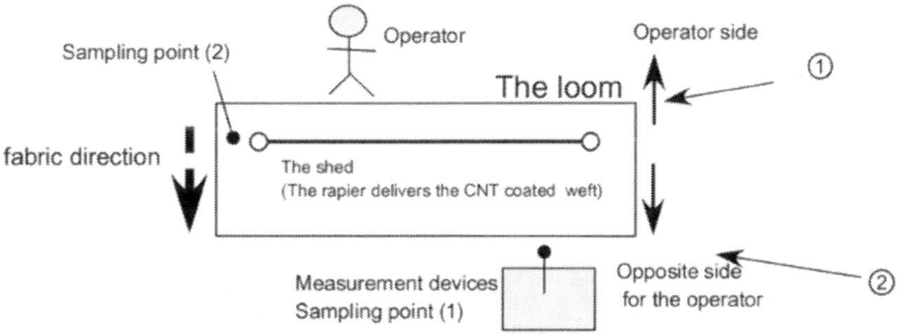


**Figure 4**. Schematic of sampling site 5 in Figure 2. (Takaya et al. [21])

## 3. Results

### *3.1. Handling MWCNT*

**Table 2** lists MWCNT number concentrations and the corresponding sampling locations. MWCNT number concentrations were detected in the handling large quantities of MWCNTs without an enclosure. When average MWCNT concentrations during a process were less than 0.015 mg/m$^3$, direct reading instruments did not detect particle release.

The highest concentrations were observed when the enclosure was not used during handling of large quantities of MWCNTs. In the facility, since packaging is normally performed automatically inside the enclosure, the operator rarely entered the facility and opened the equipment. Exposure concentration was low at 0.020 and 0.010 mg/m$^3$ by both area and personal sampling, respectively. Clearly, enclosure and automation are effective measures for handling large quantities of CNTs.

At sites 2 and 3, concentrations in the worker's personal breathing zone during handling of smaller quantities of MWCNTs (on the order of 1 kg to 10 kg) were not higher than 0.015 mg/m$^3$ in our measurements. The concentration on the windward side was less than that observed in personal sampling. According to the results of site 2, we found in weighing tasks, even a loose enclosure had a beneficial effect on exposure conditions. The efficacy of the LEV and the chamber volume affect the concentration inside the chamber. The concentration outside the chamber was low, which shows that the simplified chamber works well to maintain the surrounding environment.

### *3.2. Handling MWCNT-containing materials and solutions*

At sites 4 and 5, we observed no neat MWCNTs by scanning electron microscopy (SEM). MWCNT concentrations shown in **Table 2** are the sum of the MWCNTs, binders, and embedded fibers. Graphitized EC produced from OC cannot be perfectly distinguished from EC produced from MWCNTs; therefore, MWCNT concentrations may be overestimated at these sites. Even though the results include some types of errors, the measured MWCNT concentration is a suitable indicator because an overestimate errors on the side of safety. The cutoff diameter at site 4 is different from that at the other sites because we used a different sampler for respirable particles.

**Table 2.** MWCNT concentration (mg/m³) and [number concentration] (number/cm³)

| | Area sampling | | Personal sampling | Background | |
|---|---|---|---|---|---|
| **CNT Production** | Near | Far | | Indoor | Outdoor, or Indoor at night |
| Packaging **1** | 0.020*1-0.060*2<br>[ND-11000*3]<br>by SMPS 10–400nm<br>[ND-490*3]<br>by OPC 300–5000 nm | 0.010 – 0.060*2 | 0.010 – 0.050*2 | 0.002 – 0.010*2 | 0.001<br>[3000]<br>by SMPS 10–400nm<br>[54]<br>by OPC 300–5000 nm |
| **Dispersion** | | | | | |
| Weighing and Mixing **2** | 0.005 | 0.005 | 0.015 | 0.002 | 0.001 |
| Transferring and Molding **3** | 0.005 | 0.003 | 0.005 | 0.002 | 0.001 |
| Drying coated solution containing MWCNTs*4 **4** | 0.015 (near winder) | 0.003 | 0.016 | 0.002 | |
| **Manufacturing of CNT containing Products** | | | | | |
| Weaving of MWCNT containing fiber*4 **5** | 0.005 | 0.004 | 0.004 | | 0.001 |

*1 At the opening of the enclosure; *2 Not enclosed, not automated; *3 Background-subtracted; *4 Particulates containing CNTs and fibers were observed, but free CNTs were not observed by SEM observation; MWCNT concentrations in embedded particles.

## 4. Discussion

We proposed a method for measuring MWCNT exposure and conducted exposure assessments over the MWCNT lifecycle. As shown in **Figure 1**, we categorized the lifecycle into six stages: handling of powdered CNTs (stages 1 and 2) and other stages where exposure to released CNTs might be possible. The main routes of occupational exposure to CNTs are found in stages 1 and 2.

In stage 1, since larger quantities of MWCNTs are handled, strict control measures should be applied to lower worker exposure. Our results show that appropriate protective measures can accomplish this. Airborne MWCNT concentrations in this study were found to be in the range reported previously [12,16]. Airborne MWCNTs seem to be well controlled, at a level of a few to a few tens of micrograms per cubic meter.

Sites 2 and 3 used smaller quantities of MWCNTs. Here, airborne MWCNT concentration was less than a few tens of micrograms per cubic meter, and these concentrations were lower than the OEL proposed by Nakanishi: 0.030 mg/m$^3$. No release into the environment outside the chamber was observed. Smaller-scale handling is often performed in a fume hood, which can also reduce exposure.

Real-time number concentration monitoring can be useful for assessing MWCNT aerosol release. In the present study, outside air intruded into the facility caused sudden increases or fluctuations in the background concentration of particulate matter. In these cases, it is difficult to detect MWCNT aerosol generation by direct reading instruments.

Carbon analysis can be used to measure MWCNTs embedded or bound to organic carbonaceous materials [14] by adjusting the protocol used. In this way, MWCNTs cam be distinguished from resins and other base materials in the composite. At least at site 5, SEM analysis indicated no neat MWCNTs, although MWCNT-fiber fragments showed MWCNT extrusion on the fiber surface. Generally, neat MWCNTs are not observed during treatment of MWCNT-containing composites products [21]. Since MWCNTs and the base materials cannot be perfectly separated from each other, base materials contribute positive error to the MWCNT concentration data. Over the course of stages 3 and 4, MWCNT concentration (including the positive error) was lower than the OEL proposed by Nakanishi [1]. In vitro tests have shown that particles containing MWCNTs have a weaker effect than neat MWCNTs [22]. The facilities examined in this study are considered well controlled.

Although carbon analysis method can be applied to exposure measurement of MWCNTs embedded in composite, SEM/TEM observations are essential to confirm if neat MWCNT is released.

## 5. Conclusion

We performed exposure assessments of MWCNTs at workplaces over their lifecycle by our carbon analysis method, which uses the MWCNT size distribution to distinguish MWCNTs from background carbonaceous substances. In most facilities, MWCNT concentration is well controlled by using appropriate control measures, and workers usually wear protective equipment. Further validation of the method and further examples of its application are needed in order to reach consensus regarding exposure measurements for MWCNTs.

## 6. Acknowledgment

This research was supported in part by Project Research No. N-P25-01 of the Japan National Institute of Occupational Safety and Health. We are deeply grateful to Kurarayliving Co. Ltd. and Matsubun Textile Co. Ltd. for giving us the opportunity to conduct a field survey in a MWCNT-related textile factory.

The authors declare no conflicts of interest.

## References

[1] Nakanishi J 2011 Risk Assessment of Manufactured Nanomaterials:"Approach"-Overview of Approaches and Results- Final report issued on Aug 17, 2011. NEDO project (P06041) "Research and development of Nanoparticle characterization method." available at https://www.aist-riss.jp/projects/nedo-nanorisk/nano_rad2/download/download.cgi?ES_e

[2] National Institute for Occupational Safety and Health (NIOSH) 2010 niosh current intelligence bulletin: Occupational Exposure to Carbon Nanotubes and Nanofibers. available at http://www.cdc.gov/niosh/docket/review/docket161a/pdfs/carbonNanotubeCIB_PublicReviewOfDraft.pdf

[3] NIOSH 2003 Manual of Analytical Methods (NMAM). in: Schlecht P C, O'Connor P F (Eds.), Method 5040 update DHHS (NIOSH) Publication No. 2003-154, Third Supplement to NMAM, fourth ed., National Institute for Occupational Safety and Health, Cincinnati, OH, USA

[4] Birch M E, Cary R A. 1996 *Aerosol Sci Technol* **25** 221–4

[5] Birch M E 2002 *Appl Occup Environ Hyg* **17** 400–5

[6] Chow J C, Watson J G, Pritchett L C, Pierson W R, Frazier C A, Purcell R G 1993 *Atmos Environ* **27A** 1185–1201

[7] Myojo T, Oyabu T, Nishi K, Kadoya C, Tanaka I, Ono-Ogasawara M, Sakae H, Shirai T 2009 *J Nanoparticle Res* **11** 91–9

[8] Han J H, Lee E J, Lee J H, So K P, Lee Y H, Bae G N, Lee S-B, Ji J H, Cho M H, Yu I J 2008 *Inhal. Toxicol* **20** 741–9

[9] Maynard A D, Baron P A, Foley M, Shvedova A A, Kisin E R, Castranova V 2004 *J Toxicol Environ Health,* Part A **67** 87–107

[10] Methner M, Hodson L, Dames A, Geraci C 2010 *J Occup Environ Hyg* **7** 163–76

[11] Birch M E, Ku B K, Evans D E, Ruda-Eberenz T A 2011 *Ann Occup Hyg* **55** 1016–36

[12] Dahm M M, Evans D E, Schubauer-Berigan M K, Birch M E, Fernback J E 2012 *Ann Occup Hyg* **56** 542–556

[13] Ono-Ogasawara M, Myojo T 2013 *Adv Powder Technol* **24** 263–269

[14] Ono-Ogasawara M, Takaya M, Kubota H, Shinohara Y, Koda S, Akiba E, Tsuruoka S, Myojo T 2013 *J. Phys.: Conf Ser*, 429, 012004; http://iopscience.iop.org/1742-6596/429/1/012004.

[15] Takaya M, Serita F, Ono-Ogasawara M, Shinohara Y, Koda S 2010 *San Ei Shi* **52** 182–8 (in Japanese)

[16] Hedmer M, Isaxon C, Nilsson P T, Ludvigsson L, Messing M E, Genberg J, Skaug, V, Bohgard M, Tinnerberg H, Pagels J H 2014 *Ann Occup Hyg* **58** 355–379

[17] Schlagenhauf L, Nüesch F, Wang J 2014 *Fibers* **2** 108–127

[18] Ogura I, Kotake M, Shigeta M, Ueijima M, Saito K, Hashimoto N, Kishimoto A 2013 *J Phys: Conf Ser* **429** 012049; http://iopscience.iop.org/1742-6596/429/1/012049.

[19] Han Y M, Cao J J, Chow J C, Watson J G, An Z S, Jin Z D, Fung K, Liu S X 2007 *Chemosphere* **69** 569–74

[20] Misra C, Singh M, Shen S, Sioutas C, Hall P. 2002 *J Aerosol Sci* **33** 1027–47

[21] Takaya M, Ono-Ogasawara M, Shinohara Y, Kubota H, Tsuruoka S, Koda S 2012 *Ind Health* **50** 147–55

[22] Wohlleben W, Meier M W, Voge, S, Landsiedel R, Cox G, Hirth S, Tomovic Z 2013 *Nanoscale* **5** 369–380

# Properties of nanoparticles affecting simulation of fibrous gas filter performance

**Paolo Tronville[1], Richard Rivers[2]**

[1] Politecnico di Torino DENERG, Corso Duca degli Abruzzi, 24 10129 Turin, Italy
[2] EQS, Inc. 1262 Bassett Ave. Louisville, Kentucky 40204 USA

E-mail: paolo.tronville@polito.it

**Abstract.** Computational Fluid Dynamics (CFD) codes allow detailed simulation of the flow of gases through fibrous filter media. When the pattern of gas flow between fibers has been established, simulated particles of any desired size can be "injected" into the entering gas stream, and their paths under the influence of aerodynamic drag, Brownian motion and electrostatic forces tracked. Particles either collide with a fiber, or pass through the entire filter medium. They may bounce off the fiber surface, or adhere firmly to the surface or to particles previously captured. Simulated injection of many particles at random locations in the entering stream allows the average probability of capture to be calculated. Many particle properties must be available as parameters for the equations defining the forces on particles in the gas stream, at the moment of contact with a fiber, and after contact. Accurate values for all properties are needed, not only for predicting particle capture in actual service, but also to validate models for media geometries and computational procedures used in CFD. We present a survey of existing literature on the properties influencing nanoparticle dynamics and adhesion.

## 1. Introduction

The immense growth of nanotechnology in recent years means that nanoparticles will appear in industrial and research locations at levels which demand careful air pollution control. Vehicular exhausts pollute the air in cities and along highways with unacceptable aerosol concentrations, including nanoparticles. It is important to understand and quantify the details of nanoparticle by fibrous filters, to promote the development of effective, reliable and minimum-cost solutions to nanoparticle filter systems design.

The authors have previously discussed the effect of nanofiber additions to air filter media [1]. The present paper is concerned with the physical properties of nanoparticles which affect their capture and build-up in fibrous air filters, regardless of the characteristics of the fibers in the filter media.

Several studies found in our literature review had a recurring theme: the behavior of instruments based on "classic" concepts cannot necessarily be extended to aerosol particles with low-nanometer sizes. Flow patterns in differential mobility analyzers that appear unimportant for micrometer-sized particles may affect results for nanoparticles. Particle-charge neutralization in bipolar ion fields are not described by the same expressions for nanoparticles as for larger particles. Particle count devices may have very different count efficiencies for particle sizes above and below 100 nm. Agglomeration of fundamental nanoparticles can confuse results.

## 2. Nanoparticle properties of interest to filtration simulation

Properties of aerosol particles which must be known for CFD simulations are:

- Form and size distribution;
- Aerodynamic drag as a function of size and form;
- Density, both of simple forms and agglomerates;
- Count in a sample or sample flow;
- Bounce;
- Adhesion properties, both to filter media fibers and to other particles of their own kind;
- Diffusion parameters;
- Electrical charge distributions as a function of size and neutralizer properties;
- For liquid particles, evaporation rates.

Tests of the validity of CFD simulations are, ultimately, the accuracy with which a CFD model and calculation procedure duplicates experimental data on the filter media simulated. Incorrect property values lead to meaningless simulations.

### *2.1. Forms of nanoparticles*

Nanoparticle can have relatively simple geometries, such as spheres, cubes, cylindrical rods. Platelet forms – thin round discs or flat polygons, exist. Agglomerates with maximum dimensions below 100 nm – the usual definition of nanoparticles – are commonplace, since they are created in flames and by both spark-ignition and diesel engines. Such aggregates consist of smaller primary nanoparticles, but for filtration studies, aggregates must be considered as single, though complex, forms. Nanoparticles of carbon have some of the most exotic forms: hollow quasi-spherical "fullerenes"; single or multilayer hollow nanotubes; spirals. Gel nanoparticles can exist in spherical or ellipsoidal form.

The behavior of a non-spherical particle is often expressed in terms of the diameter of a sphere having like behavior. Unfortunately, the equivalent spherical diameter depends on the specific effect or property considered. Figure 1 shows five equivalent diameters for a relatively simple hexagonal platelet particle.

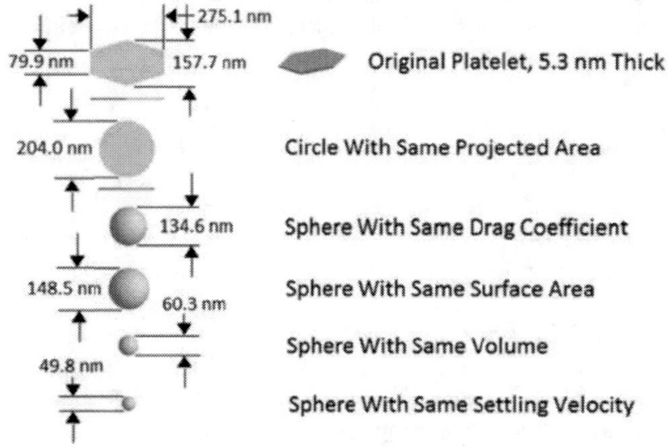


**Figure 1.** Circle and sphere equivalents to a hexagonal platelet (Adapted from [2])

Shape, size and density are especially difficult to define for an agglomerate. One needs to know the number and dimensions of its fundamental particles, often rather uniform spheres of low nanometer diameter. Other dimensions of its size and structure are needed: maximum length and width, and the *radius of gyration* of the agglomerate. With these, a quantitative measure of the agglomerate complexity, its *fractal dimension,* can be calculated. For aerosol agglomerates, this can often be determined by analysis of two-dimensional electron microscope images using computer algorithms. Rogak, Flagan and Nguyen [3] provide a description of these image-analysis procedures. An approximation to the fractal dimension is a simple function of the number of primary particles in an agglomerate and square root of the product of the particle's length and width (Sander [4]). Examples of agglomerate characterization are given in Patterson and Kraft [5]. Useful explanations of the terminology in that paper and others which make use of fractal concepts are given on two internet sites [6] and [7].

Several studies have taken a somewhat different approach to establishing the fractal dimension and fractal prefactor of aggregates. Assuming or calculating the size and size distribution of primary particles, and fractal dimension and fractal prefactor values, they generate large numbers of images of aggregates. Various models of the agglomeration physics are possible. These simulated images frequently match the appearance of TEM or SEM images of the agglomerates studied quite well, indicating that the parameter choices were appropriate. [8] and [9] discuss ways in which the agreement between simulations and measurements can be quantified, and in particular, the best means to obtain reliable fractal descriptions of 3D agglomerates from 2D images.

### *2.2. Size distributions of nanoparticles*

Many studies have shown that primary nanoparticles can exist as essentially uniform particles. More often, however, the size distributions are log-normal. The same mathematical relationships and graphical representations that are used with micrometer-scale particles apply to distributions at nanometer scale. Simple particle forms have a characteristic dimension, which for a cloud of particles can have a size-dependent distribution. Sodium chloride particles, for example are frequently cubical, characterized by the length of a side; cylindrical particles can have size distributions for both diameter and length. Agglomerates can have wide size ranges, but no easily defined size, hence no easily specified size. What can be defined and measured for agglomerates are their mobilities, aerodynamic, diffusive and electrical. Mobility is defined as: [particle velocity relative to local gas velocity] / [force on particle].

The term [force on particle] is different for aerodynamic, diffusive, and electrical mobility. These three forces and means for measuring them are discussed in sections below. Because agglomerates have no definable size, their size distributions are often stated in terms of some "equivalent mobility diameter", dependent on the method used to obtain the distribution. An alternative "size" is some measurable geometric value, obtained from SEM images. One example of SEM-measured size is the determination of the smallest rectangle able to enclose the particle. Image-analysis software packages, both commercial and open-source, are available to measure such dimensions more or less automatically, and characterize the distributions of them.

### *2.3. Aerodynamic drag of nanoparticles*

The aerodynamic drag $\mathbf{F}_{\text{drag}}$ of a sphere with low Reynolds number moving in a gas stream is given by Stokes law with the Cunningham correction:

$$\mathbf{F}_{\text{drag}} = 3\pi\mu d_p\,(\boldsymbol{u}_{\text{g}} - \boldsymbol{u}_{\text{p}}) \cdot (m_p\, C_c)^{-1}$$

Where $\mu$ = gas dynamic viscosity; $d_p$= sphere diameter; $\boldsymbol{u}_{\text{g}}$ = gas velocity; $\boldsymbol{u}_{\text{p}}$ = particle velocity; $m_p$ = particle mass; and $C_c$ = Cunningham's correction for slip at the particle surface, a function of gas type, temperature and pressure. With $\lambda$ = gas mean-free-path and Knudsen Number defined as

$$Kn = 2\lambda(\, d_p\,)^{-1}$$

$$C_c = 1 + Kn(A + B\,\exp(-C_C /\, Kn))$$

This classic relationship has been shown [10] to hold for particles as small as 0.012 times the mean free path of the gas molecules. [10] provides *A*, *B*, and *C* for dry air. Since the expression for $C_c$ is a function of Knudsen Number, the values of *A*, *B*, and *C* should be applicable to any gas and set of test conditions for which the mean-free path of the gas and the particle diameter are known.

Tammet [11] demonstrates that modifications to the expression for the Cunningham correction in [10] are needed for spheres with diameters below 4 nm. His correction is added to the values of $d_p$ wherever they appear in the Cunningham correction, including the values for *Kn*, and also in the above expression for aerodynamic drag $\mathbf{F}_{drag}$.

If an airflow containing aerosol particles is passed through a properly dimensioned sequence of circular orifices, the aerosol particles are confined to a narrow beam, which expands in the region downstream of the last orifice. A typical pattern of flow through such an "aerodynamic lens" is shown in figure 2, from [12]. CFD calculations compared to the measured angular dispersion of the aerosol beams and the fraction of particles passing completely through the lens provide sensitive tests of both the expression for aerodynamic drag and the parameters values used in the CFD calculations.

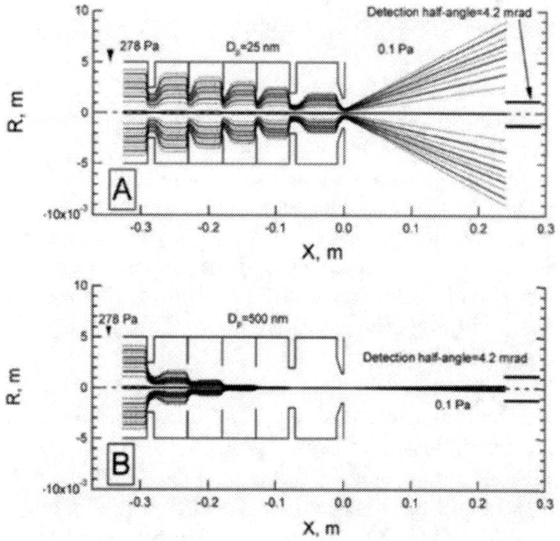


**Figure 2.** CFD-Simulated particle paths through an aerodynamic lens.
A: 25 nm diameter particle;
B: 500 nm diameter particle
( from Zhang [12 ])

It is, however, important to include the effect of Brownian motion in the simulation. Reference [13] discusses the combined effects of aerodynamic drag and Brownian motion on aerodynamic lens performance for nanoparticles. The net result of these studies appears to be that the theory-based expressions for aerodynamic drag and Brownian motion are reliable, provided "slip" of gas molecules at the particle surface is properly included (also cf. Lewis [14]).

Some particles approaching an aerodynamic lens will strike the surface surrounding the entrance orifice. Other particles will strike the walls of the chambers and the orifices inside the lens. The fraction of entering particles which reaches the lens outlet can be calculated by CFD, and compared to measured values. Liu et.al. [15] reporting such comparisons, conclude that the agreement between simulation and measurement validates the parameter values and equations used for simulation. Sorensen [14] reviews the extensive literature on determination of the "mobility radius" of agglomerates, which is the radius of a sphere having the same aerodynamic drag as the agglomerate. He covers the entire range of particle Knudsen Numbers, hence nanoparticles.

### *2.4. Density of nanoparticles and agglomerates*

Density is mass divided by volume. For analyses of particle capture in fibrous filtration "volume" may not be the volume of the fundamental material making up the particle, but the volume which defines its aerodynamic drag. For relatively simple nanoparticle forms, it may be possible to determine particle volume from SEM images, but masses cannot be determined gravimetrically.

The mass of aerosol particles can be measured with a single device, the Aerosol Particle Mass Analyzer (APM) making use of two properties of the particle simultaneously. The APM injects particles into the gap between two rotating cylinders. The gas in this gap rotates, imparting a centripetal force on the particle, proportional to its mass. This force is balanced by an electrostatic field established between the cylinders acting on a known charge on the particle. Figure 3a shows the design of the design of the APM, and figure 3b the equations which allow the calculation of the mass of the particle.

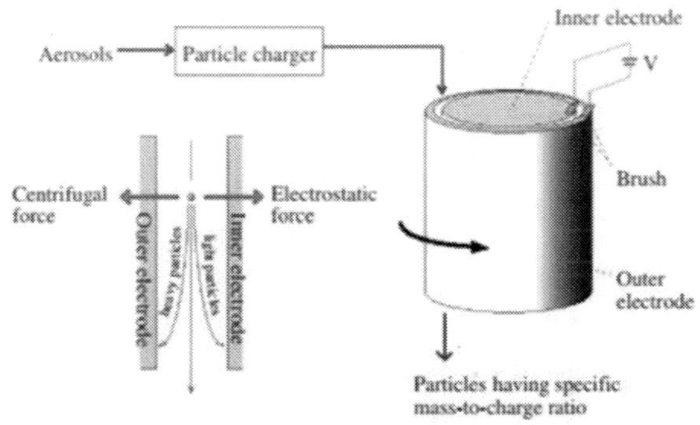


**Figure 3a.** Aerosol Particle Mass Analyzer Concept (from Kanomax APM Operation Manual)

$$\mathrm{m}r\omega = q\frac{V}{r\cdot\ln(r_0/r_i)}$$

where $m$ = particle mass
$\omega$ = angular rotation speed
$r$ = particle location relative to axis of rotation
$q$ = particle charge
$r_i$ = radius of inner electrode
$r_o$ = radius of inner electrode
$V$ = applied voltage

**Figure 3b.** APM Force Balance Equation (after Kanomax APM Operation Manual)

The density of nanoparticles can also be determined by pairs of instruments which allow extraction of the density value. A common pair is the differential mobility analyzer (DMA) and the vacuum aerodynamic particle sizer. An extensive review of non-spherical aerosol particle density measurement methods is given in [15].

### *2.5. Bounce properties of nanoparticles*

Most early studies of particle bounce were made for the configurations present in single impactors and cascade impactors. A few metal-fiber filter forms exist, and for these the impactor studies are relevant. The impactor substrates studied rarely bear any resemblance to the common fibers in fibrous filter media, which are now formed of glasses, ceramics, synthetic organic polymers or natural organic materials like cotton (cellulose) or animal hairs (proteins). In addition, impactor substrates are often coated with greases or oils, or wetted, specifically to eliminate particle bounce. Such coatings are rare in fibrous filtration.

Several studies have attempted to quantify particle bounce from the fundamentals of elastic-body mechanics. [16], [17], [18] and [19] are among those who have taken this approach. Sato, Chen and Pui [20] extend these concepts to nanoparticles, and consider adhesion and removal effects as well.

### *2.6. Adhesion properties of nanoparticles*

In general, adhesion forces binding particles to surfaces become stronger as particle size decreases, the forces of removal by aerodynamic drag become smaller for decreasing size, and bombardment by other particles becomes less probable. Thus re-entrainment of nanoparticles from filter fibers is minimal. Still, re-entrainment can occur. The theoretical aspects of adhesion forces are discussed in [21] and [22]. Extensive treatments of many aspects of particle adhesion are given in [23]. Remarkable adhesion effects resulting from high-energy collisions of nanoparticles with surfaces are described in [24]. A series of slides illustrating the complexity of particle adhesion is given at the website [25].

### *2.7. Diffusion of nanoparticles in filtration*

In filtration, the diffusion process of interest is Brownian motion, the erratic path taken by particles as a result of unbalanced impacts from the molecules of the gas surrounding the particle. The (vector) acceleration of a particle can be defined as the sum of the (vector) aerodynamic drag force per unit mass and the (vector) Brownian force per unit mass:

$$\frac{d^2\mathbf{V}}{dt^2} = \mathbf{F}_D + \mathbf{F}_B$$

The aerodynamic drag on the particle is the "classic" definition described above, with the Cunningham correction $C_C$ using a value of $Kn$ appropriate for the nanoparticle size. To simulate Brownian motion, one must create a series of randomly directed forces at random time intervals. The power spectrum of these forces must have a Gaussian distribution, and must have an average power level $S_{nn}$ given by

$$S_{nn} = kTC_C / (3\pi\mu d_P)$$

The time of application of each Brownian impact is randomized, as is its spatial direction. This procedure produces the erratic pattern characteristic of Brownian motion, superimposed on the general direction of particle motion resulting from $\mathbf{F}_D$.

### *2.8. Nanoparticle charging and neutralization*

The unipolar charging of nanoparticles has been studied extensively, chiefly in connection with electrostatic precipitators and to support the interpretation of data from instruments that make use of charging to measure particle size and density. Charged aerosols are more easily captured by fibrous filters, whether or not they have electret properties or externally applied electrostatic fields. But charging aerosols to improve fibrous filter efficiency is dangerous, for if the capture efficiency somehow declines, the charged particles pass into the downstream space, and the charged particles precipitate rapidly on surfaces in the space, even forming visible patterns.

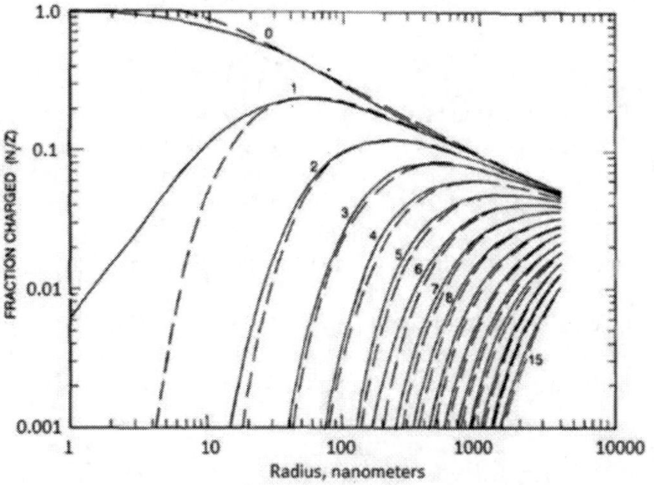


**Figure 4.** Probability of $N$ elementary charges per particle on an aerosol at Boltzmann equilibrium, i.e. "neutralized". (From Hoppel and Frick [27]. Dashed curves follow Fuchs [26])

Unipolar charging is used to neutralize aerosols which carry high unipolar charges as a result of their generation method. Bipolar charging - the exposure of an aerosol to a cloud of mixed positive and negative ions - is important to aerosol experimentation in many ways. Knowledge of charge distributions on aerosols is also critical for simulation of particle capture in fibrous filtration. The probable number of charges on each diameter of particle needs to be known. An electrostatically "neutral" aerosol cloud is not charge-free; each individual particle in the cloud carries some number of elementary charges, ranging from zero to $N$ elementary charges, either positive or negative. For an aged aerosol, the Boltzmann equilibrium distribution will be approached. Aerosols freshly generated in the laboratory are passed through clouds of mixed positive and negative ions to reach equilibrium quickly. The ion-production devices of such neutralizers can depend on radioactive sources (Kr-85 or Po-210), soft X-rays, or high-voltage corona discharge to produce the bipolar ion clouds needed. Covert, Wiedensohler and Russell [26] critique neutralizing devices. Like so many things in aerosol experimentation, care in choosing a neutralizer adequate for the conditions of an experiment is essential.

Fuchs [27] developed the basic form of the distribution of charges on aerosols exposed to a balanced bipolar ion cloud (one with equal numbers of positive and negative ions). Hoppel and Frick [28] extended the analysis to nanometer-size particles, and provided plots of the Boltzmann equilibrium distribution of charge on particles with radii 1 nm to 4μm. Stommel and Riebel [29] found errors in these earlier works, and provide corrections. The limited results of their corrections deviate slightly from the plot from Hoppel and Frick (figure 4).

### *2.9. Evaporation of liquid nanoparticles*

The literature on nanoparticle evaporation is chiefly related to high-temperature evaporation of metal particles. Li and Davis [30] conducted experiments on evaporation of dibutyl-phthalate particle evaporation in air at pressures decreasing from 13.7 kPa to 0.016 kPa. Their results are expressed as evaporation rates as a function of Knudsen Number, so may be applicable to nanoparticles at normal pressures. Their experimental conditions involved values of $Kn$ as high as 2. Sutter et.al. [31] studied the evaporation of n-hexadecane particles with mass-mean diameters as small as 1.5 μm actually on filter fibers. Evaporation rates were substantially lower than predicted by traditional mass-transfer theory (Fick's Law). The mass of nanoparticle aerosols is very small, so evaporation may not result in appreciable health hazards.

### *2.10. Counting of nanoparticles for filtration studies*

Light-scattering aerosol spectrometers, widely used for counting particles of micrometer dimensions, have lower detection limits of about 90 nm, hence limited application in nanoparticle studies.

Nanoparticles can be examined by a transmission electron microscope (TEM) or scanning electron microscope (SEM), and the number of particles counted. Nuclear track-etched polycarbonate membrane filters provide a relatively featureless image background, allowing some automation of the counting process when SEM images are obtained. Preparation of nanoparticle samples for these imaging processes is more difficult than for light microscopy; the samples must be given a thin conductive coating, usually of a gold/palladium alloy, in an ion sputtering device.

Image-analysis computer programs are available to assist the particle counting process, including an open-source one (fraclac, an add-on for imagej). Electron microscope imaging is essentially the reference method for evaluating all other devices used to study aerosol particle geometry. It avoids the assumptions about particle charging, aerodynamic drag and diffusion effects necessary for interpreting the results of data from instruments such as the MOUDI, ELPI, DMA, SMPS, and APM.

The usual device for nanoparticle counting in gas flows is the condensation particle counter (CPC), which increases the size of individual particles by condensing a vapor onto them, then counting the enlarged particles by light-scattering. While the counting efficiency of a CPC is not 100% for all size particles, the efficiency values are reasonably stable and predictable, and useful counts can be obtained for particles below 10 nm diameter. The accuracy of results decreases as the size detected decreases.

Particle capture efficiency measurements for filter media require simultaneous upstream and downstream sampling, or very stable aerosol generation. In addition, it is essential that the sampling systems upstream and downstream, and the two flow rates, be as nearly identical as possible, so that particle losses in the two sampling systems are the same.

Journal of Physics: Conference Series **617** (2015) 012010 doi:10.1088/1742-6596/617/1/012010

## 3. Generation of nanoparticles for filtration studies

Biskos et.al.[32] provide descriptions of the many ways available for nanoparticle generation. They discuss the potential of several methods to yield compact aerosol forms, rather than agglomerates, and provide an extensive bibliography for the details of various methods. Here we comment on some of the available generation methods.

Tiwari, Fields and Marr [33] dispersed dry nanoparticle powders using the generator described in [34]. The nanoparticle powders must be produced in some way which provides high yields of particles with defined properties. Powders of many substances are commercially available.

Liquid dispersions of nanoparticles can be diluted with ultrapure water, aerosolized by spraying, then passed through an evaporator to create individual spherical nanoparticles. Dispersions of polystyrene latex, gold and silver with certified diameters and standard deviations are available commercially with diameters as small as 1 nm. Raabe [35] gives the dilution needed to yield a given fraction of single particles in each spray droplet, depending on the droplet size distribution. Sheehan et.al. [36] generated nanoparticles by creating solution droplets with a Collison atomizer, stripping the larger particles with a cyclone, then evaporating the droplet water. Using an electrospray, instead of more conventional atomizers, gives smaller droplets, increasing the chance of dispersing individual particles. Electrosprayed droplets can also be solutions which evaporate to yield individual nanoparticles whose shape depends on the nature of the solute [37]. Figure 5 shows such a generator, which incorporates a bipolar ion source to neutralize the aerosol, which would otherwise be highly charged.

A popular means of generating liquid nanoparticles is to evaporate a liquid, then condense its vapor. Metals, such as gold and silver, can also be vaporized if high enough temperatures are available. Ji et.al. [38] describe such a generator. A widely-used means to reach temperatures high enough to vaporize metals is spark-discharge, Meuller et.al. [39] review controlled spark-discharge nanoparticle generation.

Flames can of course produce carbon nanoparticle agglomerates in the laboratory, as they do in the outdoor environment. They are also produce occasional fullerene-structure particles. For filter media testing, steady aerosol generation rates with substantial particle counts are needed. Means to generate all fullerene forms – single – and multiple-wall nanotubes, "buckyballs", spirals – have all been developed. The methods are employed for carbon nanoparticle generation, and many other materials, include laser ablation, spark generation; and furnace, flame, plasma and laser-stimulated reactors. These last four methods allow the creation of nanoparticles of selected chemical composition, especially organic compounds and metal oxides. Flows of *precursor compounds* are provided, either in solution or as gases, then passed into zones where energy is injected. The addition of substantial energy produces the reactions needed to create the desired compounds. A cooling zone is provided; there the hot vapors condense into particles. Sometimes an additional hot zone is provided for melting or *sintering* any agglomerates present into compact particles. Figure 6 shows a generator using a flame to oxidize a feed flow of liquid aerosol precursor particles. The mix of combustion and inert gases, and the form of the flame, provide conditions which optimize the yield of nanoparticles, either in primary form or as agglomerates as desired [40].

Iida et.al. [41] generated particles with diameters as small as 300 nm using an ink-jet generator; it seems possible to do the same with vibrating orifice generators, such as described by Mitchell, Snelling and Stone [42].

Wang and Tronville examined nanoparticle generation methods for their suitability to adaption as standardized test methods [43].

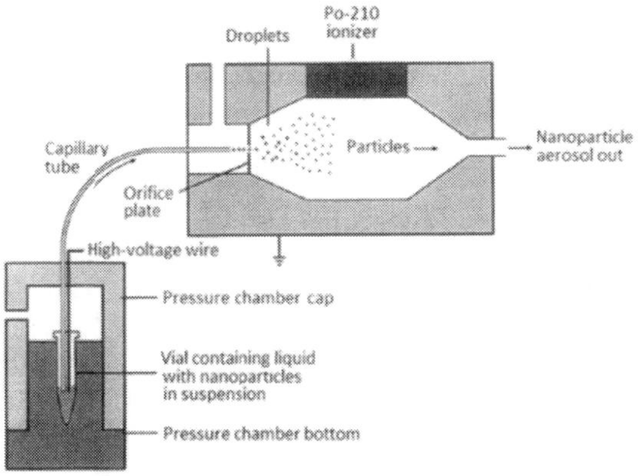


**Figure 5.** Electrospray nanoparticle generator (modified from TSI catalog illustration)

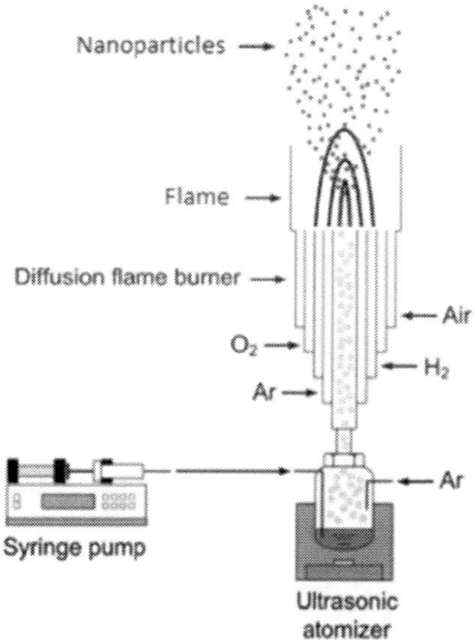


**Figure 6.** Flame spray pyrolysis generator (modified from Cho et al. [38])

## References

[1] Zhou B, Tronville P, Rivers R 2013 *HVAC&R Research* **19** 503–512
[2] Gallego-Urrea J A, Hammes J, Cornelis G and Hassellov M 2014 *J. Nanopartical Res.* **16** 2383–2400
[3] Rogak S N, Flagan R C and Nguyen H V 1993 *Aerosol Sci. & Tech.* **18** 25–47
[4] Sander M, Patterson R I A, Raj A and Kraft M 2009 On the fractal dimension of soot particles Cambridge Centre for Computational Chem. Eng. *Preprint* 80
[5] Patterson R I A and Kraft M 2007 *Combustion and Flame* **151** 160–172
[6] fractalfoundation.org
[7] imagej.nih.gov/ij/plugins/fraclac
[8] Chakrabarty R K, Garro M A, Garro R A, Chancellor S, Moosmütter H and Herald C M 2011 *Aerosol Sci. & Tech.* **45** 75–80
[9] Chakrabarty R K, Garro M A, Garro R A, Chancellor S, Moosmütter H and Herald C M 2011 *Aerosol Sci. & Tech.* **45** 903–908
[10] Kim J H, Mulholland G W, Kukuck S R and Pui D Y H 2005 *J. Res.National Inst. Standards and Tech.* **110** 31–54
[11] Tammet H 2012 *Aerosol Sci. & Tech.* **46** i–iv
[12] Zhang X F, Smith K A, Worsnop D R, Jiminez J L, Jayne J T, Kolb C E, Morris J and Davidovits P 2004 *Aerosol Sci. & Tech.* **38** 619–638
[13] Wang X L, Kruis F E and McMurry P H 2005 *Aerosol Sci. & Tech.* **39** 624–636
[14] Lewis E R 2010 *J. Aerosol Sci.* **41** 418–25
[15] Liu P S K et.al. 2007 *Aerosol Sci. & Tech.* **41** 721–733
[16] Sorensen C M 2011 *Aerosol Sci. & Tech.* **45** 765–779
[17] DeCarlo P F, Slowik J G, Worsnop D R, Davidovits P and Jiminez J L 2004 *Aerosol Sci. & Tech.* **38** 1185–1205
[18] Dahneke B 1975 *J. Colloid Interface Sci.* **51** 58–65
[17] Tsai C J, Pui D Y H and Liu B D Y H 1990 *Aerosol Sci. & Tech.* **12** 497–507
[18] Dunn P F, Brach R M and Janson G G 1996 *Aerosol Sci. & Tech.* **25** 445–465

[19] Brach R M and Dunn P F 1998 *Aerosol Sci. & Tech.* **29** 379–388
[20] Sato S, Chen D R and D.Y.H. Pui D Y H 2007 *Aerosol and Air Quality Res.* **7** 278–303
[21] Carrillo J Y, Raphael E and Dobrynin A V 2010 *Langmuir* **26** 12973–12979
[22] Thomas J 2003 Mechanics of nanoparticle adhesion – a continuum approach. *Particles on Surfaces* ed K.L. Mittal (New York: Marcel Dekker) Chap. 8
[23] Mittal K L and Jaiswal R 2015 *Particle Adhesion and Removal* (New York: Wiley)
[24] Suri M and Dumitrica T 2008 *Phys. Rev. B* **78** 081405
[25] web2.clarkson.edu/projects/crcd/me537/downloads/slides/Rimai.pdf
[26] Covert D, Wiedensohler A and Russell L 1997 *Aerosol Sci. & Tech.* **27** 206–214
[27] Fuchs N 1963 *Geofis. Pura Appl.* **56** 185–192.
[28] Hoppel W and Frick G 1990 *Aerosol Sci. & Tech.* **12** 471–496
[29] Stommel Y G and Riebel U 2007 *Aerosol Sci. & Tech.* **41** 840–847
[30] Li W G and Davis E J 1996 *Aerosol Sci. & Tech.* **25** 11–21
[31] Sutter B, Bémer D, Appert-Collin J C, Thomas D and Midoux N 2010 *Aerosol Sci. & Tech.* **44** 394–404
[32] Biskos G, Vons, V, Yurteri C U and Schmidt-Ott A 2008 *KONA Powder & Particle J.* **26** 13–35
[33] Tiwari A J, Fields C G and Maur L C 2013 *Aerosol Sci. & Tech.* **37** 1267–1275
[34] Tang P, Fletcher D F, Chan H K and Raper J A 2008 *Powder Tech.* **187** 27–36
[35] Raabe O G 1968 *Amer. Indust. Hygiene Assoc. J.* **29** 439-443
[36] Sheehan M J, Peters T M, Cena L, O'Shaughnessy P T and Gussman R A 2009 *Aerosol Sci. & Tech.* **43** 1091–1098
[37] Ji J H, Bae G U, Yun S H, Jung J H, Noh A S and Kim S S 2007 *Aerosol Sci. & Tech.* **41** 786–793
[38] Meuller B O, Messing M E, Engberg D L, Jansson A M, Johansson L I M, Norten S M, Tureson N and Deppart K 2012 *Aerosol Sci. & Tech.* **46** 1256–1270
[39] Cho K, Chang H K, Kil D S, Park J H, Jang H D and Sohn H Y 2009 *Aerosol Sci. & Tech.* **43** 911–920
[40] Chen D, Pui D H Y and Kaufman S I 1995 *J. Aerosol Sci.* **26** 963–977
[41] Iida K, Sakurai H, Saito K and Ehara K 2014 *Aerosol Sci. & Tech.* **48** 789–802
[42] Mitchell J P, Snelling K W and Stone R L 1987 *J. Aerosol Sci.* **18** 231–243
[43] Wang J and Tronville 2014 *J. Nanopart. Res.* **16** 2417

# Detailed characterization of welding fumes in personal exposure samples

**B Quémerais[1], James Mino, M R Amin, H Golshahi and H Izadi**
Division of Preventive Medicine, University of Alberta, 107 University Campus NW, Edmonton, AB T6G 2S2, Canada

quemerai@ualberta.ca

**Abstract**. The objective of the project was to develop a method allowing for detailed characterization of welding particles including particle number concentration, size distribution, surface chemistry and chemical composition of individual particles, as well as metal concentration of various welding fumes in personal exposure samples using regular sampling equipment. A sample strategy was developed to evaluate the variation of the collection methods on mass concentration. Samples were collected with various samplers and filters at two different locations using our collection system. The first location was using a robotic welding system while the second was manual welding. Collected samples were analysed for mass concentration using gravimetry and metal concentration using ICP/OES. More advanced analysis was performed on selected filters using X-Ray Photoelectron Spectroscopy to determine surface composition of the particles, and X-Ray Diffraction to determine chemical composition of the fumes. Results showed that the robotic system had a lot of variation in space when the collection system was located close to the weld. Collection efficiency was found to be quite variable depending upon the type of filter. As well, metal concentrations in blank filters were dependent upon the type of filter with MCE presenting with the highest blank values. Results obtained with the XRD and XPS systems showed that it was possible to analyse a small of powdered welding fume sample but results on filters were not conclusive.

## 1. Introduction

It is well-known that exposure to welding fumes, especially in poorly ventilated areas, causes adverse pulmonary health effects. Metal fume fever, presenting with flu-like symptoms, is among the most common occupational illnesses in welders [1]. Other diseases include siderosis, which relates to the accumulation of iron oxide particles in the lungs [2, 3]. In addition, welders have an increased risk of developing chronic obstructive pulmonary disease (COPD) as well as occupational asthma and pneumonia [4, 5, 6]. Long-term exposure to high concentrations of welding fumes is associated with increased incidence of interstitial pulmonary fibrosis, where permanent tissue damage and scarring occurs in the lungs. This illness contributes to poor quality of life and premature death [7, 8]. The International Agency for Research on Cancer has classified welding fumes as possible carcinogens [9]. It is believed that lung exposure to metal oxide particles produced by the welding process induce pro-inflammatory cytokine production and oxidative stress. Indeed, increased levels of cytokines have been detected in the blood of welders exposed to zinc oxide welding fumes in controlled settings [1].

[1] To whom any correspondence should be addressed

It is well-known that the welding process produces particles ranging from few nanometers to few micrometers [10, 11, 12, 13, 14, 15]. Very detailed characterization of fumes have been done in controlled settings using scanning or transmission electron microscopy techniques coupled with energy dispersive spectroscopy (EDS), X-ray fluorescence (XRF), and X-ray photoelectron spectrometry (XPS) [16, 17, 18, 19, 20]. The main compound in welding fumes is iron oxide but manganese is present in significant concentrations. Carbon, oxygen, potassium, silica, and fluoride have also been found in particles produced by welding techniques using a flux [17, 18, 19]. It has been suggested that, in some cases, a core-shell structure is present for welding particles [18]. In addition, fume generation rate may vary with the welding technique [20]. The aim of the study was to develop a sampling and analytical technique, using common filters and samplers, to performed detailed characterization of welding fumes in personal exposure samples. It was important to find a filter suitable for the determination of both mass and metal concentration in one sample. It was also important to evaluate the efficiency of collection of various samplers.

## 2. Materials and methods

### *2.1. Sample collection*

Welding fumes were collected at two different locations: the Canadian Centre for Welding and Joining (CCWJ), Department of Chemical Engineering, University of Alberta, and the Northern Alberta Institute of Technology (NAIT).

Since the aim of the project was also to test various samplers, a collection system was built to allow for multiple sample collection at once. Sampling assembly was based on a design from Zugasti et al. [21]. The assembly was composed of two vertical fiberglass rods inserted in two support stands and two horizontal fiberglass rods. Samplers were mounted as close as possible to each other on two separate rods. Five samplers were mounted per horizontal rod for a total of ten samplers, and the distance between the rods was kept at four to five inches. Personal sampling pumps Universal PCXR4 (SKC, Eighty Four, Pennsylvania) were used to collect samples. Pumps were calibrated prior to and after sampling and the average flow was used to calculate the sampling volume. All pumps were calibrated to run at a flow of 2 L/min since it is the recommended flow for the IOM sampler. For each sampling day, particle number concentrations were determined using the P-Trak (TSI, Shoreview, Minnesota). The P-Trak was located just next to the sampling assembly. All samplers were cleaned using a 2% detergent solution (Citranox, Alconox, White Plains, New York) prior to be used to avoid contamination of samples.

At both locations, welding was performed on carbon steel (Wilkinson Steel and Metals, Surrey, British, Columbia). Metal concentration from the steel was obtained from the Material Safety Data Sheet and was 91-99 % iron, < 2% manganese, < 0.1% chromium, < 1% nickel, and < 1% copper.

#### *2.1.1. CCWJ*

Welding fumes were generated using an Arc Mate 100iC robot (Fanuc Robotics, Rochester Hills, Michigan). The power source was a Power Wave 455M/STT (Lincoln Electric, Toronto, Ontario) and the feeder was a 4R 220 Autodrive (Lincoln Electric, Toronto, Ontario). All welding fume generation was performed using a program developed for gas metal arc welding (GMAW) with a wire feed speed of 150 in/min and a voltage of 20 V. The wire was a 0.045 SuperArc L-56 (ER70S-6, Lincoln Electric, Toronto, Ontario) with a diameter of 0.045 in. The travel speed of the robot was 6.8 in/min and welding lasted about 10 min per experiment. Plates were grinded to remove rust prior to being used, and a new plate was used for each run. Sample collection was performed at CCWJ on 28 October and 7 November 2014.

The sampler assembly was put on a table located along the plate being welded, with the lowest rod mounted twenty centimeters higher than the plate to simulate the breathing zone of a welder. Local exhaust ventilation was installed 30 in behind the samplers at an angle of 45° (Figure 1).

**Figure 1.** Sampling layout at CCWJ

On October 28th, three samplers were tested: the conductive plastic IOM sampler (SKC, Eighty Four, Pennsylvania), 37 mm total dust cassettes (SKC, Eighty Four, Pennsylvania), and a 47 mm Teflon PFA filter holder (SKC, Eighty Four, Pennsylvania). The testing was carried out in three runs with three samplers of each type attached to the rod for each run (Figure 1) for a total of 9 samples per sampler. Location of each sampler on the rod was rotated between runs to get all samplers in all positions. Sampling was carried out using 0.4 µm polycarbonate membranes (SKC, Eighty Four, Pennsylvania). On November 7th, sampling was carried out using only the IOM and the total dust cassettes (SKC, Eighty Four, Pennsylvania) in three different runs. For each run five of each sampler was tested. In the first run samplers were loaded with a 5 µm PVC filter (SKC, Eighty Four, Pennsylvania), and in runs 2 and 3 samplers were loaded with 0.4 µm polycarbonate membranes (SKC, Eighty Four, Pennsylvania) in order to verify the collection efficiency of different filters.

### *2.1.2. NAIT*

Sample collection was performed while students were welding on December 3rd, 2014. Samples were collected while 3rd year students were working on a pipe using shielded metal arc welding (SMAW) using 0.0125 in diameter E6010 (Lincoln Electric, Toronto, Ontario) and 0.09375 in diameter E7018 electrodes (Blueshield, Edmonton, Alberta). Afternoon classes start at 11 am and generally end around 3 pm, therefore sampling was carried out for 2h45.

The sampler assembly was mounted on a mobile cart located 1.2 m from the weld with the lowest rod mounted 20 cm higher than the weld height to simulate breathing zone of the welder. Local exhaust ventilation was located behind the samplers (Figure 2).

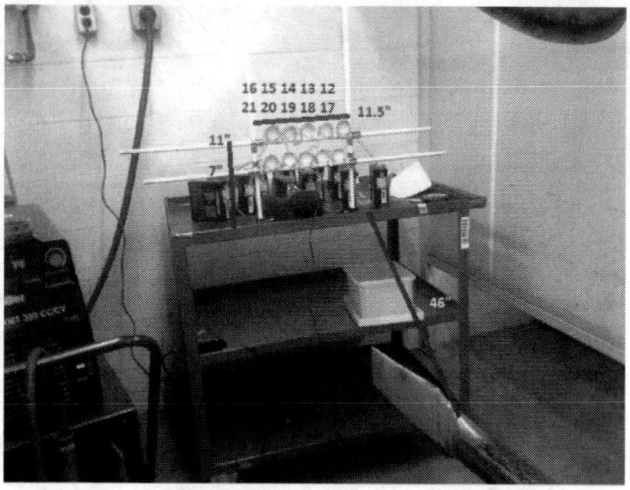


**Figure 2.** Sampling layout at NAIT

Sampling was carried out using total dust cassettes (SKC, Eighty Four, Pennsylvania) loaded with 5 μm PVC filters (SKC, Eighty Four, Pennsylvania) and in total ten samples were collected. This sampling was performed to verify sampling variation for each sampler as results from CCWJ showed a lot of variation within samplers.

### *2.2. Analysis*

All samples were analysed for mass concentration and blank filters were analysed for metal concentration. In addition, two bulk welding fume samples were collected at NAIT on the ground after class and sieved down to a fraction lower than 106 μm. These samples were used to develop analytical methods, mainly XRD and XPS. One of the samples was collected after shielded metal arc welding (SMAW) and the other one was collected after flux core arc welding (FCAW).

#### *2.2.1. Mass concentration*

All membranes were weighed pre and post-sampling to calculate the mass on the filter. Weighing was performed in a temperature and humidity-controlled room at the Department of Civil and Environmental Engineering (University of Alberta, Edmonton, Alberta). All filters were equilibrated with the room humidity for 24 hours prior to be weighed. Since indoor air is very dry in winter in Edmonton, static charging is an issue for weighing filters. All filters were passed through and antistatic device (Mettler Toledo, Mississauga, Ontario) prior to be weighed. All weighing was performed on a microbalance (Sartorius, Goettingen, Germany). To account for variation in weighing, blanks filters were weighed as samples for each sampling day.

#### *2.2.2. Metal concentration*

Metal analysis were performed on an ICP/OES iCAP 6000 (Thermo Fisher Scientific, Cambridge, UK) at the Natural Resources Analytical Laboratory (Faculty of Agricultural, Life and Environmental Sciences, University of Alberta, Edmonton, Alberta). The laboratory uses an Yttrium internal standard to correct for matrix interference and the software corrects for iron interference. All blank filters were digested using a 4:1 $HNO^3$:HCl solution in a microwave oven [22] and diluted to 25 mL with deionized water prior to analysis. The SMAW bulk sample was analysed as well to verify the suitability of the method. The analytical run included digestion blanks, instruments blanks, and an external reference check (200 ppb per metal). Samples were analysed for Al, Ba, Be, Ca, Cd, Co, Cr, Cu, Fe, K, Mg, Mn, Mo, Ni, Pb, Si, V, and Zn. These metals were selected since they may be present in the steel or in the electrodes.

*2.2.3. XRD and XPS analysis*
X-Ray Diffraction analysis (XRD) was performed on a Rigaku XRD Ultima IV (Rigaku, Tokyo, Japan) at the Department of Chemical Engineering (University of Alberta, Edmonton, Alberta) on both bulk samples. X-Ray Photoelectron Spectroscopy (XPS) was performed on an AXIS 165 (Kratos, Manchester, United Kingdom) using a monochromatized Al Kα source (hv=1486.6 ev) at an emission current of 12mA and HT=14kV. The XPS is located the Department of Physics (University of Alberta, Edmonton, Alberta) and analysis was done only on the FCAW bulk sample. In all cases analyses were performed directly ion the powered sample.

*2.3. Statistical analysis*
All statistical tests were performed on SPSS Statistics software version 22 (IBM, Armonk, New York). Mann-Whitney U test was used to compare samplers or filters. Averages and standard deviations were calculated from Excel (Microsoft, Redmond, Washington).

## 3. Results and discussion

*3.1. Mass concentration*
Blanks filters for all day of analysis showed differences between pre and post weighing from 0.001 to 0.002 mg. Therefore weighing was considered stable between pre and post weighing days.

Results from CCWJ showed a lot of variation for both days of analysis within and in between samplers (Figures 3 and 4) and results from November 7th are shown in Table 1.

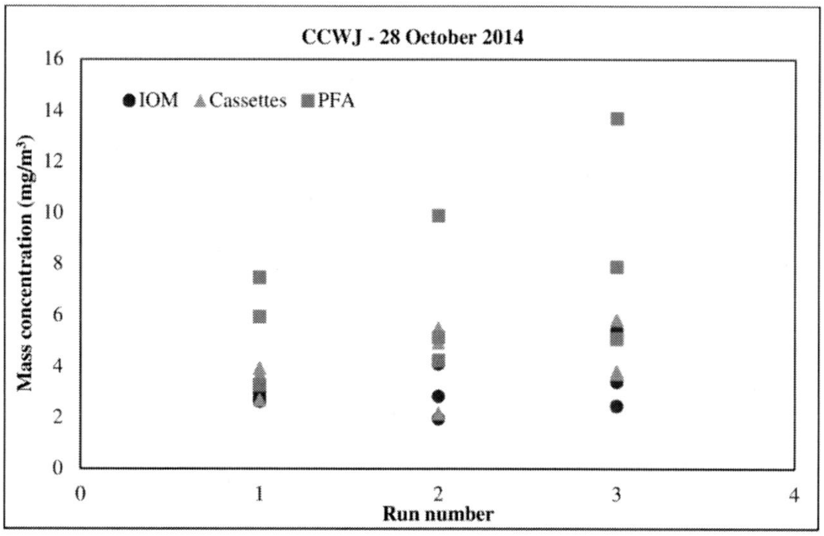


**Figure 3.** Comparison of three samplers loaded with 0.4 μm polycarbonate membranes

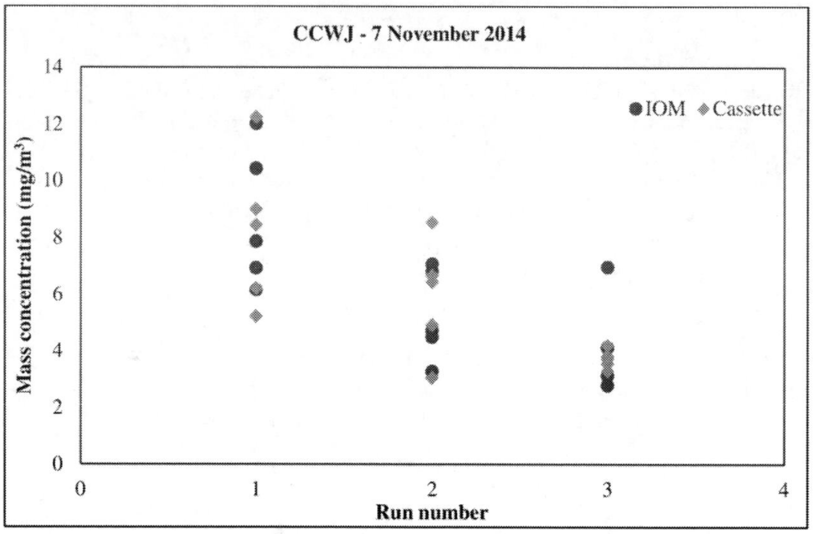


**Figure 4.** Comparison of the IOM sampler and the total dust cassette loaded with PVC filters (run 1) and polycarbonate filters (runs 2 and 3)

The results show that there was a lot of variation when sampling at CCWJ as compared to what was obtained in the study done by Zugasti et al. were they reported coefficient of variation close to 4% [21]. In fact, coefficients of variations were 30-35% for each sampler except for the PFA sampler that showed a coefficient of variation of 46%.

Average results for each samplers obtained on October 28th were respectively 3.16 mg/m$^3$ (standard deviation = 1.03 mg/m$^3$, n = 9) for the IOM sampler, 4.01 mg/m$^3$ (standard deviation = 1.21 mg/m$^3$, n = 9) for the total dust cassette, and 6.94 mg/m$^3$ (standard deviation = 3.24 mg/m$^3$, n = 9) for the PFA filter holder. Mann-Withney U tests were run to verify if results were significantly different between samplers. Results from the statistical tests showed that the IOM was not significantly different than the cassette (U = 23; Z = -1.545; p = 0.122), the IOM was different than the PFA filter holder (U = 6; Z = -3.046; p = 0.002), and the total dust cassette was different than the PFA (U = 14; Z = -2.340; p = 0.019). Since the PFA filter holder gave higher mass concentrations than the IOM sampler or total dust cassette, it was decided not to use it in further testing.

**Table 1.** Results from November 7th 2014 using PVC and polycarbonate filters

| | | IOM sampler | Total dust cassette |
|---|---|---|---|
| PV filter | Average (mg/m$^3$) | 8.7 | 8.2 |
| | Standard deviation (mg/m$^3$) | 2.5 | 2.7 |
| | Coefficient of variation (%) | 26 | 33 |
| Polycarbonate membrane | Average (mg/m$^3$) | 4.8 | 4.9 |
| | Standard deviation (mg/m$^3$) | 1.6 | 2 |
| | Coefficient of variation (%) | 34 | 39 |

In the experiment from 7 November, results with both PVC filters and polycarbonate filters showed no difference between the IOM sampler and the total dust cassette. However a difference was noted between the two types of filters with averages for PVC and polycarbonate respectively at 8.45 and 4.79 mg/m$^3$. This difference was found to be significant when using the Mann-Whitney U test (U =

23; Z = -3.388; p = 0.001). Therefore the polycarbonate filters are not suitable to determine accurately the mass concentration in air samples since they underestimate the results.

Since results obtained from CCWJ were very variable, another test was performed at NAIT on December 3rd to verify the variation in mass concentration. Results are shown in table 2.

**Table 2.** Results obtained at NAIT with total dust cassettes and PVC filters 5 µm

| Sample number | Mass concentration ($mg/m^3$) |
|---|---|
| 1 | 0.438 |
| 2 | 0.478 |
| 3 | 0.487 |
| 4 | 0.510 |
| 5 | 0.528 |
| 6 | 0.438 |
| 7 | 0.487 |
| 8 | 0.535 |
| 9 | 0.494 |
| 10 | 0.540 |
| Average | 0.493 |
| Standard deviation | 0.036 |

Results obtained at NAIT show much better reproducibility of the sampling and analytical method with a coefficient of variation of 7%, which was closer to what had been previously published [21]. Various reasons can explain the variability obtained when sampling at CCWJ. The samplers were mounted very close to the area being welded. It is possible the welding process produces a lot of spatial variations which would have been recorded by our sampling design. In addition, sampling at NAIT was carried out for a longer time therefore any error in the timing would result in a lower error on the calculation of the total volume sampled.

### *3.2. Metal concentration*

Instrument blank was negligible for all metals analysed. External reference check gave average recovery values from 82% (silica) to 102% (zinc). Results for digestion blanks and various filters are shown in Table 3.

**Table 3.** Metal results for acid blanks and selected filters

| **Filter type** | **Al (ng)** | **Ba (ng)** | **Be (ng)** | **Ca (ng)** | **Cd (ng)** | **Co (ng)** | **Cr (ng)** | **Cu (ng)** | **Fe (ng)** |
|---|---|---|---|---|---|---|---|---|---|
| Digestion Blank | 169 | 23 | ND | 752 | ND | ND | ND | 196 | 134 |
| MCE[a] | 988 | 113 | ND | 47291 | ND | 1 | 357 | 305 | 1730 |
| PVC | ND | 29 | ND | 907 | ND | ND | 41 | ND | 72 |
| Fluoropore[b] | ND | 28 | ND | ND | ND | ND | 7 | ND | 99 |
| Omnipore[c] | ND | 18 | ND | 583 | ND | ND | 20 | ND | ND |
| Polycarbonate | ND | 23 | ND | 494 | ND | ND | 19 | ND | 159 |

| **Filter type** | **K (ng)** | **Mg (ng)** | **Mn (ng)** | **Mo (ng)** | **Ni (ng)** | **Pb (ng)** | **Si (ng)** | **V (ng)** | **Zn (ng)** |
|---|---|---|---|---|---|---|---|---|---|
| Digestion Blank | ND | 61 | 4 | ND | ND | 2 | 365 | ND | 105 |
| MCE[a] | 1685 | 11614 | 81 | 2 | 20 | 13 | | 24 | |
| PVC | 1180 | 56 | 2 | ND | 27 | ND | 1546 | ND | 75 |
| Fluoropore[b] | 600 | ND | 1 | ND | 2 | ND | 246 | 8 | 3 |
| Omnipore[c] | ND | 52 | ND | ND | ND | ND | 368 | ND | 28 |
| Polycarbonate | ND | 11 | 4 | ND | 74 | ND | 334 | ND | 16 |

[a] Mixed cellulose ester
[b] Hydrophobic PTFE membranes (Millipore, Billerica, Massachusetts)
[c] Hydrophilic PTFE membranes (Millipore, Billerica, Massachusetts)

Both calcium and potassium gave blank values quite high in the microgram range. MCE filters gave blank values higher than the digestion blanks for Al, Cr, Cu, Fe, Mn, Ni, Pb, and V. Since these metals are of interest in welding fumes, MCE filters were not used in our experiments. For elements other than Ca and K, blank values for filters were similar to digestion blanks suggesting than most of the contamination was coming from the acid digestion. It is unknown why polycarbonate membranes gave a higher value for nickel as compared to PTFE membranes. The Omnipore hydrophilic PTFE membranes (Millipore, Billerica, Massachusetts) showed to be suitable for metal analysis. Further investigation will determined if the membrane is also suitable for the determination mass concentration in samples.

Detection limit was calculated as three times the standard deviation of the digestion blank [23] and quantification limit was calculated as five times the detection limit. Detection limits and quantification limits are given in Table 4. These values are considered sufficient for the analysis of welding fumes.

**Table 4.** Detection limits and quantification limits for selected metals

| | **Al (µg)** | **Cd (ng)** | **Co (ng)** | **Cr (ng)** | **Cu (µg)** | **Fe (µg)** | **Mn (ng)** | **Mo (ng)** | **Ni (ng)** | **Pb (ng)** | **V (ng)** | **Zn (µg)** |
|---|---|---|---|---|---|---|---|---|---|---|---|---|
| Detection limit | 1.7 | 1 | 11 | 19 | 1.3 | 0.5 | 9 | 9 | 18 | 52 | 10 | 0.4 |
| Quantification limit | 8.9 | 6 | 56 | 94 | 6.6 | 2.3 | 46 | 47 | 88 | 261 | 48 | 1.8 |

Further work will involve the analysis of real welding fume samples using various instruments to confirm measurements.

### *3.3. XRD and XPS analysis*

Results for XRD showed that both FCAW and SMAW bulk sample contained iron and manganese oxides. Silica was also found in both fumes while titanium was found only in the SMAW sample. An example of graph is given in Figure 5.

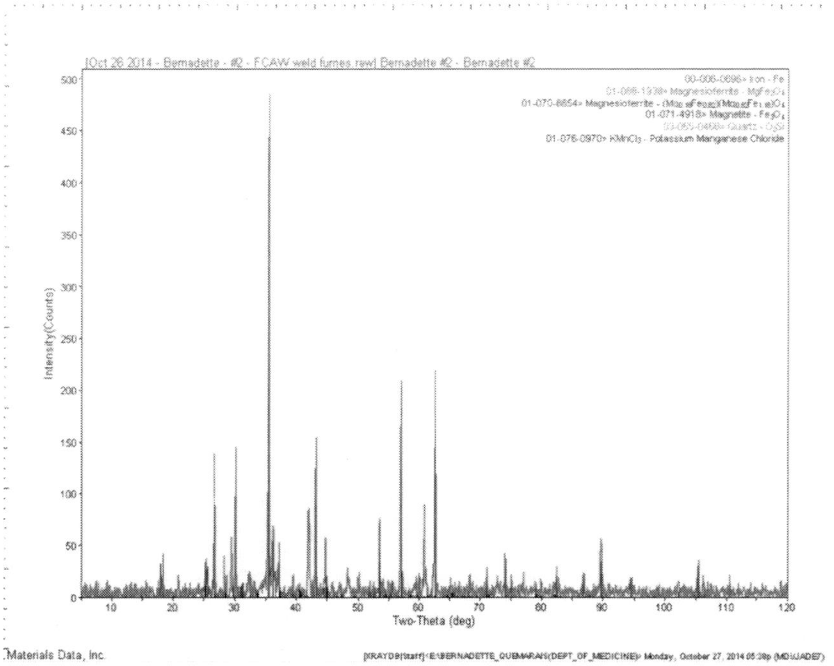


**Figure 5.** XRD analysis on FCAW bulk sample.

Results for XPS are shown in figure 6. The surface chemistry of the FCAW samples shows a lot of different chemical species due to the presence of a flux on the FCAW electrode.

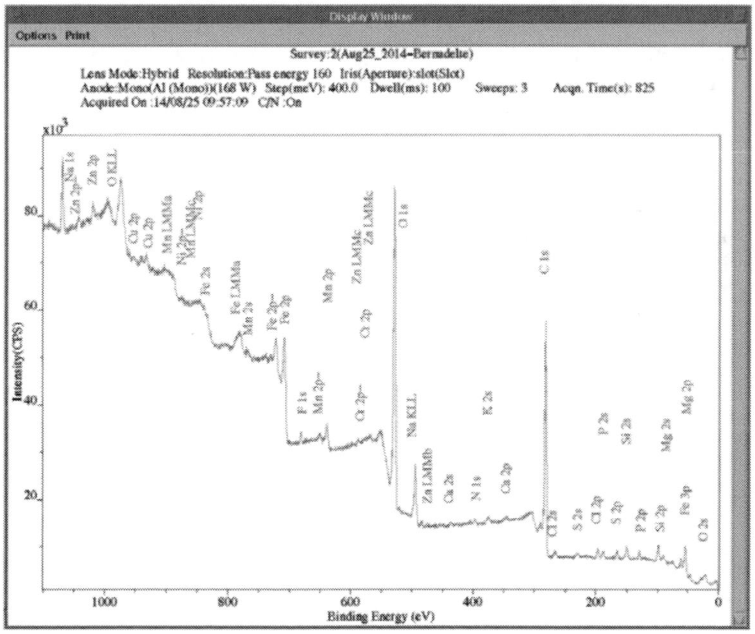


**Figure 6.** Analysis of a FCAW bulk sample using X-Ray Photoelectron Spectroscopy

Results obtained with XRD and XPS showed that it is possible to perform this type of analysis on small welding fume bulk samples. Unfortunately analysis on filters was not successful because the filters presented some white surfaces. Further investigations are necessary to improve this type of analysis.

## Acknowledgements

We want to thank Marcio Correa de Carvalho from CCWJ for his help with the robotic system, and Marcelo Mena from NAIT his help in organizing the sampling at NAIT. We also want to thank Candis Scott for her valuable help in the laboratory and in the field. This work was supported by a research contract from the Government of Alberta, Jobs, Skills, Training, and Labour, Occupational Disease Prevention Unit.

## References

[1] Blanc P D *et al* 1993 *American Review of Respiratory Disease* **147** 134
[2] Doig A T and McLaughlin A I G 1936 *The Lancet* **227** 771
[3] Antonini J M *et al* 2003 *American Journal of Industrial Medicine* **43** 350
[4] Wang Z P *et al* 1994 *American Journal of Industrial Medicine* **26** 741
[5] Beach J R *et al* 1996 *American Journal of Respiratory and Critical Care Medicine* **154** 1394
[6] Coggon D *et al* 1994 *The Lancet* **344** 41
[7] Buerke U *et al* 2002 *American Journal of Industrial Medicine* **41** 259
[8] Simon-Deckers A *et al* 2012 Role of metal oxide nanoparticle in welder's lung injury Nanosafe 2012 Grenoble France
[9] International Agency for Research on C 1990 *Iarc monographs on the evaluation of carcinogenic risks to humans - chromium, nickel and welding* IARC (Geneva) Volume 49
[10] Debia M *et al* 2013 *Journal of Occupational and Environmental Hygiene*
[11] Lehnert M *et al* 2012 *The Annals of Occupational Hygiene* **56** 557
[12] Dasch J and D'Arcy J 2008 *Journal of Occupational and Environmental Hygiene* **5** 444
[13] Pfefferkorn F E *et al* 2010 *The Annals of Occupational Hygiene* **54** 486
[14] Moroni B and Viti C 2009 *Journal of Aerosol Science* **40** 938
[15] Zimmer A T and Biswas P 2001 *Journal of Aerosol Science* **32** 993
[16] Jenkins N T and Eagar T W 2005 *Welding Journal (Miami, Fla)* **84** 87
[17] Sowards J W *et al* 2008 *Welding Journal (Miami, Fla)* **87** 106
[18] Sowards J W *et al* 2010 *Welding Journal (Miami, Fla)* **89** 82s
[19] Sowards J W *et al* 2008 *Welding Journal (Miami, Fla)* **87** 76
[20] Fuglsang K *et al* 2011 Measurement of ultrafine particles in emissions from welding processes 16th International Conference on Joining Materials Elsinore, Denmark
[21] Zugasti A *et al* 2012 *Journal of Environmental Monitoring* **14** 375
[22] International Standard Organization 2012 *Workplace air — determination of metals and metalloids in airborne particulate matter by inductively coupled plasma atomic emission spectrometry — part 2: Sample preparation* ISO (Geneva Switzerland) pp 54
[23] Howard A G and Statham P J 1997 *Inorganic trace analysis: Philosophy and practice* John Wiley & Sons (Chichester Sussex England) pp 182

# Design of an exposure chamber to test samplers used in the evaluation of personal exposure to nanoparticles

**R Amin[1], H Izadi and B Quémerais**

Division of Preventive Medicine, University of Alberta, 107 University Campus NW, Edmonton, AB T6G 2S2, Canada

mdruhul@ualberta.ca

**Abstract.**

The aim of this study was to design a laboratory size exposure chamber for the testing of samplers used to collect personal exposure samples for nanoparticles. A polyethylene cylindrical container with a diameter of 42 cm and height of 60 cm was used as the testing chamber. The chamber was divided into 2 parts by an aluminium honey comb. Particles generated using a 1 jet Collison nebulizer (BGI) operating at a flow rate of 4L/min were inserted into the chamber via a tube located near to the top of the chamber. A heater was inserted just after the nebulizer to avoid condensation of water in the tubing, and dilution air, running at 10L/min was inserted just after the heater. As particle charge can dramatically affect sampling a particle neutralizer was attached to the generation system so as to neutralize the particles before they enter the chamber. A diffusion dryer was used to remove any water from the air stream prior to enter the chamber. A fan was used to mix and distribute the generated particles. After generation and mixing, the particles passed through the aluminium honeycomb which is essential to eliminate any turbulent or unwanted air flow. Six sampling ports along with a pressure gauge were placed on the walls 15 cm from the bottom of the chamber. The pressure gauge was added to ensure the desired pressure is achieved during sampling. The sampling ports allowed for the connection of five samplers and sampling pumps as well as the connection of an ultrafine particle counter. The exposure chamber was developed to assess various samplers for carbon nanotubes and cellulose nanocrystals. Results showed that the chamber was working properly and that mixing was sufficiently uniform to test samplers.

## 1. Introduction

The Nanotechnology Consumer Products Inventory lists over 1,600 consumer products containing nanomaterials [1]. In 2008, there were about 400,000 workers worldwide exposed to nanomaterials. This number is estimated to grow up to 2 million workers in 2015 [2]. Carbon nanotubes, in particular, have become a concern since they are believed to have asbestos-like fiber toxicity leading to mesothelioma and pulmonary fibrosis [3]. Recent animal studies have shown that exposure to carbon nanotubes lead to changes in mesothelial tissue [3, 4, 5]. However, there are no standardized methods for the sampling and analysis of nanomaterials and only few recommendations for occupational

[1] To whom any correspondence should be addressed

exposure limits are available [6]. In an effort to protect their workforce in the lack of methodology and regulation, the National Institute for Occupational Safety and Health (NIOSH) wrote a comprehensive report in which they recommend the use of NIOSH methods 5040 and 7402 to assess exposure to carbon nanotubes. In addition, NIOSH recommends to use an exposure limit of 1 $\mu g/m^3$ for elemental carbon in the respirable fraction, providing that carbon nanotubes and nanofibers are predominant in the sample [7]. The Government of Alberta, Canada, is concerned about the exposure to carbon nanotubes and has asked the University of Alberta to verify if it is practical to routinely use these methods to assess workers exposure to carbon nanotubes.

Various samplers are available for the collection of respirable particles. However, in North America, the IOM sampler coupled with polyurethane foam (Institute of Occupational Medicine, Edinburg, Scotland), and the cyclone coupled with a total dust cassette are by far the samplers the most commonly used to collect respirable aerosols. It was decided to test both the IOM and the cyclone to evaluate the difference between these samplers, and their efficiency in collecting carbon nanotubes.

Unfortunately there was no known workplace using or producing carbon nanotubes. It was then decided to build an exposure chamber to test our samplers. This paper describes the design of the exposure chamber.

## 2. Materials and methods

### *2.1. Exposure chamber system*

The exposure chamber system was composed of an aerosol generator system, an aerosol neutralizer, and the exposure chamber itself. The whole system was based on designs previously published [8, 9, 10].

*2.1.1. Aerosol generation.* Aerosol generation was performed using a 1 jet Collison nebulizer (BGI, Waltham, Maryland)) followed by a heater to avoid condensation of water inside the tubing. A solution of carbon nanotubes in purified water (2.5 g/L) was prepared directly in the nebulizer jar and placed in an ultrasonic bath for 15 minutes. The gas pressure used was 50 psi, which produces a flow of approximately 4 L/min at the outlet of the nebulizer. To balance the inlet flow to the outlet flow inside the exposure chamber, purified compressed air (ultra-zero air, Praxair, Mississauga, Ontario), was added just after the heater at a flow of approximately 10 L/min.

*2.1.2. Diffusion dryer.* In order to remove water from the aerosol particles, a diffusion dryer filled with indicating silica gel was installed between the aerosol generating system and the aerosol neutralizer.

*2.1.3. Aerosol neutralizer.* Since the indoor air in the location where experiments are conducted (Edmonton, Canada) is extremely dry in winter (approximately 20% relative humidity), static charging is a significant problem. An EAN 581 electrostatic aerosol neutralizer (Topas, Dresden, Germany) was used to eliminate the effects of static charging and to balance the positive and negative charges inside the chamber.

*2.1.4. Exposure chamber.* A 60 cm height and 46 cm outside diameter polyethylene tank was used as the testing chamber. The aerosol inlet was inserted on top of the chamber along with a fan to promote mixing of the particles. An aluminium honeycomb (5 cm height) was inserted into the chamber 10 cm from the top to eliminate unwanted turbulence inside the chamber. Six ports and a pressure gauge were placed at 15 cm from the bottom of the chamber. The ports allowed for the connection of five samplers and a P-Trak (TSI, Shoreview, Minnesota). The pressure gauge was used to verify the pressure inside the chamber since the chamber had to be run at atmospheric pressure. Layout of the chamber is shown in Figure 1 and scheme of the whole system is shown in Figure 2.

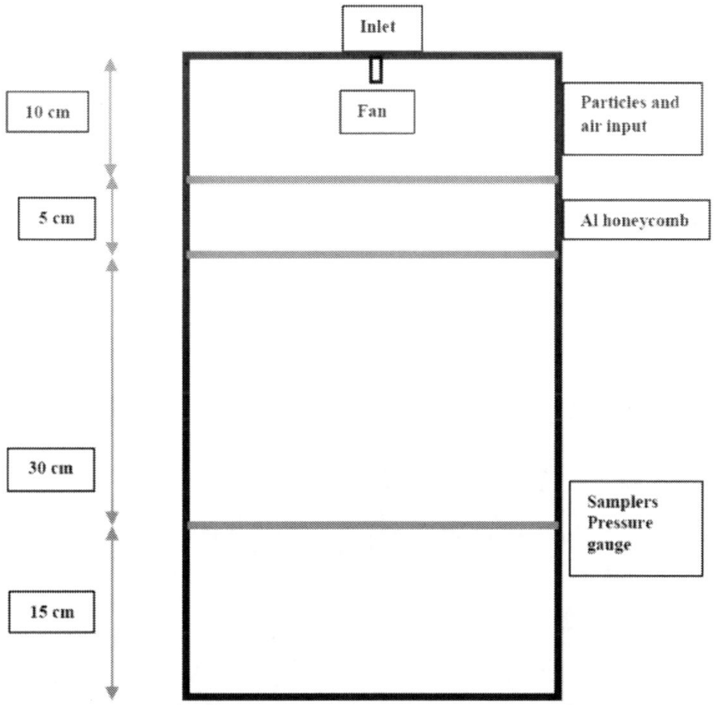


**Figure 1.** Layout of the exposure chamber

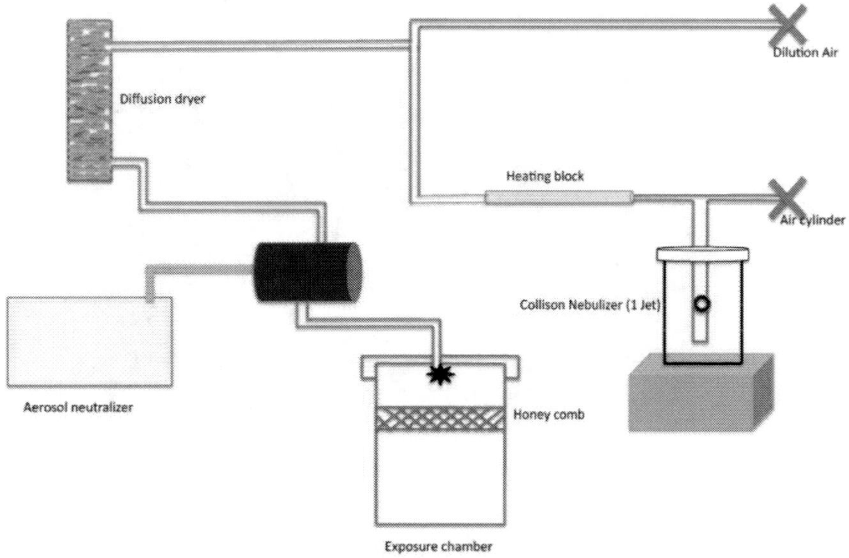


**Figure 2.** Design of the overall system

### *2.2. Testing of samplers*

Testing was performed using two types of samplers: 1) the IOM aerosol sampler (conductive plastic SKC, Eighty Four, Pennsylvania), and 2) a total dust cassette connected to a cyclone (GS3, SKC, Eighty Four, Pennsylvania). The IOM sampler was equipped with polyurethane foam to collect only the respirable fraction. The IOM sampler runs at 2 L/min while the cassette/cyclone assembly runs at 2.75 L/min. Preliminary testing was performed using GLA-5000 PVC filters 25 mm and 37 mm diameter, 5 μm porosity (SKC, Eighty Four, Pennsylvania). The chamber was operated 4 times at a carbon nanotube concentration of 2.5 g/L. In each run, 2 samplers of one type and 3 samplers of the other type were connected to the sampling ports and the P-Trak was attached to the 6$^{th}$ port just to control the particle count. In total ten filters were collected with the IOM sampler and ten filters were collected with the cassette/cyclone assembly. Pumps were calibrated prior to and after sampling and average flow was used. PVC filters were weighed before and after sampling in a temperature and humidity-controlled room. Filters were allowed to equilibrate in the room for 24 hours before being weighed. The sampling set-up is shown in Figure 3.

**Figure 3.** Sampling set-up.

## 3. Results and discussion

Preliminary testing showed that the mixing inside the chamber was not working properly as a visual difference was noted between filters (Figure 4).

**Figure 4.** Preliminary testing performed using IOM samplers. Sample on the left shows fewer particles on the filter than the other four samples.

It was found that there were leaks at this particular port. All the ports were sealed to avoid potential leaks using silicone paste. In the following test, filters did not show any visual difference in particulate loading (Figure 5) therefore both samplers were tested and particulate concentration was measured.

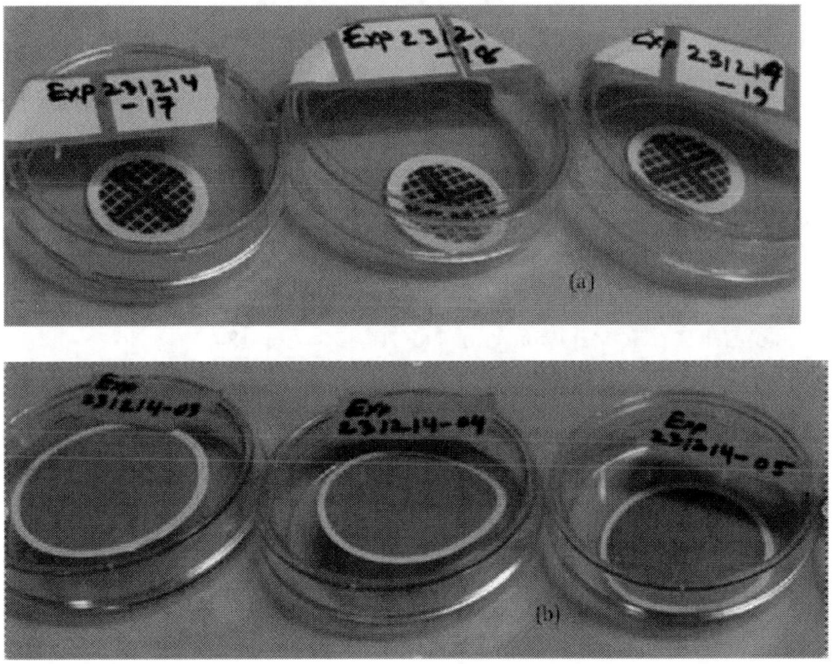


**Figure 5.** Loading on filters for a) IOM sampler, and b) cassette/cyclone assembly

For the cassette/cyclone assembly, the mean mass concentration was 0.81 $mg/m^3$ (standard deviation: 0.14 $mg/m^3$, n = 10). The IOM sampler gave a mean mass concentration of 0.83 $mg/m^3$ (standard deviation: 0.19 $mg/m^3$, n = 10). Average concentrations were similar for both samplers but the IOM sampler showed a higher variation than the cassette/cyclone assembly with respective coefficients of variation of 22.5% and 17%. Number concentration of particles obtained from the P-Trak showed a fast increase from approximately 800 part/mL for the baseline up to 80,000 part/mL during the experiment. The particles concentration stayed stable during the experiments and went back to baseline rapidly at the end of the experiment.

The results show some variation in the concentration of particles within the exposure chamber probably due to the presence of turbulence inside the chamber. However, the variation was considered to be acceptable to test methods and samplers for the collection of nanoparticles. It is interesting to note that, for carbon nanotubes, both types of samplers collect the same amount of particles, therefore they are both suitable to use for the collection and analysis of carbon nanotubes. Further work will include the testing of samplers for the analysis of elemental carbon in carbon nanotubes, testing of the chamber and method development for cellulose nanocrystals and titanium dioxide nanoparticles.

**Acknowledgments**
We want to thank Andrew Martin from Mechanical Engineering and Sam Grazioni at the Department of Medicine workshop for their help in designing and building the exposure chamber. We also want to thank Candis Scott for her valuable help in the laboratory. This work was supported by a research contract from the Government of Alberta, Jobs, Skills, Training, and Labour, Occupational Disease Prevention Unit.

**References**

[1] Lowell A 2013 [Accessed January 19th 2015] *Inventory finds increase in consumer products containing nanoscale materials* Available from: http://www.wilsoncenter.org/article/inventory-finds-increase-consumer-products-containing-nanoscale-materials
[2] Roco M C *et al* 2010 *Nanotechnology research directions for societal needs in 2020 - retrospective and outlook* WTEC (Arlington VA) pp 614
[3] Poland C A *et al* 2008 *Nat Nano* **3** 423
[4] Xu J *et al* 2012 *Cancer Science* **103** 2045
[5] Chen T *et al* 2014 *Toxicology letters* **226** 150
[6] Van Broekhuizen P *et al* 2012 *Annals of Occupational Hygiene* **56** 515
[7] National Institute for Occupational Safety and Health 2013 *Occupational exposure to carbon nanotubes and nanofibers* National Institute for Occupational Safety and Health (Cincinnati Ohio) DHHS (NIOSH) Publication No. 2013–145 pp 184
[8] Noël A *et al* 2013 *Journal of Occupational and Environmental Hygiene* **10** 86
[9] Feather G A and Chen B T 2003 *Aerosol Science and Technology* **37** 261
[10] Golshahi L *et al* 2011 *Journal of Aerosol Science* **42** 474

# Multi-Instrument Manager Tool for Data Acquisition and Merging of Optical and Electrical Mobility Size Distributions

**Torsten Tritscher[1], Amine Koched[2], Hee-Siew Han[3], Eric Filimundi[2], Tim Johnson[3], Sherrie Elzey[3], Aaron Avenido[3], Carsten Kykal[1] and Oliver F. Bischof[1]**

[1] TSI GmbH, Particle Instruments, Neuköllner Str. 4, 52068 Aachen, Germany

[2] TSI France Incorporated, Technopôle de Château-Gombert, 13382 Marseille, France

[3] TSI Incorporated, 500 Cardigan Road, Shoreview, MN 55126, USA

E-mail: torsten.tritscher@tsi.com

**Abstract**. Electrical mobility classification (EC) followed by Condensation Particle Counter (CPC) detection is the technique combined in Scanning Mobility Particle Sizers (SMPS) to retrieve nanoparticle size distributions in the range from 2.5 nm to 1 μm. The detectable size range of SMPS systems can be extended by the addition of an Optical Particle Sizer (OPS) that covers larger sizes from 300 nm to 10 μm. This optical sizing method reports an optical equivalent diameter, which is often different from the electrical mobility diameter measured by the standard SMPS technique. Multi-Instrument Manager (MIM™) software developed by TSI incorporates algorithms that facilitate merging SMPS data sets with data based on optical equivalent diameter to compile single, wide-range size distributions. Here we present MIM 2.0, the next-generation of the data merging tool that offers many advanced features for data merging and post-processing. MIM 2.0 allows direct data acquisition with OPS and NanoScan SMPS instruments to retrieve real-time particle size distributions from 10 nm to 10 μm, which we show in a case study at a fireplace. The merged data can be adjusted using one of the merging options, which automatically determines an overall aerosol effective refractive index. As a result an indirect and average characterization of aerosol optical and shape properties is possible. The merging tool allows several pre-settings, data averaging and adjustments, as well as the export of data sets and fitted graphs. MIM 2.0 also features several post-processing options for SMPS data and differences can be visualized in a multi-peak sample over a narrow size range.

**Keywords:** SMPS, OPS, NanoScan SMPS, wide-range particle measurement, data post-processing

## 1. Introduction

The comprehensive measurement of number concentrations and size distributions of aerosols in the sub-micrometer range has received much consideration during the past years. Assessing possible exposure to nanomaterial used in or added to products but also the investigation of airborne nanoparticles in general requires suitable and sensitive instruments. Toxicology studies reported a high rate of pulmonary deposition of nanoparticles, and also show their ability to travel from lung to systemic sites (Oberdörster et al., 2005). Additionally a high inflammation potential of nanoparticles was reported by other studies. Nanoparticles have a relative large surface area per mass which can be seen as interaction interface. Therefore it has been concluded that particles in the nanometer range are more biologically active (Gurr et al., 2005).

Measuring exposure to nanoparticles has been conducted in academic research using high-end instrumentation in the laboratory. With the need to move these measurements out of the laboratory into the field to study indoor air quality and occupational health and safety, more compact and easy to use instrumentation is required to characterize the aerosol.
Commonly the technique of choice is electrical mobility classification followed by Condensation Particle Counter (CPC) detection. Both are typically combined in a Scanning Mobility Particle Sizer (SMPS) to retrieve nanoparticle size distributions in the range from 2.5 nm to 1 μm. During the last NanoSafe Conference we introduced the NanoScan SMPS (TSI model 3910, Tritscher et al., 2012), which is a novel, portable nanoparticle sizing and counting instrument that is capable of covering the range from 10 to 420 nm. Its straightforward portability is a key to help in source identification measurements.

The detection size range of SMPS systems can be extended by the addition of an Optical Particle Sizer (OPS, TSI model 3330) that covers larger particle sizes from 300 nm to 10 μm and is also a portable and compact device. Its optical sizing method reports an optical equivalent diameter, which is often different from the electrical mobility diameter measured by the standard SMPS technique. Discrepancies between different measurement principles as well as the challenges in combining size spectral data were the motivation to develop a small but powerful data merging tool. Multi-Instrument Manager (MIM™) software was introduced by TSI to integrate algorithms that facilitate the merging of SMPS data based on electrical mobility diameter with data dependent on an optical equivalent diameter into one single, wide-range data set. Here we present the functionality of MIM 2.0, the next-generation of the data merging tool and give some data merging, data acquisition and SMPS post-processing examples.

## 2. Method and Instrumentation

The Multi-Instrument Manager (MIM™) software is a MATLAB-based tool that allows reviewing, averaging, merging and post-processing data from TSI SMPS spectrometers equipped with the previous 3080 and the new 3082 Electrostatic Classifier generations, NanoScan SMPS (TSI model 3910) and/or OPS (TSI model 3330) instruments. MIM software was introduced by TSI after the original development of dedicated algorithms (Han et al., 2011) that facilitate merging SMPS data based on electrical mobility diameter with data based on optical equivalent diameter to compile a single, wide-range data set. MIM 2.0 features increased channel resolution of up to 128 size channels per decade of recorded SMPS raw data files, comparison of SMPS correction steps and combination of multiple units. One of the key features when merging OPS with electrical mobility distributions is that the software takes aerosol optical properties into account by automatically determining an overall aerosol effective refractive index that was used to adjust the merged data. Thus an indirect and average characterization of aerosol optical and shape properties is possible.

In Figure 1 we show an example for measurements from SMPS and OPS with the resulting composite fit. The merging procedure is heavily dependent on the particle type, concentration and distribution.

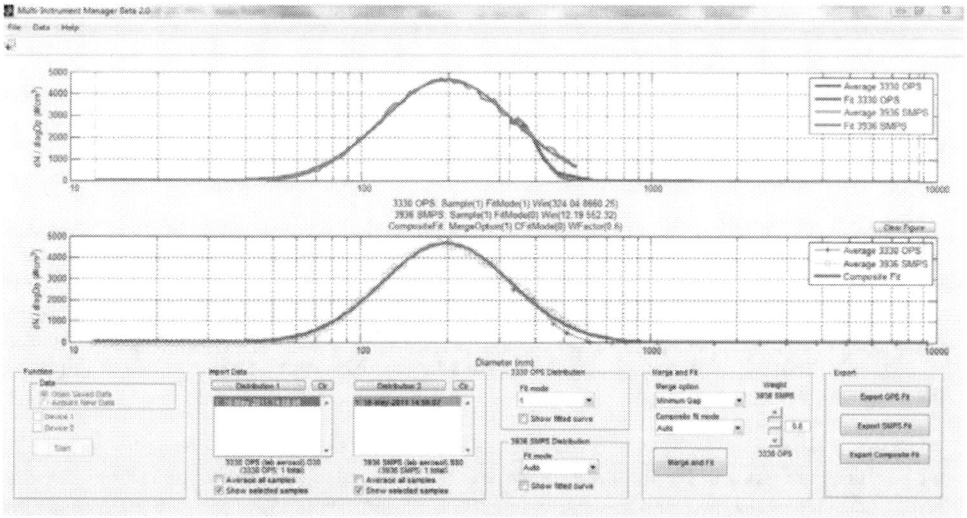


**Figure 1:** Size distribution of ambient air aerosol from SMPS and OPS measurement displayed in the MIM 2.0 view. The top panel shows the individual size distributions as well as their average and the lower graph shows the merged curve as a result in red.

Data merging is done is by combining two size distribution measurements that have been uploaded, which can also be the averages of several size distributions. The Merge and Fit panel (data control area, see Figure 1) allows parameterization of the merge algorithm. The merge option selection panel includes "No Correction", "Minimum Gap" and "Calibration Data". The software's "Minimum Gap" option will seek to minimize the gap (global minimum) between the SMPS and the OPS distributions by varying the real part of the refractive index and shape correction factor of the OPS data. Typical values displayed for the real part of refractive index are between 1.3 to 1.8, and for the shape correction factor 0.8 to 1.05. The optimal refractive index found by the algorithm will provide the "best fit" to the data but is not necessarily equivalent to the bulk refractive index of the aerosol of interest. However, in many cases it can help to provide further insight into the characteristics of the optical aerosol properties.

For the "Calibration Data" option, the algorithm seeks to minimize the difference between the calibration data and reference monodisperse sizes by varying the shape factor as well as the real and imaginary part of the refractive index. This option is only possible if the user created a calibration data set earlier on. As the OPS is calibrated with polystyrene latex (PSL) reference particles, the optically determined particle diameters from the OPS may differ from physical particle diameters due to material properties such as refractive index, morphology and shape. By using an Electrostatic Classifier to supply monodisperse aerosols to the OPS, the OPS can be calibrated to measure mobility diameter. Particles used for such a calibration should be the same or as a minimum comparable to the particles sampled as size distributions that should be merged.

Additional settings like fit mode selection (e.g. auto/1/2/3) help to find the best fitting curve and the weighting option allows prioritizing data either based on SMPS/NanoScan SMPS or OPS in the overlapping region. All merged size distributions and data sets as well as fitted graphs can be

Journal of Physics: Conference Series **617** (2015) 012013 doi:10.1088/1742-6596/617/1/012013

exported. MIM 2.0 permits several pre-settings, data averaging and adjustments. More details on the different options and settings can be found in the MIM user guide.

## 3 Results and Discussion

### 3.1. Multi-Modal Laboratory Aerosol

Data merging of uni-modal size distributions as shown in Figure 1 is typically not very challenging for any fitting procedure. Therefore a laboratory aerosol was generated from two different aerosol sources and subsequently measured. We used Arizona Road Dust in a Small Scale Powder Disperser (SSPD, TSI model 3433) and a sucrose solution in a constant output atomizer (TSI model 3076) with diffusion dryer. NanoScan SMPS and OPS were used to detect and determine the size distribution of the mixed lab aerosol. The two instruments sampled the laboratory aerosol from a mixing chamber (blue "bean pot" as shown in Figure 2).

The aerosol generation was largely stable over a time span of several minutes, which is displayed in the several thin curves in the top panel of Figure 3. The average size distributions from NanoScan SMPS and OPS were used for the data merging but the same is possible also for individual scans. Figure 3 shows the data merging result using the Minimum Gap option in the lower panel and suggested a three modal composite fit. The software tool can easily deal with multi-modal distribution as these results demonstrate.

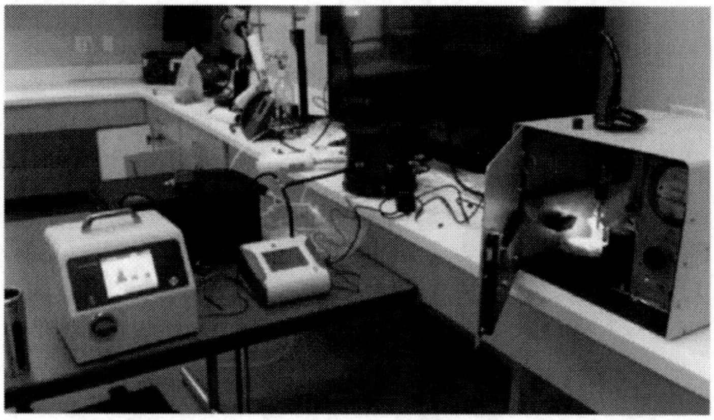

**Figure 2:** Setup with NanoScan SMPS and OPS for laboratory-generated bi-modal aerosols. A constant output atomizer and a Small Scale Powder Disperser (SSPD) were used for the primary generation and the aerosol was subsequently mixed inside the blue mixing manifold.

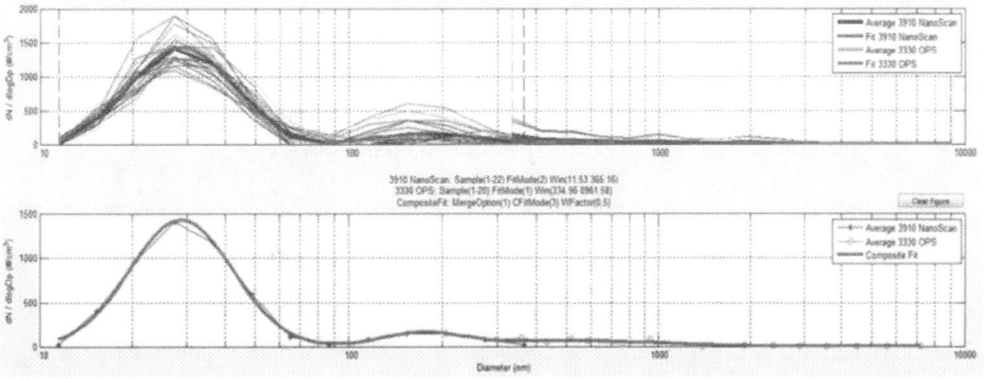

**Figure 3:** Multimodal size distributions of a laboratory aerosol displayed in the MIM 2.0 view. Minimum gap merging and three modal fitting were pre-selected. The red curve in the lower panel shows the resulting fit.

### 3.2. Case Study Fireplace

The capabilities of MIM were further tested for a real-world aerosol that requires a wide measurement size range. In a brief case study we collected data with NanoScan SMPS and OPS instruments from a wood-burning fireplace. The fireplace was located in the living room of a domestic building. The ambient air in the room where the fireplace is located was monitored for two hours, with the instruments set to 1 min sample time per run. Compressed wood was used as fuel, which was again added after one hour of fireplace operation. The details of the actual measurement setup are considered to be of secondary importance here as the main purpose of this work is not the evaluation of the burning behavior but collecting exemplary data for use with the new MIM 2.0 software.

Figure 4 summarizes the results during the two hour measurement period complemented with photos of the corresponding burning phases. The top panels show 1 min size distribution scans and their average value and the red curves in the lower panels present the merged size distribution curves, respectively. The size distribution shown ranges from 10 nm to 10 µm. A nucleation mode develops during the measurement and after an hour forms a stable mode around 20-30 nm, while the accumulation mode at around 100 nm decreases relatively and shifts slowly to size larger than 100 nm. For a more comprehensive study that describes the complete development of the bimodal size distribution of the fireplace emission, smaller average intervals should be chosen as some information gets lost, especially in the beginning when size distributions vary rapidly. However, here we focus exclusively on the MIM processing option and using average values before merging is one possible approach if not every single size distribution should be merged individually.

Journal of Physics: Conference Series **617** (2015) 012013 doi:10.1088/1742-6596/617/1/012013

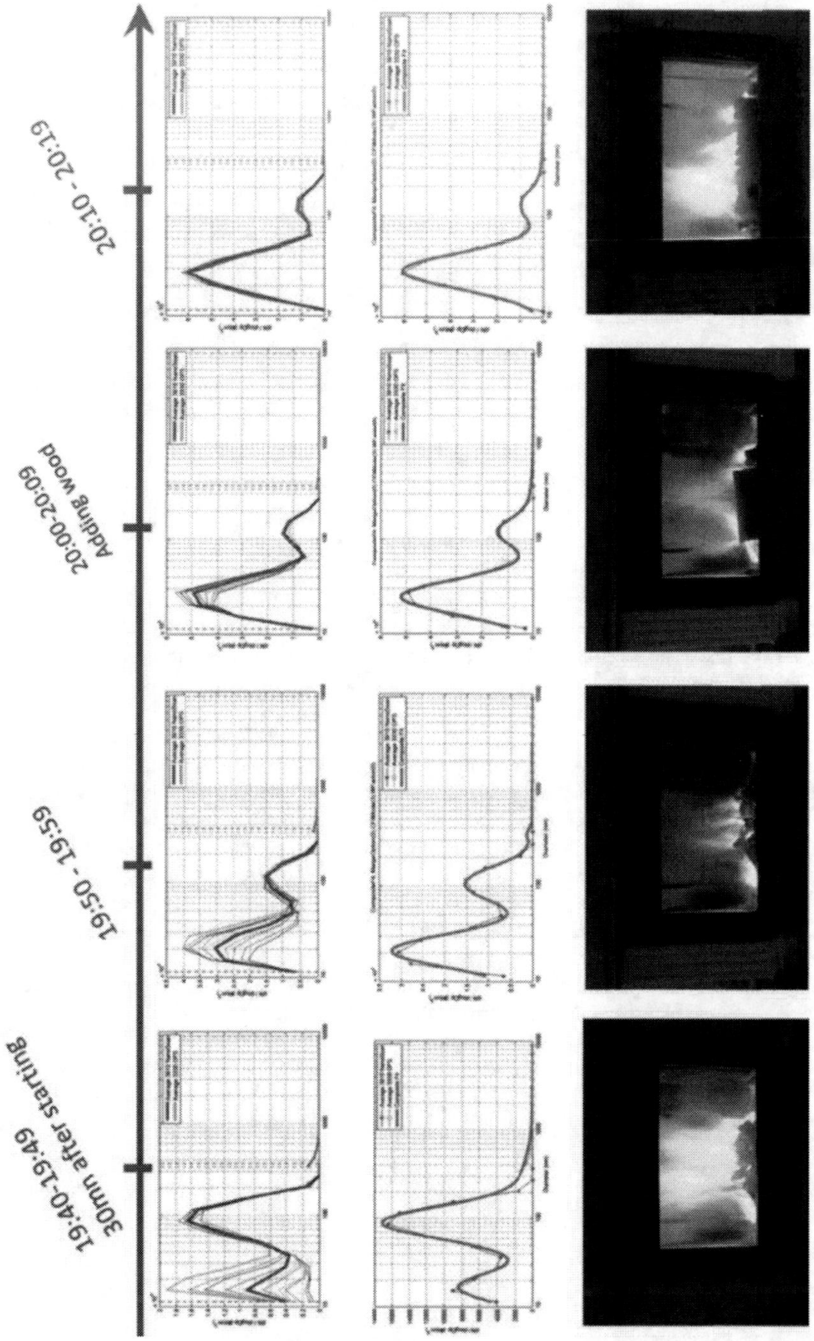


**Figure 4:** Development of measured aerosol size distribution during a fireplace burning cycle. The upper panels show the individual measured size distributions for NanoScan SMPS and OPS and their average (bold curve). The merging results of the bimodal size distributions are shown in the middle panel and on the bottom photographs that illustrate the corresponding burning phases.

### 3.3. Post-processing of SMPS Data

Other than the actual merging of measurement data, MIM 2.0 allows also post-processing electrical mobility size distribution data recorded with the high-resolution component SMPS. Data from the recently-introduced SMPS, TSI model 3938 (Tröstl et al., 2014) that uses a newly-developed Electrostatic Classifier (TSI model 3082) as well as most data from SMPS based on the 3080 model Electrostatic Classifier can be loaded for re-processing.

There are several configuration options within the SMPS Settings window for the data inversion that is applied. The main panels for modifications are Classifier Platform, Time Delay, Correction and Channel Resolution. Three drop-down menus exist within the Classifier Platform panel to modify the data corrections for impactor size, DMA high voltage polarity and neutralizer type that were used when collecting the data. The Time Delay panel allows for adjustment of the amount of time it takes for an aerosol to travel from the outlet of the classifier DMA to the CPC detector.
Figure 5 gives an example of how the Correction panel in the SMPS Settings window of MIM can be used to apply physical property adjustments. In this example a multi-peak particle size distribution in the low diameter range sub-20 nm is shown without corrections (blue curve). The same size distribution with multiple charge and diffusion loss corrections applied can be directly compared to the uncorrected curve in the same graph. Accounting for diffusion losses results in a higher concentration, while the multi-charge correction only has a negligible effect in this example due to the small, completely single charged particle diameters (Figure 5).

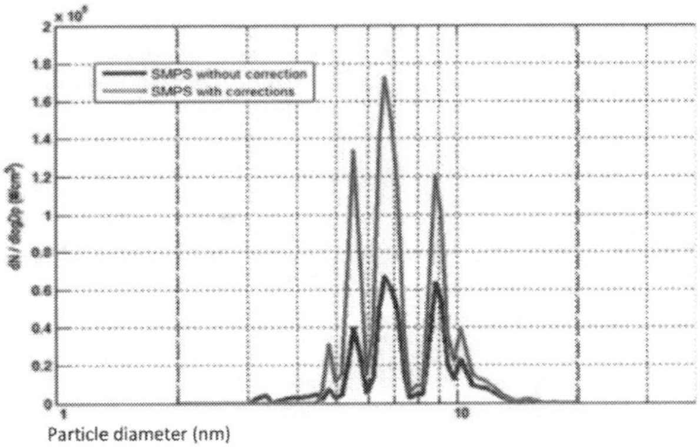


**Figure 5:** Direct comparison of SMPS size distribution with and without corrections for diffusion losses and multiple charging in MIM.

A second SMPS post-processing example is shown in Figure 6 for the Channel Resolution panel. The MIM software enables increasing the displayed SMPS size resolution from 64 channels per decade to 128 channels per decade by reanalyzing the raw measurement data, which are typically not accessible. Results from a multi-peak scan in the lower nanometer range displayed with 64 channels per decade and 128 channels per decade are shown in Figure 6. This higher resolution can be useful in cases like the selected example but there are many size distributions which do not necessarily benefit from an increased channel resolution. We also want to point out that it is only possible to reprocess the data for higher size resolution if the SMPS spectrometer settings during the collection of the raw data were set accordingly.

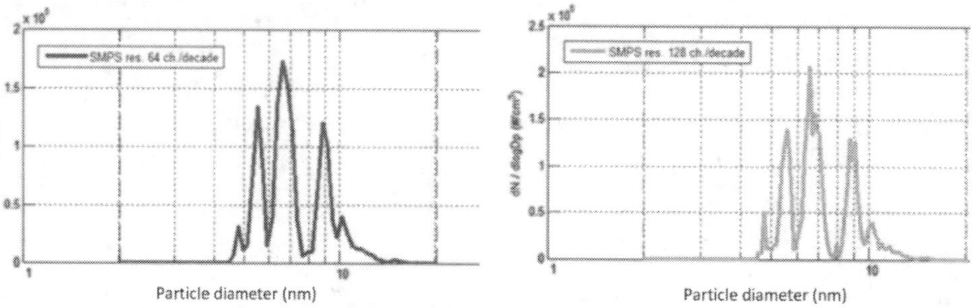


**Figure 6:** Multi-peak scan in the lower nanometer size range of the SMPS with different size resolution. The standard resolution is 64 channels per decade (blue). The increased resolution with 128 channels per decade (green) can be reprocessed from raw data.

## 4. Conclusions

In this paper we describe the new Multi-Instrument Manager (MIM 2.0) software which is an easy to use tool to merge and post-process data from commonly used TSI SMPS, NanoScan SMPS and OPS instruments. Merging of size distributions from electrical mobility and optical diameter based instruments can be done in three different ways and by selecting several fitting options with the objective of retrieving one single size distribution over a wide particle size range. Case studies for laboratory and fireplace aerosols conducted with NanoScan SMPS and OPS instruments demonstrated that curve fit distributions with up to three modes of lognormal distribution from nano- and micrometer sizing instruments can be merged well up to three modes of lognormal distribution. Also online data acquisition over a wide range as well as data processing from two instruments of the same type are now possible.

In addition we have shown that MIM is capable of doubling the size resolution of existing SMPS data to 128 channels per decade. Recorded SMPS data sets can be reprocessed to take into account different post-processing settings and the applied corrections can be displayed for direct comparison. This feature provides SMPS users better control and more options to evaluate and process size distribution data sets. While MIM software does not have the purpose to replace instrument measurement software such as the Aerosol Instrument Manager (AIM, TSI Inc.), it is an additional tool that helps to develop a better understanding of aerosol size distribution data over a wide size range.

## 5. References

Gurr, J.R., Wang, A.S.S., Chenb, C-H. and K.Y. Jan, Ultrafine titanium dioxide particles in the absence of photoactivation can induce oxidative damage to human bronchial epithelial cells. Toxicology, 213: 66-73, 2005.

Han, H. S., Sreenath, A., Birkeland, N. T. and G. J. Chancellor, 7th Asian Aerosol Conference, pp. Paper Number ID153, Xi'an, China, 2011.

Oberdörster G, Oberdörster E and J. Oberdörster, Nanotoxicology: an emerging discipline evolving from studies of ultrafine particles. Environ Health Perspect, 11: 823-839, 2005.

Tritscher, T., Beeston, M., Zerrath, A.F., Elzey, S., Krinke, T.J., Filimundi, E. and O.F. Bischof, NanoScan SMPS – A novel, portable nanoparticle sizing and counting instrument. Journal of Physics (IOP): Conference Series 429, 012061, 2013.

Tröstl, J., Tritscher, T., Bischof, O.F., Horn, H.-G., Krinke, T.J., Baltensperger, U., Gysel, M., Fast and Accurate Measurement in the Sub-20 nm Range using a Scanning Mobility Particle Sizer. Journal of Aerosol Science, submitted December 2014.

## 6. Supplementary comment

SMPS/NanoScan SMPS/OPS customers can download MIM 2.0 software free of charge from the following website: www.tsi.com/MIM .

Journal of Physics: Conference Series **617** (2015) 012014 doi:10.1088/1742-6596/617/1/012014

# Quantitative measurement of carbon nanotubes released from their composites using thermal carbon analysis

**I Ogura[1,2], M Shigeta[2], M Kotake[2], M Uejima[2] and K Honda[1,2]**

[1] National Institute of Advanced Industrial Science and Technology (AIST), 16-1 Onogawa, Tsukuba, Ibaraki 305-8569, Japan

[2] Technology Research Association for Single Wall Carbon Nanotubes (TASC), 1-1-1 Higashi, Tsukuba, Ibaraki, 305-8565, Japan

E-mail: i-ogura@aist.go.jp

**Abstract**. The ability of thermal carbon analysis to determine CNTs was evaluated in the presence of a polymer (Polystyrene, PS). Samples placed in an Au (Pt) foil boat were measured using a thermal-carbon analyzer, and the results were compared with gravimetric measurements of sample masses obtained using an ultra-microbalance. First, debris from the polymer without CNTs (i.e., PS debris) was analyzed. The amount of PS debris detected in the organic carbon (OC) fraction was found to be in good agreement with the gravimetrically measured mass of the PS debris, while the amount of pyrolytically generated carbon soot detected in the elemental carbon (EC) fraction was negligible. Next, single-wall CNT (AIST/TASC Super-Growth) powder was analyzed, and the amount of the CNT powder detected in the EC fraction was found to be 95–96% of the gravimetrically measured mass of the CNT powder. Subsequently, a mixture of the PS debris and the CNT powder was analyzed, and the amounts of detected OC and EC were found to be comparable to the gravimetrically measured masses of the PS debris and the CNT powder, respectively. Finally, debris from 5 wt% CNT–PS composites was analyzed, and amounts of OC and EC detected were found to be approximately comparable to the estimated masses of the PS and the CNTs in the debris of CNT–PS composite, respectively. The results therefore indicate thermal carbon analysis is capable of determining CNTs in the presence of PS.

## 1. Introduction

Carbon nanotubes (CNTs) have unique properties, and therefore, their use as a filler material in composites is considered promising. In the future, CNT composites are expected to be used in a wide range of industrial applications and consumer products. However, there are concerns regarding the potential impact of CNTs on the health of workers and users, and in this respect, the evaluation and control of CNT exposure is required.

Thermal carbon analysis, such as Method 5040 of the National Institute for Occupational Safety and Health (NIOSH), is often used as a quantitative measurement of CNTs in the air [1]. This is a method to determine the fractional content of organic carbon (OC) and elemental carbon (EC). The method used for analysis is as follows: a sample collected with a quartz-fiber filter is heated in stages in a helium atmosphere to vaporize OC, after which the EC is burned by heating in stages in the presence of oxygen. The vaporized or burned carbon is then completely oxidized to $CO_2$ with a catalyst. Furthermore, by reducing $CO_2$ to $CH_4$ with another catalyst, $CH_4$ is detected using a flame-ionization detector. Using this method, CNTs are detected within the EC fraction. This method is

Journal of Physics: Conference Series **617** (2015) 012014 doi:10.1088/1742-6596/617/1/012014

currently one of the most reliable existing quantitative methods for measuring CNTs in the air. However, when a large quantity of other carbon material exists, this technique may be inapplicable.

When CNTs are used in a mixed state with a polymer as a composite material, particles of mixed CNTs and debris from the polymer itself can be released during mechanical and abrasive processing. Unattached (free) CNTs can also be released, but they may be in the minority. The harmful effects of mixed CNTs have not yet been fully evaluated, although Wohlleben et al. (2011) indicated that they are smaller than those of free CNTs [2]. However, it is considered that under the present circumstances, any exposure to mixed CNTs (as well as to free CNTs) should be measured and controlled.

In this study, we evaluated the capability of using thermal carbon analysis to determine CNTs in the presence of a polymer (Polystyrene, PS).

## 2. Methods

We prepared four test samples: **PS**: ground PS debris; **CNT**: CNT powder (AIST/TASC Super-growth single-wall CNTs with a tube diameter of approximately 3 nm and a carbon purity of 99.9%); **PS & CNT**: mixture of PS debris and CNT powder; and **CNT–PS composite**: ground debris of CNT–PS composites (approximately 5wt% CNTs).

Masses of OC, EC, and total carbon (TC) in each sample placed in an Au (or Pt) foil boat were measured using a thermal-carbon analyzer (CAA-202M-D, Sunset Laboratory Inc., USA), and the results were compared with the gravimetric measurement of the sample mass obtained using an ultra-microbalance. We used two different temperature-step protocols based on the NIOSH Method 5040 [3] and the IMPROVE (Interagency Monitoring of Protected Visual Environments) method [4], but with slight modifications (Tables 1 and 2). The optical pyrolysis correction for pyrolytically generated carbon soot from OC during analysis was not used.

**Table 1.** NIOSH 5040 method-based protocol.

| | Duration (sec) | Temperature (°C) | Gas |
|---|---|---|---|
| OC | 80 | 310 | He |
| OC | 80 | 475 | He |
| OC | 80 | 615 | He |
| OC | ***240*** | 870 | He |
| OC | 45 | 550 | He |
| EC | 45 | 550 | 2%$O_2$/He |
| EC | 45 | 625 | 2%$O_2$/He |
| EC | 45 | 700 | 2%$O_2$/He |
| EC | 45 | 775 | 2%$O_2$/He |
| EC | ***300*** | 870 | 2%$O_2$/He |
| EC | ***120*** | ***900*** | 2%$O_2$/He |

**Table 2.** IMPROVE method-based protocol.

| | Duration (sec) | Temperature (°C) | Gas |
|---|---|---|---|
| OC | 180 | 120 | He |
| OC | 180 | 250 | He |
| OC | ***540*** | 450 | He |
| OC | ***540*** | 550 | He |
| EC | 300 | 550 | 2%$O_2$/He |
| EC | 210 | 700 | 2%$O_2$/He |
| EC | ***540*** | ***870*** | 2%$O_2$/He |
| EC | ***120*** | ***900*** | 2%$O_2$/He |

Journal of Physics: Conference Series **617** (2015) 012014 doi:10.1088/1742-6596/617/1/012014

## 3. Results and discussion

The thermograms of the samples obtained using the thermal-carbon analyzer with two different temperature-step protocols are shown in figure 1, and the results are summarized below.

### *3.1. PS debris (Table 3)*

The amount of PS debris detected in the OC fraction was in good agreement with the gravimetrically measured mass of the PS debris, and the amount of pyrolytically generated carbon soot detected in the EC fraction was negligible.

### *3.2. CNT powder (Table 4)*

The amount of CNT powder detected in the EC fraction was 95–96% of the gravimetrically measured mass of the CNT powder, while the amount of TC was in good agreement with the gravimetrically measured mass of the CNT powder. Thee detected OC could be derived from the volatile gas adsorbed on the surface of the CNTs.

### *3.3. PS debris & CNT powder (Table 5)*

The amounts of detected OC and EC were comparable to the gravimetrically measured masses of the PS debris and the CNT powder, respectively. Thus, even when a mixture of the PS debris and the CNT powder was analyzed, the amounts were be distinct from each other.

### *3.4. CNT–PS composite (Table 6)*

The amounts of detected OC and EC were approximately comparable to the estimated masses of the PS and the CNTs in the debris of CNT–PS composites, respectively. Although the thermogram (figure 1) indicated that CNTs in the composites were burned at a lower temperature than that of the as-produced CNT powder, it was considered that the PS and CNTs were distinct from each other.

## 4. Conclusions

Our results indicate that thermal carbon analysis can be used to determine CNTs in the presence of PS.

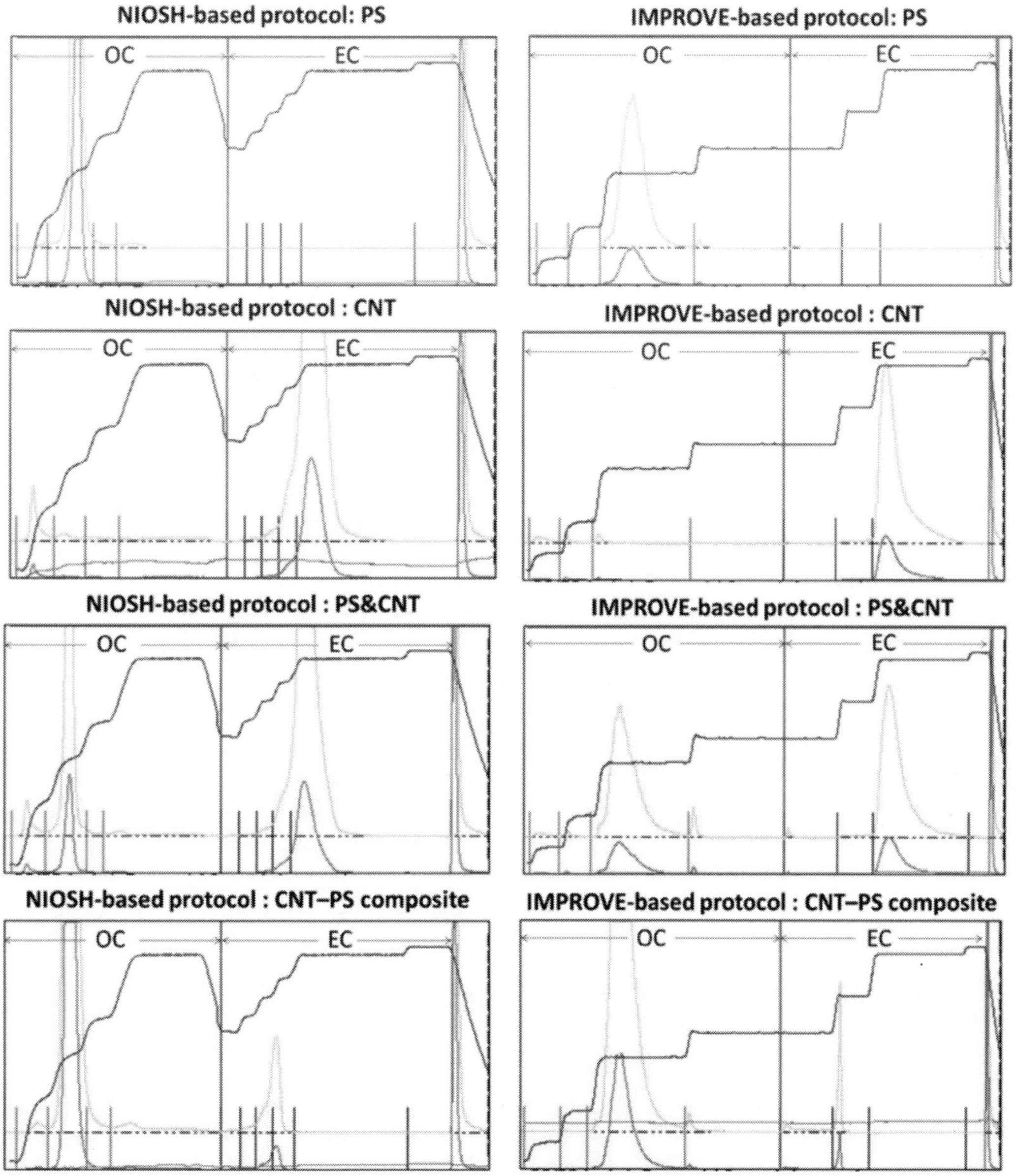


**Figure 1.** Thermograms produced using thermal-carbon analyzer.

Blue line: Temperature; green and purple lines: Flame ionization detector response

**Table 3.** Results of PS.

| | $n$ | OC / 0.92×$M_{PS}$ | EC / 0.92×$M_{PS}$ | TC / 0.92×$M_{PS}$ |
|---|---|---|---|---|
| NIOSH-based protocol | 8 | 1.02±0.05 | 0.0014±0.0009 | 1.02±0.05 |
| IMPROVE-based protocol | 4 | 0.99±0.04 | 0.0043±0.0044 | 1.00±0.04 |

**$M_{PS}$**: Mass of PS debris; **0.92**: Carbon content in PS

**Table 4.** Results of CNT.

| | $n$ | OC / $M_{CNT}$ | EC / $M_{CNT}$ | TC / $M_{CNT}$ |
|---|---|---|---|---|
| NIOSH-based protocol | 9 | 0.073±0.008 | 0.95±0.02 | 1.02±0.02 |
| IMPROVE-based protocol | 5 | 0.028±0.016 | 0.96±0.03 | 0.98±0.02 |

**$M_{CNT}$**: Mass of CNT powder

**Table 5.** Results of PS & CNT.

| | $n$ | OC / 0.92×$M_{PS}$ | EC / $M_{CNT}$ | TC / 0.92×$M_{PS}$+$M_{CNT}$ |
|---|---|---|---|---|
| NIOSH-based protocol | 9 | 1.06±0.07 | 0.96±0.02 | 1.00±0.03 |
| IMPROVE-based protocol | 11 | 0.97±0.03 | 0.97±0.03 | 0.96±0.01 |

**$M_{PS}$**: Mass of PS debris; **$M_{CNT}$**: Mass of CNT powder; **0.92**: Carbon content in PS

**Table 6.** Results of CNT–PS composite.

| | $n$ | OC / 0.92×0.95×$M_{COMP}$ | EC / 0.05×$M_{COMP}$ | TC / (0.92×0.95+0.05)×$M_{COMP}$ |
|---|---|---|---|---|
| NIOSH-based protocol | 12 | 1.09±0.06 | 0.80±0.08 | 1.07±0.05 |
| IMPROVE-based protocol | 8 | 0.96±0.03 | 0.93±0.03 | 0.96±0.03 |

**$M_{COMP}$**: Mass of CNT–PS composite; **0.92**: Carbon content in PS;
**0.05**: CNT content in composite (approximately); **0.95**: PS content in composite (approximately)

**Acknowledgment**

This work is based on results obtained from a project commissioned by the New Energy and Industrial Technology Development Organization (NEDO).

**References**

[1] NIOSH (National Institute for Occupational Safety and Health) 2013 NIOSH Current intelligence bulletin 65: Occupational exposure to carbon nanotubes and nanofibers. http://www.cdc.gov/niosh/docs/2013-145/

[2] Wohlleben W, Brill S, Meier M W, Mertler M, Cox G, Hirth S, von Vacano B, Strauss V, Treumann S, Wiench K, Ma-Hock L and Landsiedel R 2011 On the lifecycle of nanocomposites: comparing released fragments and their in-vivo hazards from three release mechanisms and four nanocomposites *Small* **7** 2384–2395

[3] NIOSH (National Institute for Occupational Safety and Health) 2003 Method 5040 Issue 3, Diesel particulate matter (as Elemental Carbon). NIOSH Manual of Analytical Methods (NMAM; 4th ed.)

[4] Ono-Ogasawara M, Takaya M, Kubota H, Shinohara Y, Koda S, Akiba E, Tsuruoka S and Myojo T 2013 Approach to the exposure assessment of MWCNT by considering size distribution and oxidation temperature of elemental carbon *J. Phys.: Conf. Ser.* **429** 012004

Journal of Physics: Conference Series **617** (2015) 012015 doi:10.1088/1742-6596/617/1/012015

# Preliminary evaluation of the encapsulation of new anti-diabetic sulphonylhydrazone and antitumor *N*-acylhydrazone derivatives using PLGA nanoparticles

**F N Costa[1], A L Ibiapino[1], L P de Figueiredo[1], E J Barreiro[2,3], L M Lima[2,3], D N do Amaral[2], C E de Castro[1], F C Giacomelli[1] and F F Ferreira[1]**

[1] Centro de Ciências Naturais e Humanas (CCNH), Universidade Federal do ABC (UFABC), Santo André, SP, Brazil.

[2] LASSBio, Institute of Biomedical Sciences, Federal University of Rio de Janeiro (UFRJ), Rio de Janeiro, RJ, Brazil.

[3] Graduate Program of Chemistry, Institute of Chemistry, Federal University of Rio de Janeiro (UFRJ), Rio de Janeiro, RJ, Brazil.

E-mail: fabio.furlan@ufabc.edu.br

**Abstract.** It has been demonstrated the feasibly of using PLGA nanoparticles to promote the encapsulation of novel anti-diabetic sulphonylhydrazone and antitumor *N*-acylhydrazone derivatives. The motivation is to further demonstrate the possibility of long-term release of anti-diabetic as well as higher accumulation of the antitumor derivative by using the nanotechnology-based production. The produced nanoparticles were obtained by the nanoprecipitation method, which revealed to be effective in the encapsulation of the bioactive compounds. The determined sizes were in the range of ~100 nm, which are supposed to be suitable for both potential applications. The preliminary experimental data demonstrated the formation of stable nanosystems and further experiments are underway in order to determine the loading content, encapsulation efficiency and release profile of the hydrophobic bioactive compounds.

## 1. Introduction

Diabetes mellitus and cancer are prevalent global diseases that demand high economic and social costs [1, 2]. The development of diabetes mellitus (DM) has increased enormously during the last years and estimations report that the number of cases may reach 300 million by the year 2025 [3]. In turn, according to the World Health Organization (WHO), the different types of cancer figure among the leading causes of morbidity and mortality worldwide, with approximately 14 million new cases and 8.2 million cancer-related deaths in 2012. The number of new cases is expected to rise by about 70% over the next two decades [4]. Moreover, type 2 diabetes and cancer share many risk factors including aging, familiar history, obesity, physical inactivity, poor diet, alcohol consumption and smoking [5].

DM is characterized by hyperglycemia as a result of defects in insulin secretion, insulin action or both [6]. Consistently, high blood glucose levels can lead to serious diseases affecting the heart and blood vessels, eyes, kidneys, and nerves. These complications are the main causes of morbidity and mortality in diabetic patients.

The three main types of diabetes are type 1 diabetes, type 2 diabetes and gestational diabetes, being type 2 diabetes the most common one. It usually occurs in adults, however, it has been increasingly seen in children and adolescents [7]. Although the body is able to produce insulin, in type 2 diabetes the amount of insulin may not be sufficient or the body cannot respond to its effects, leading to an accumulation of glucose in the blood. Currently available hypoglycemic agents include sulfonylureas, biguanides, thiazolidinediones, α-glycosidase inhibitors and dipeptidyl peptidase-4 inhibitors. These drugs are widely used to control hyperglycemia, but fail to alter the course of diabetic complications and have limited use due to undesirable side effects and high rates of secondary failure [8].

Cancer is a generic term for a large group of diseases that can affect any part of the body and can be defined as the rapid and unplanned growth of abnormal cells that can invade other parts (healthy) of the body, thus spreading to other organs. The latter process is referred to as metastasis - the major cause of cancer death. For males and females combined, the most common types of cancer that lead to death are: lung, liver, stomach, colorectal, breast, oesophageal, among others [4, 9].

Many types of cancer can be prevented, reduced and/or controlled mainly if early detected but the treatment is still a major challenge. Among the drugs that are in clinical trials and have been approved in the United States we can quote: imatinib mesylate (Gleevec®, Novartis) - that is used for gastrointestinal stromal tumor and chronic myeloid leukemia, gefitinib (Iressa®, AstraZeneca & Teva) used for non-small-cell lung cancer, rituximab (Rituxan®, Biogen Idec & Genentech) used in the treatment of B-cell non-Hodgkin's lymphoma and B-cell leukemia and trastuzumab (Herceptin®, Genentech) used in the therapy of breast cancer [10]. It is therefore crucial and strategic the effort to identify new drug candidates that are more potent, selective and less toxic in the therapeutic arsenal against cancer.

The significance of these diseases and the great challenge to discover new therapeutic agents for the treatment of DM, their complications and for treatment of cancer were the major motivation of this work, which aims the development of a nanosystem of two new bioactive compounds - LASSBio-1773 and LASSBio-1735 - which have been found to promote hypoglycemic and antiproliferative activities, respectively. The polymeric nanoparticles (NPs) used in drug-delivery systems have been the subject of extensive research to promote benefits such as protection of encapsulated substances against degradation, reduced side-effects, higher therapeutic efficacy, sustained drug release, specific accumulation, high cellular internalization and long blood circulation half-life [11-14]. In this work we describe preliminary results on the preparation and characterization of LASSBio-1773- and LASSBio-1735- encapsulated into polymeric nanoparticles. These compounds were planned and synthesized in the Laboratory for the Evaluation and Synthesis of Bioactive Substances (LASSBio®) of the Federal University of Rio de Janeiro (UFRJ) and, subsequently, the nanoparticles were prepared at the Laboratory of Crystallography and Structural Characterization of Materials (LC*CEM*) of the Federal University of ABC (UFABC).

## 2. Experimental

### 2.1 Materials and reagents

LASSBio-1471, a sulphonylhydrazone derivative, which was reported as a PPARγ ligand, was shown to be effective in the treatment of diabetic neuropathy in streptozotocin-injected rats [8]. Aiming to optimize the hypoglycemic properties of this prototype, the medicinal chemistry approach of structural modifications using molecular homologation, ring opening and molecular simplification strategies, as displayed in Figure 1, was employed. Then, a new sulphonylhydrazonic compound, LASSBio-1773, was designed. Its activity was evaluated using the murine model of streptozotocin-induced hyperglycemia. One dose of 55 $\mu mol\ kg^{-1}$ of LASSBio-1773 compound was capable to reduce the blood glucose levels (reduction of 53% in 120 minutes).

LASSBio-1773 is also a therapeutic candidate for the treatment of diabetic neuropathy, which is one of the complications resulting from type 2 diabetes. Also, it was able to reduce the thermal hyperalgesia and mechanical allodynia in a diabetic neuropathy model. These results suggest that the presence of the dimetoxyl group, the ester group and the aromatic ring attached to the imine are required for hypoglycemic effect observed in this model [8, 15].

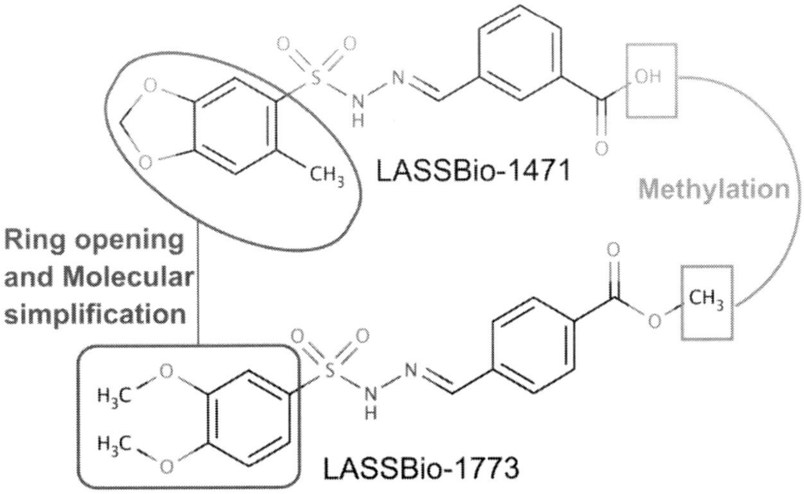


Figure 1. Medicinal chemistry strategies used in LASSBio-1773 synthesis.

The *N*-acylhydrazone derivative - LASSBio-1735 - was designed from LASSBio-1586, a simple antitumor drug candidate, a Combretastatin A4 (CA-4) analogue, using the *N*-methylation strategy (see Figure 2). Among the known classes of anticancer agents, the microtubule-targeted antimitotic drugs are considered to be the most important ones. CA-4, which is a natural stilbene isolated from *Combretum caffrum*, is a microtubule-destabilizing agent that binds to the colchicine domain on $\beta$-tubulin and exhibits low toxicity. LASSBio-1586 was capable of inhibiting microtubule polymerization and displayed a broad *in vitro* and *in vivo* antiproliferative profile, as well as a better selectivity index than the prototype CA-4, indicating improved selective cytotoxicity toward cancer cells. LASSBio-1735 compound, in turn, presented *in vitro* antiproliferative activity against HL-60 (human leukemia), SF-295 (human glioblastoma), MDA-MB435 (melanoma), and HCT-8 (ileocecal adenocarcinoma) tumor cells and was compared with the data from LASSBio-1586 and CA-4. The homologous compound, LASSBio-1735, was well tolerated, exhibiting a slight increase in the cytotoxic potency against HL-60 ($IC_{50}$ = 0.03 μM) and HCT-8 ($IC_{50}$ = 0.54 μM) tumor cell lines relative to LASSBio-1586 compound [16].

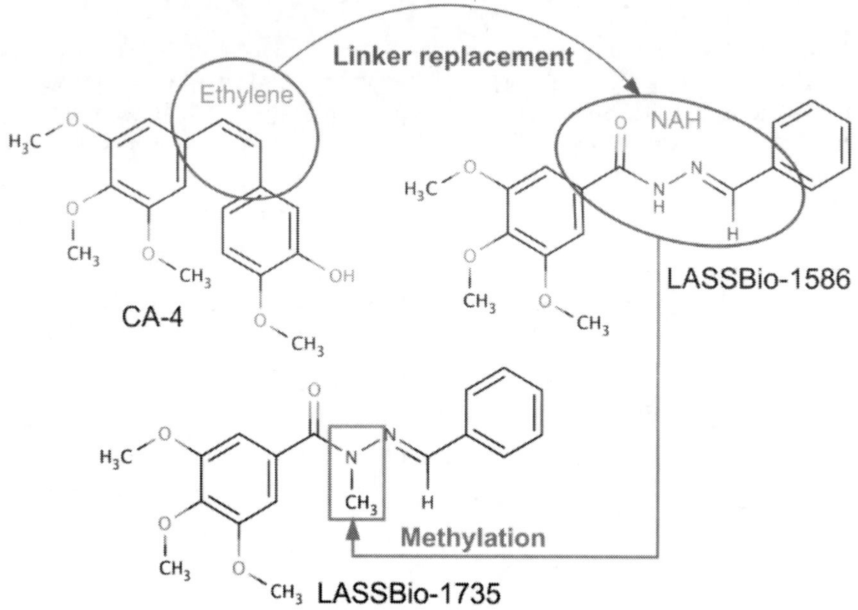


Figure 2. Synthetic strategy used in the production of LASSBio-1735 from LASSBio-1586, an analogue of Combretastatin A4.

In the synthetic procedures of these compounds, it has been produced solid-phase samples with particles of micrometric sizes. Both LASSBio-1735 and LASSBio-1773 were obtained as white crystalline solids [15, 16]. The FDA approved poly(lactic-*co*-glycolic acid), PLGA ($M_w$ = 50.000-75.000 g $mol^{-1}$; L/G 85:15), was purchased from Sigma Aldrich and used as received. The solvent employed in the preparation of the NPs (acetone) was of analytical grade and was also used as received. The water was pretreated with the Milli-Q® Plus System.

## 2.2 Preparation of the nanoparticles

The PLGA NPs were prepared by the nanoprecipitation approach, as displayed in Figure 3 [17], by using a syringe at a flow rate of 2 mL $min^{-1}$. The PLGA copolymer (5.0 mg) was firstly completely dissolved in acetone (1.0 mL) at room temperature and subsequently the organic solution was injected into the aqueous phase under stirring. The organic solvent was further removed by dialysis. The final aqueous volume after the nanoprecipitation procedure and solvent elimination was set to 5.0 mL. The LASSBio-1773 and LASSBio-1735 PLGA-loaded NPs were similarly prepared. In such case, 0.5 mg of each bioactive compound was also dissolved in acetone along to the copolymer. The prepared NPs were used immediately or stored at 4 °C.

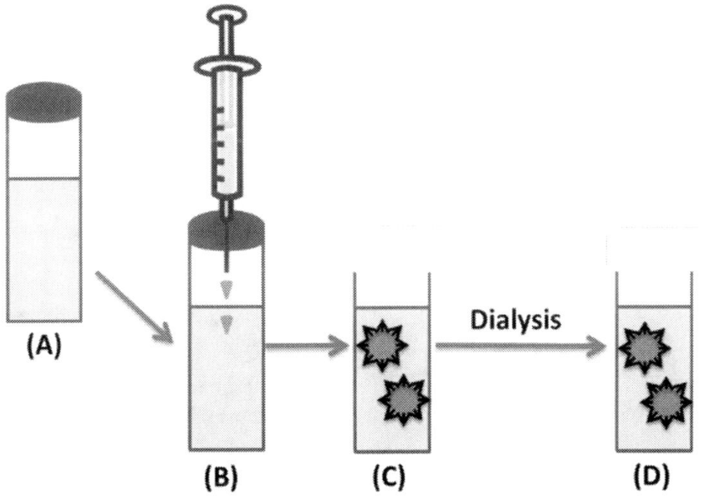


Figure 3. Schematic representation of the preparation of PLGA NPs: (A) organic (polymer or polymer/drug) solution, (B) nanoprecipitation followed by (C) solvent diffusion and (D) solvent elimination *via* membrane dialysis.

## 2.3 Characterization of the nanoparticles

The preliminary characterization of the NPs was performed by means of electrophoretic light scattering (ELS), dynamic light scattering (DLS) and static light scattering (SLS).

*2.3.1 Electrophoretic light scattering (ELS).* The ELS measurements were performed in order to determine the average zeta potential (ζ) of the nanoparticles using the Zetasizer NanoZS equipment (Malvern Instruments, UK). It measures the electrophoretic mobility ($U_E$) of the nanoparticles and converts the value to a ζ potential (mV) through the Henry's equation:

$$U_E = \frac{2\varepsilon\zeta f(ka)}{3\eta} \quad (1)$$

where $\varepsilon$ is the dielectric constant of the medium and $\eta$ is the viscosity. The parameter $f(ka)$ is the Henry's function, which has been calculated through the Smoluchowski approximation, $f(ka) = 1.5$ [14, 18]. Each ζ potential value reported herein is an average of 10 independent measurements with a repeatability of ±2%.

*2.3.2 Dynamic light scattering (DLS).* The DLS measurements were performed using an ALV/CGS-3 compact goniometer system consisting of a 22 mW HeNe linearly polarized laser operating at a wavelength of λ = 633 nm, an ALV 7004 digital correlator and a pair of avalanche photodiodes operating in the pseudo-cross-correlation mode. The samples were loaded in 10-mm-diameter glass cells and maintained at a constant temperature of (25 ± 1) °C. The autocorrelation functions reported herein are based on 3 independent runs with 10 s counting time. The data were collected and further averaged by using the ALV Correlator Control software, and they were analysed by using the Cumulant method according to Equation 2:

$$\ln g_1(t) = \ln C - \Gamma t + \frac{\mu_2}{2} t^2 \cdots \tag{2}$$

where the parameters Γ (relaxation frequency) and $\mu_2$ (second-order cumulant) could be determined. The hydrodynamic radius ($R_H$) of the nanoparticles was determined by using the Stokes-Einstein relation as:

$$R_H = \frac{k_B T q^2}{6\pi\eta\Gamma} \tag{3}$$

where $k_B$ is the Boltzmann constant, $T$ is the absolute temperature, $q$ is the scattering vector, $\eta$ is the viscosity of the solvent and Γ is the mean relaxation frequency related to the diffusion of the nanoparticles. The polydispersity (PDI) of the samples was computed as PDI $= \mu_2 \Gamma^{-2}$ [19, 20].

*2.3.2 Static light scattering* (*SLS*). The SLS measurements were also performed using the ALV/CGS-3 compact goniometer system. In the SLS mode, the scattering angle was varied from 30° to 150° with an angular step of 5°. At each angle, the light-scattered intensity was measured in triplicate and the average values were reported. The measured scattered intensity $I(q)$ can be related to the radius of gyration ($R_G$) of the nanoparticles through the Guinier approximation:

$$\ln I(q) = \ln \underset{q\to 0}{I} - \frac{q^2 R_G^2}{3} \tag{4}$$

## 3. Results and discussion

Light scattering techniques were employed in order to analyze the stability of the promising bioactive compounds - LASSBio-1773 and LASSBio-1735 - after encapsulation into PLGA nanoparticles. Firstly, we carried out the zeta potential measurements. The magnitude of the zeta potential is an important physicochemical parameter that is related to the potential stability of nanoparticles. Large positive or negative zeta potential means that the particles tend to repel each other, preventing further aggregation/flocculation. Particles with zeta potentials more than ±30 mV are normally considered stable [18]. The zeta potential values determined for LASSBio-1773-loaded, LASSBio-1735-loaded and unloaded PLGA NPs were respectively -40 mV, -36 mV and -38 mV. These values shed some light on the stability of the formed NPs and we can also infer that these nanoparticles are stable and these results seem to indicate that the bioactive compounds were preferably encapsulated rather than adsorbed, as the zeta potential remains the same.

The DLS data were used to determine the mean hydrodynamic radius ($R_H$) and the polydispersity of the systems. Figure 4 displays the autocorrelation function for drug-free and drug loaded PLGA NPs. These curves suggest a monomodal distribution ($\mu_2 \Gamma^{-2} \leq 0.16$), which is related to the presence of polymeric nanoparticles and indicate the criteria for good quality (homogeneous) sample preparation. The mean values for both $R_H$ and the PDI ($\mu_2 \Gamma^{-2}$) are given in Table 1.

Journal of Physics: Conference Series **617** (2015) 012015 doi:10.1088/1742-6596/617/1/012015

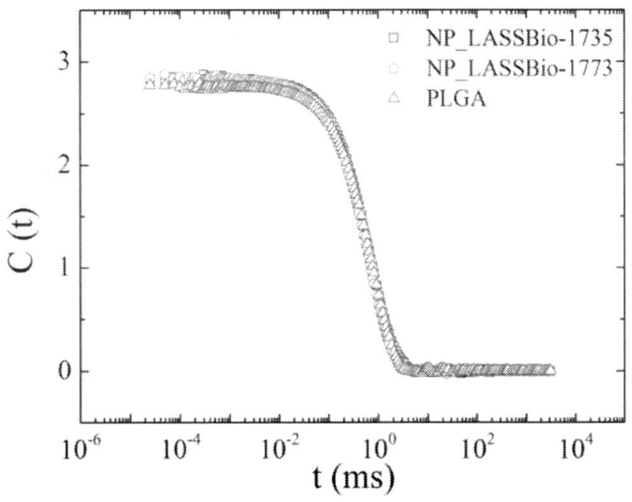


Figure 4. Autocorrelation functions measured at a scattering angle of 90° for LASSBio-1773-loaded (red circles), LASSBio-1735-loaded (black squares) and drug-free PLGA (blue triangles) NPs.

The variation of the scattered intensity with the scattering angle was plotted to measure the mean static particle size ($R_G$). A decrease in $R_G$ as a function of the presence of both sulphonylhydrazone and *N*-acylhydrazone derivatives was observed.

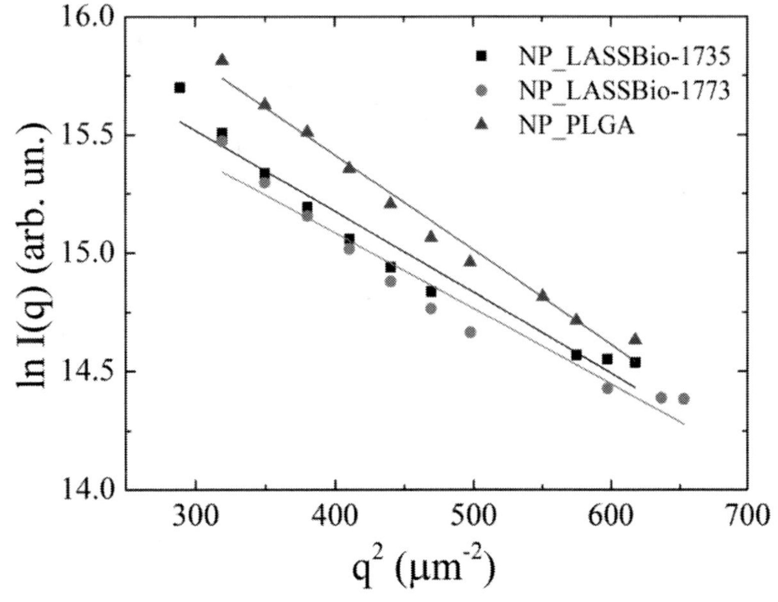


Figure 5. Guinier plot – Angular variation of the scattering intensity as a function of $q^2$ for LASSBio-1735 and LASSBio-1773 compounds.

It is well established that the ratio $\rho = R_G/R_H$ is a characteristic parameter related to the conformation of polymer chains and self-assembled macromolecular objects in solution. The $\rho$-value of spherical objects is dependent on the inner structure and compactness, being close to 0.775 for very compact spheres, $\rho \sim 0.8$–$0.9$ for block copolymer micelles due to solvation phenomena and $\rho \sim 1.0$ for hollow spheres and vesicles [12, 13, 21-23]. The $R_G/R_H$ ratios (Table 1) suggest that these nanoparticles behave as compact spheres. The reduction of $R_G$, $R_H$ and $R_G/R_H$ as the probes are loaded is probably explained by strong hydrophobic interactions between PLGA and the bioactive compounds LASSBio-1773 and LASSBio-1735.

Table 1. Characteristics of the nanoparticles obtained by light scattering experiments. $R_H$ is the hydrodynamics radius, PDI indicates the polydispersity index ($\mu^2 \Gamma^{-2}$), $\zeta$ is the zeta potential, $R_G$ is the radius of gyration and $R_H/R_G$ ratio indicates the conformation characteristic of the NPs.

| NPs | $R_H$ (nm) | PDI | $\zeta$(mV) | $R_G$ (nm) | $R_G/R_H$ |
|---|---|---|---|---|---|
| PLGA | 132.5 | 0.03 | -38(3) | 103.4 | 0.78 |
| LASSBio-1773 | 128.7 | 0.16 | -40(3) | 97.9 | 0.76 |
| LASSBio-1735 | 123.7 | 0.13 | -36(3) | 96.5 | 0.78 |

Finally, it is important to highlight that further experiments are underway in order to determine the loading content and encapsulation efficiency of the developed nanosystems, which will then be followed by investigations regarding the release profile of the hydrophobic bioactive compounds.

## 4. Conclusions

In this work we demonstrated the feasibility of using PLGA to promote the encapsulation of new anti-diabetic sulphonylhydrazone and antitumor *N*-acylhydrazone derivatives. The produced polymeric nanoparticles were obtained by the nanoprecipitation method, which revealed to be effective in the encapsulation of these bioactive compounds. We also verified that both systems are highly hydrophobic. Our preliminary experimental data demonstrated the formation of nanosystems with promising features that should be further investigated in more details.

**Acknowledgments**

The authors thank the financial support of CAPES (PROAP and postdoctoral fellowship PNPD), the Graduate Program in Nanosciences and Advanced Materials (PPG-Nano) of UFABC, the National Council for Scientific and Technological Development (CNPq, grant nr. 305186/2012-4), INCT-INOFAR (BR, # 573.564/2008-6), FAPERJ and FAPESP (2012/14087-8).

**References**

[1] Nugent R 2008 *Ann. N.Y. Acad. Sci.* **1136** 70
[2] Shi Y and Hu F B 2014 *The Lancet* **383** 1947
[3] King H, Aubert R and Herman W 1998 *Diabetes Care* **21** 1414
[4] Organization W H, *World Cancer Report 2014.* 2014: International Agency for Research on Cancer (IARC).
[5] Giovannucci E, Harlan D M, Archer M C, Bergenstal R M, Gapstur S M, Habel L A, Pollak M, Regensteiner J G and Yee D 2010 *Diabetes care* **33** 1674
[6] American Diabetes Association 2014 *Diabetes Care* **37** S81

[7] International Diabetes Federation 2013 IDF DIABETES ATLAS **1** 6th ed. (Brussels International Diabetes Federation) 155
[8] Zapata-Sudo G, Lima L, Pereira S, Trachez M, da Costa F, Souza B, Monteiro C, Romeiro N, D'Andrea E, Sudo R and Barreiro E 2012 *Curr. Top. Med. Chem.* **12** 2037
[9] Ma X and Yu H 2006 *Yale J. Biol. Med.* **79** 85
[10] Narang A S and Desai D S, *Pharmaceutical Perspectives of Cancer Therapeutics*. Anticancer Drug Development - Unique Aspects of Pharmaceutical Development. 2009: Springer Science Business Media.
[11] de Oliveira A M, Jäger E, Jäger A, Stepánek P and Giacomelli F C 2013 *Colloids Surf. A* **436** 1092
[12] Jäger A, Gromadzki D, Jäger E, Giacomelli F C, Kozlowska A, Kobera L, Brus J, Říhová B, El Fray M, Ulbrich K and Štěpánek P 2012 *Soft Matter* **8** 4343
[13] Li M, Jiang M, Zhu L and Wu C 1997 *Macromolecules* **30** 2201
[14] Jiang J, Oberdörster G and Biswas P 2008 *J. Nanopart. Res.* **11** 77
[15] Da Costa F, *Novos candidatos a protótipos de fármacos hipoglicemiantes com atividade anti-inflamatória: LASSBio-1773 e LASSBio-1774*, in *Instituto de Ciências Biomédicas*. 2013, Federal University of Rio de Janeiro: Rio de Janeiro. p. 126.
[16] do Amaral D, Cavalcanti B, Bezerra D, Ferreira P, Castro R, Sabino J, Machado C, Chammas R, Pessoa C, Sant'Anna C, Barreiro E and Lima L 2014 *Plos One* **9** e85380
[17] Barichello J M, Morishita M, Takayama K and Nagai T 1999 *Drug Dev. Ind. Pharm.* **25** 471
[18] Hunter R J, *Zeta Potential in Colloidal Science, Principles and Applications*. 3rd ed. Vol. 1. 1988, London: Academic Press Ltd.
[19] Stepánek P, *Data Analysis in Dynamic Light Scattering*, in *Dynamic Light Scattering: The Method and Some Applications*, Brown W., Editor. 1993, Oxford University Press: Oxford. p. 735.
[20] Stepánek P and Konak C 1984 *Adv. Colloid Interface Sci.* **21** 195
[21] Burchard W, Kajiwara K and Nerger D 1982 *J. Polym. Sci.* **20** 157
[22] Giacomelli F C, Riegel I C, Petzhold C L, da Silveira N P and Stepánek P 2009 *Langmuir* **25** 731
[23] Sedlák M and Koňák C e 2009 *Macromolecules* **42** 7430

Journal of Physics: Conference Series **617** (2015) 012016 doi:10.1088/1742-6596/617/1/012016

# What is the impact of Silicon Carbide nanoparticles to the mineral composition of rat lungs? A PIXE-µPIXE comparative study

**O Lozano[1,†], J.L. Colaux[2], J. Laloy[3], JM Dogné[3], and S Lucas[1]**

[1] Research Centre for the Physics of Matter and Radiation (PMR), Namur Nanosafety Center (NNC), NAmur Research Institute for LIfe Sciences (NARILIS), University of Namur (UNamur), Rue de Bruxelles 61, B-5000 Namur, Belgium
[2] University of Surrey, Ion Beam Centre, Guildford GU2 7XH, England
[3] Department of Pharmacy, Namur Nanosafety Center (NNC), NAmur Research Institute for LIfe Sciences (NARILIS), University of Namur (UNamur), Rue de Bruxelles 61, B-5000 Namur, Belgium
[†] Corresponding author's e-mail: omar.lozanogarcia@unamur.be

**Abstract**. The exposure to nanomaterials can yield changes in the mineral composition of tissues which may have long term health repercussions. In this study, the changes in mineral composition of rat lungs, exposed to a nanoaerosol of silicon carbide (SiC), has been studied by means of global and local ion beam probes with the Particle-Induced X-ray Emission (PIXE) technique, measuring the whole lung contents and selected areas where SiC was found, respectively. It was found that from a global perspective there is a small decrease in the mineral contents (phosphorous, sulphur, chlorine and potassium) of the lung except for Ca, while locally these mineral contents tend fluctuate.

## 1. Introduction

Nanosafety and nanotoxicology studies focus mainly in the impact of nanomaterials (NMs) through the analysis of biomarkers and, when possible, the quantification of the NM dose within the target such as tissues or cells [1-3]. One aspect often overlooked in these studies is the fact that the presence of an external agent in the media environment can cause alterations to the same media which may result in an increase or decrease of the media contents, such as proteins or minerals, and hence have an impact on the composition of the target. For example, recent *in vivo* studies with silicon carbide (SiC) and titanium carbide (TiC) NMs in oral administration have shown a perturbation of the mineral absorption within the gastrointestinal tract [1, 4].

In this study the impact of a SiC nanoaerosol on the composition of rat lungs is investigated. SiC was selected as a model NM due to its extensive industrial usage [5]. The mineral composition of rat lungs exposed to SiC was quantified by Particle-Induced X-ray Emission (PIXE) from two perspectives: using a broad beam (PIXE analysis) to quantify the lung overall mineral contents, and using a micro beam (µPIXE analysis) to quantify the lung mineral contents in areas around SiC deposition.

## 2. Materials and methods

### *2.1. Animals*

Nulliparous and non-pregnant female rat Sprague-Dawley (8 weeks, 190-200 g; supplier: Charles River Laboratories) were selected for this study. Each one was housed in ventilated cages and was acclimatized 2 weeks previous to the exposure. Experiments were performed with the agreement of the Committee on the Ethics of Animal Experiments of the University of Namur (protocol "FUNDP10/128 DO inhalation", approved in April 2010).

Rats were sacrificed at 0, 3, 7, or 28 days after exposure, and their lungs were excised after a heart lavage.

### 2.2. *Exposure*

Silicon carbide (SiC) NPs were bought from Io-Li-Tec. No trace of endotoxins was found and it was used without further treatment for nanoaerosol production.

Rats were exposed in a Whole-Body Exposure Model based on the Organisation for Economic Co-operation and Development (OECD) guideline 403, as described by Laloy et al. [6]. Rats were exposed 6 h a day during 5 consecutive days to a SiC nanoaerosol diluted at $2.5x10^5$ particles/$cm^3$. The control group was exposed to the same filtered air used to produce the nanoaerosol.

### 2.3. *Sample preparation*

For broad beam measurement, rat lungs were dried 48 hours at 37 °C and prepared into pellets following a well-established protocol for high volume analysis [2].

For micro beam measurement, rat lungs were frozen in liquid nitrogen and sections of 6 x 6 $mm^2$ and 10 µm thick were obtained with a cryostat microtome. The sections were placed on polypropylene films fixed to aluminium holders and were left to dry at ambient conditions without any further treatment.

### 2.4. *Ion beam analysis*

Samples were chemically quantified with a broad beam (few $mm^2$) at the Laboratory of Analyses by Nuclear Reaction (LARN) of the Physics of Matter and Radiation Unit of the University of Namur, and with a micro beam (few $\mu m^2$) at the Surrey Ion Beam Centre. The Particle-Induced X-ray Emission (PIXE) and Rutherford Back-Scattering (RBS) techniques were used to measure heavy elements (i.e. Si & Ca) and light elements (i.e. C & O), respectively. The principles of PIXE and RBS have been described elsewhere [2]. The techniques PIXE and RBS are called as such for broad beam analysis, while the "µ" prefix is added for micro beam analysis, hence µPIXE and µRBS. RBS (µRBS) data was used to adjust the matrix for PIXE (µPIXE) analysis. The measurement geometry for broad beam has been described previously [1], and used a 2.0 or 2.5 MeV proton beam in this study. The geometry for the micro beam has been described previously [7], and used a 2.5 MeV proton beam in this study.

### 2.5. *Statistical analysis*

Data was analyzed using the Holm–Sidak method (two way ANOVA). The statistical significance was compared between the control and exposed samples, dividing the significant results in $p<0.05$, $p<0.01$, and $p<0.001$.

Three samples (n=3) per exposure condition were measured with broad beam. Eight (n=8) scans were done per exposure condition and four (n=4) scans were done per control condition were measured with micro beam.

Mineral concentrations are presented as the weighted mean concentration with error bars representing 1 standard deviation of the weighted mean. Ratios between exposed and control samples use an error bar taking into account the propagation of errors from both samples.

## 3. Results and discussion

The mineral concentration found in the rat lungs is presented in Table 1 for PIXE measurements and Table 2 for μPIXE measurements. In the case of PIXE measurement the mineral concentration change is easier to observe for rats sacrificed at day 28: phosphorous (P), sulfur (S), chloride (Cl) and potassium (K) show a lower content with respect to their respective controls; while Ca showed a statistically significant increase with respect to the control. The results are summarized in Figure 1a.

**Table 1. Elemental analysis of trace minerals in lungs measured by PIXE. The second column gives the dry lung weight. Statistically significance exposed versus control group: * indicates p<0.05, *** indicates p<0.001. MDL: Minimum level of detection.**

| Controlgroup | | Elemental analysis (wt. ppm, weighted mean ± SD) | | | | | |
|---|---|---|---|---|---|---|---|
| Sacrifice day | Weight (mg, mean ± SD) | Si | P | S | Cl | K | Ca |
| 0 | 183 ± 22 | < MDL | 21,368 ± 469 | 13,021 ± 514 | 20,471 ± 1,141 | 15,825 ± 704 | 601 ± 58 |
| 3 | 190 ± 37 | < MDL | 21,118 ± 620 | 12,701 ± 487 | 27,027 ± 5,909 | 16,033 ± 1,350 | 621 ± 97 |
| 7 | 220 ± 24 | < MDL | 22,822 ± 885 | 14,578 ± 429 | 20,813 ± 917 | 17,928 ± 656 | 680 ± 68 |
| 28 | 212 ± 14 | < MDL | 25,319 ± 2,183 | 15,632 ± 967 | 25,456 ± 3,698 | 17,184 ± 1,472 | 851 ± 106 |
| Exposed group | | | | | | | |
| Sacrifice day | | Si | P | S | Cl | K | Ca |
| 0 | 233 ± 32 | 5,039 ± 789 *** | 20,330 ± 281 | 12,745 ± 440 | 23,059 ± 4,367 | 14,206 ± 1,686 | 654 ± 101 |
| 3 | 209 ± 17 | 5,216 ± 867 *** | 20,231 ± 633 | 12,865 ± 534 | 16,742 ± 1,090 | 16,592 ± 863 | 553 ± 89 |
| 7 | 166 ± 27 | 3,766 ± 207 *** | 21,685 ± 1,611 | 13,398 ± 1,092 | 22,497 ± 2,923 | 16,882 ± 1,951 | 667 ± 26 |
| 28 | 291 ± 96 | 40 ± 10 * | 20,048 ± 3,348 | 12,163 ± 2,020 | 17,623 ± 3,522 | 14,364 ± 3,266 | 1,264 ± 251 * |

Table 2. **Elemental analysis of trace minerals in lungs measured by μPIXE. The second column gives the dry lung weight. Statistically significance exposed versus control group: * indicates p<0.05, *** indicates p<0.001.**

| Control group | | Elemental analysis (wt. ppm, weighted mean ± SD) | | | | | |
|---|---|---|---|---|---|---|---|
| Sacrifice day | Weight (mg, mean ± SD) | Si | P | S | Cl | K | Ca |
| 0 | 183 ± 22 | 188 ± 29 | 10,092 ± 450 | 6,400 ± 293 | 9,630 ± 498 | 9,919 ± 476 | 211 ± 9 |
| 3 | 190 ± 37 | 174 ± 61 | 7,419 ± 84 | 4,679 ± 83 | 8,014 ± 172 | 7,498 ± 85 | 147 ± 8 |
| 7 | 220 ± 24 | 104 ± 10 | 7,067 ± 293 | 5,088 ± 210 | 5,701 ± 232 | 7,401 ± 315 | 241 ± 12 |
| 28 | 212 ± 14 | 82 ± 9 | 6,994 ± 155 | 4,391 ± 63 | 6,724 ± 114 | 6,826 ± 118 | 142 ± 10 |
| Exposed group | | | | | | | |

| Sacrifice day | | Si | P | S | Cl | K | Ca |
|---|---|---|---|---|---|---|---|
| 0 | 233 ± 32 | 4,393 ± 143 *** | 9,603 ± 224 | 5,254 ± 143 * | 9,041 ± 335 | 9,429 ± 235 | 220 ± 15 |
| 3 | 209 ± 17 | 3,569 ± 164 *** | 8,215 ± 243 | 5,064 ± 148 | 7,542 ± 284 | 7,995 ± 240 | 209 ± 4 |
| 7 | 166 ± 27 | 1,903 ± 220 *** | 5,210 ± 468 * | 3,282 ± 272 * | 22,146 ± 3,317 * | 4,043 ± 467 * | 397 ± 163 |
| 28 | 291 ± 96 | 732 ± 33 *** | 8,520 ± 155 * | 5,787 ± 104 * | 9,001 ± 198 | 8,282 ± 140 * | 224 ± 10 |

Mineral concentration by μPIXE showed a marked, and in some cases a statistically significant, increase in all elements with respect to the controls for days 3 and 28; with a marked decrease for P, S and K and a marked increase in Cl and Ca at day 7. The results are summarized in Figure 1b. Of the measured elements, Cl presents the highest errors bars and it is also known as a common contaminant elements for PIXE, hence its results may not be representative without further detailed measurements.

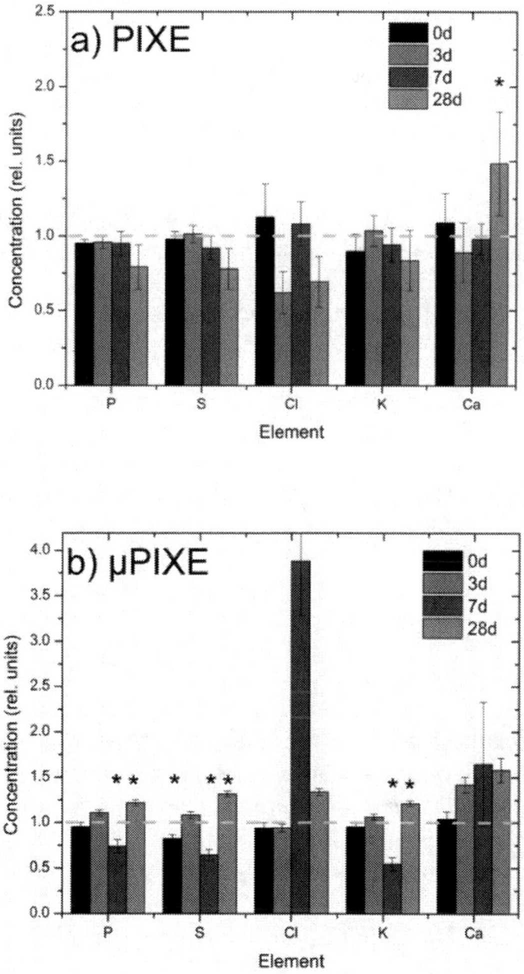

**Figure 1. Control normalized mineral concentration of rat lungs. Statistically significance exposed versus control group: * indicates p<0.05. See *Statistical analysis* subsection for details on the error bar calculation.**

The study of mineral composition variations is motivated by extracting more knowledge about the possible changes to key minerals such as Ca and P, whose ions control many of the cellular processes [8, 9], or electrical impulse conduction from ions of K in the neuron function [10], the regulation of osmotic pressue and acid-base balance by Cl [11], and the role of disulfide bonds in protein assemble and structure [12].

There is a difference between the statistically significant mineral changes in terms of concentration and days found in PIXE and μPIXE measurements. This difference reflects the type of sample being measured and hence the point of view obtained by each type of measurement: with PIXE the measured sample represents the whole lung, pulverized and made into a pellet, and therefore it reflects a global value on the total lung mineral composition. Taking into account that SiC likely remains inside the lung as observed due to the lack of impact on plasma analysis and the low levels of induced inflammation [6], most of the mineral changes, if any, should occur in the lung alveolar sac and as such any mineral composition should be relatively less evident than in the vicinity of the NM localization. On the other hand, μPIXE measurements were done on sections where SiC was found and as such it reflects a local perspective of the exposure, hence reflecting more statistical significant changes in the mineral composition. Indeed this is the case for the mineral concentration in P, S and K. From both perspectives it is judged that locally the alveolar sacs, due to the limited inflammation process, experience a depletion of minerals (7 days) which then overcompensates with higher concentrations (28 days), while as a whole the lung is effectively experiencing a decrease in minerals.

## 4. Summary

The mineral concentration of rat lungs exposed to an acute SiC nanoaerosol was investigated by PIXE and studied from the global lung mineral composition and local lung sections exposed directly to SiC, using a broad beam and a micro beam, respectively. While globally there is a slight decrease in mineral composition, local lung sections show a fluctuation of minerals.

**Acknowledgements.** This work was financially supported by: The European Community's Seventh Framework Programme (FP7/2007-2013) under grant agreement n° 227012 (Integrating activity SPIRIT - Support of Public and Industrial Research Using Ion Beam Technology), and from Service Public de Wallonie (SPW) – Direction générale opérationnelle – Economie, Emploi et Recherche (DGO6), Département des Programmes de Recherche under grants agreements n° 516252 (Nanotoxico Project, SPW/FUNDP research convention) and n° 1317938 (Complément FP7).

## References

1. Lozano, O., et al., *Effects of SiC nanoparticles orally administered in a rat model: biodistribution, toxicity and elemental composition changes in feces and organs.* Toxicol Appl Pharmacol, 2012. **264**(2): p. 232-245. doi: 10.1016/j.taap.2012.08.004
2. Lozano, O., et al., *Development of a PIXE analysis method for the determination of the biopersistence of SiC and TiC nanoparticles in rat lungs.* Nanotoxicology, 2012. **6** (3): p. 263-271 (doi:10.3109/17435390.2011.572301).
3. Lozano, O., et al., *How does the deposited dose of oxide nanomaterials evolve in an in vitro assay?* J Phys Conf Series, 2013. **429**: p. 012013.
4. Laloy, J., et al., *Can TiC nanoparticles produce toxicity in oral administrationto rats?* Toxicology Reports, 2014. **1**: p. 172-187.

5. SCP. *Silicon Carbide Products*. [cited 2014 January 8, 2014]; Available from: http://www.scprobond.com/.
6. Laloy, J., et al., *Acute inflammatory response in rats after exposure to silicon carbide nanoaerosol in a whole-body exposure model.* J Nanopar Res. Submitted, 2014.
7. Ugarte, M., G. Grime, and N. Osborne, *Distribution of trace elements in the mammalian retina and cornea by use of particle-induced X-ray emission (PIXE): localisation of zinc does not correlate with that of metallothioneins.* Metallomics, 2014. **6**: p. 274-278.
8. Clapham, D.E., *Calcium Signaling.* Cell, 2007. **131**(6): p. 1047-1058.
9. Saris, N.E., et al., *Magnesium: An update on physiological, clinical and analytical aspects.* Clin Chim Acta, 2000. **294**(1-2): p. 1-26.
10. Campbell, N., *Biology*1987, Menlo Park, California: Benjamin/Cummings Pub. Co. .
11. Nations, F.a.A.O.o.t.U. *6. ESSENTIAL NUTRIENTS - MINERALS.* Nov. 30, 2014]; Available from: http://www.fao.org/docrep/field/003/ab470e/ab470e06.htm.
12. Nelson, D.L. and M.M. Cox, *Lehninger, Principles of Biochemistry (3rd ed.)*2000: New York: Worth Publishing.

# The impact of instilled carbide nanoparticles on rat lungs: an *in vivo* perspective on acute intratracheal instillation

**O Lozano[1,†], D Lison[2], V. Escamilla-Rivera[3], J Mejía[1], O Toussaint[4], JM Dogné[5], and S Lucas[1]**

[1] Research Centre for the Physics of Matter and Radiation (PMR), Namur Nanosafety Center (NNC), NAmur Research Institute for LIfe Sciences (NARILIS), University of Namur (UNamur), Namur, Belgium
[2] Louvain centre for Toxicology and Applied Pharmacology, Université Catholique de Louvain, Brussels, Belgium.
[3] Department of Toxicology, Center of Research and Advanced Studies of IPN (CINVESTAV-IPN), Mexico City, D.F., Mexico. Department
[4] Laboratory of Biochemistry and Cellular Biology (URBC), Namur Nanosafety Center (NNC), NAmur Research Institute for LIfe Sciences (NARILIS), University of Namur (UNamur), Namur, Belgium
[5] Department of Pharmacy, Namur Nanosafety Center (NNC), NAmur Research Institute for LIfe Sciences (NARILIS), University of Namur (UNamur), Namur, Belgium
[†] Corresponding author's e-mail: omar.lozanogarcia@unamur.be

**Abstract.** In order to study a scenario of acute high concentration exposure via the pulmonary pathway of silicon carbide and titanium carbide nanoparticles, female Wistar rats were administered by intratracheal instillation doses of 0.5 and 5 mg/rat of each nanomaterial. Inflammatory parameters were studied: protein concentration, lactate dehydrogenase activity, total cell count and differentiated cell count (macrophages, neutrophils, oesonophils, lymphocytes). The genotoxicity potential was assessed by the formation of micronuclei from pneumocytes type II. It was found that silicon carbide nanoparticles induce an inflammatory response and a dose dependent genotoxicity, although the genotoxicity levels are comparably lower to the inflammatory response.

## 1. Introduction

The principal route of exposure to nanomaterials (NMs) is the pulmonary pathway [1]. Carbide nanomaterials, like Silicon Carbide (SiC) and Titanium Carbide (TiC) are widely used in several industries requiring high performance products [2, 3], yet they have not been widely studied in comparison of other material families such as their oxide versions, silica and titania. Given the exposure hazard of such materials at the nanoscale could be due to wear, breakdown or decommissioning, our research group has dedicated several efforts to map the impact of exposure on rats to such NMs in terms: toxicity[4] and biopersistence [5] by whole body exposure (WBE), biopersistence from acute intratracheal instillation [6], oral administration in acute and subacute exposures [7, 8], and *in vitro* [9]. An unexplored subject has been the toxicological impact of SiC and TiC nanoparticles (NPs) via intratracheal instillation at doses simulating an acute high concentration exposure scenario. This study presents addresses such question using rats, focusing on the inflammatory parameters from broncheoalveolar lavages (BAL) and the genotoxicity potential of SiC and TiC NPs.

## 2. Materials and methods

### *2.1. Nanoparticles*
SiC and TiC NPs were provided by Sirris (www.sirris.be). They were heated at 200 °C during 2h to eliminate possible endotoxin traces. Their hydrodynamic diamters are 31 and 25 nm for SiC and TiC NPs, respectively. Their full physicochemical characterization has been reported elsewhere [6].

### *2.2. Animals*
Wistar female rats, 200-220 g, were used in this study.

### *2.3. Exposure protocol*
The NPs were prepared in suspensions with a physiologically sterile solution (0.9% NaCl) with 1 % Tween 20 and sonicated. Suspensions were prepared in two doses: 0.5 (low dose) and 5 (high dose) mg/rat. The exposure was acute, delivering the suspensions, 300 μL/rat, to the rat lung via intratracheal instillation. During this procedure rats were anesthetized, exposing surgically the trachea. After the exposure procedure the incision was closed with a suture point.

### *2.4. Genotoxicity studies*
The genotoxicity potential of SiC and TiC NPs was studied *in vivo* in the rats pneumocytes type II. The inflammatory response was assessed 3 days after instillation by sacrificing the rats and measuring different parameters in a broncheoalveolar lavage (BAL) with a saline solution: total protein concentration, enzymatic activity of lactate dehydrogenase (LDH), and the total and differential number of cells. Pneumocytes type II were recovered after several BAL to eliminate inflammatory cells and a lung perfusion to eliminate the erythrocytes. Lung enzymatic digestion was carried out by an elastase solution flowing through the lung trachea for 30 minutes at 37 °C. Then the lung was cut in small sections and the cell suspension was filtered and then incubated in Petri boxes covered with rat IgG for 1 hour at 37 °C. The cells possessing the Fc receptor (macrophages) will adhere while the pneumocytes type II will rest in suspension and were recovered for culture. After two days, the culture chambers are washed and fixed with methanol. The plates are colored with orange acridine and counted with a fluorescent microscope.

The negative control was NaCl and the positive control was WC-Co (94:6 mass ratio, 5 mg) which generates micronuclei [10].

BAL was centrifuged, the supernatant was used to quantify the total proteins and LDH activity, and the cellular sediment was resuspended in NaCl for the cellular count. Protein concentration was measured by spectrometry after complexation with pyrogallol red-molybdate. LDH activity was measured by spectrometry after reduction of the $NAD^+$ catalysed by the enzyme. Total cell counting was done with a Neubauer plate and differential cells were counted from colored cytospins with Diff-Quick [11].

## 3. Results and discussion

### *3.1. BAL analysis*
The protein concentration reflects the permeability of the broncheo-alveolar barrier. It is observed that for the high dose of SiC there is an elevated, even if not significant, concentration of proteins when compared to those of TiC. LDH activity, a marker of cytotoxicity, is found the highest for the high dose of SiC, noting that it is statistically significant with a value higher than even the positive control. The LDH activity of the low dose of SiC and the high dose of TiC is close to the positive control. The

total cell number follow the same trend as the LDC activity: high dose of SiC has the highest statistically significant number of cells, higher than the positive control; with both SiC low dose and TiC high dose having similar levels close to the positive control.

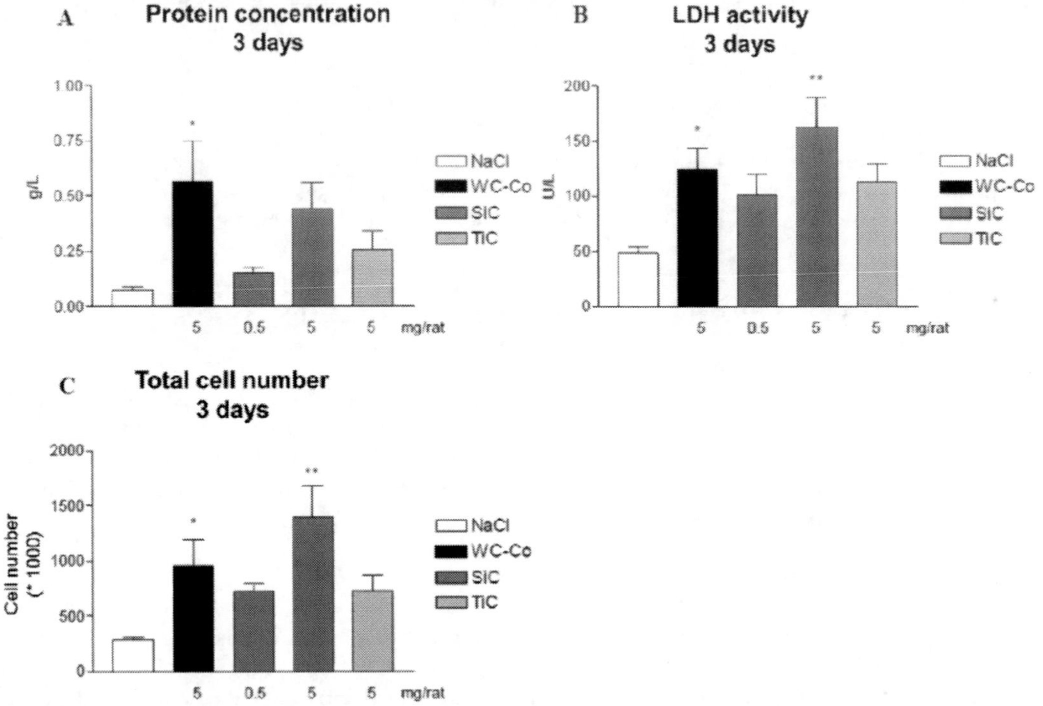


**Figure 1**. Inflammatory parameters in BAL after exposure. A) Total protein count. B) LDH activity. C) Cell total number. Nar represent the mean ± SEM (n = 4-6). Statistical significance, between negative control (NaCl) and exposed sample, evaluated by Student-Newman-Keuls test with * for $p<0.5$ and ** for $p<0.01$.

The cell count for macrophages and neutrophils has a significant increase for the high dose of SiC NPs, and lymphocytes had an increase for the low dose of SiC NPs. Eosinophils had similar cell counts for all doses, although showed no significant difference versus the negative control. Exposure to TiC NP showed higher levels (not significant) of macrophages and neutrophils with respect to the negative control.

*3.2. Pneumocyte type II analysis*

The number of micronuclei formed from pneumocytes type II for SiC and TiC NPs exposure were not significantly different from those formed on the negative control. In fact there seems to be a similar genotoxicity potential for the high dose of SiC and TiC NP, even if the high dose of SiC generated cytotoxicity and a higher presence of macrophages and neutrophils, while TiC did not generate a neutrophils response. It is also contrasting that a similar level of micronuclei is generated for both nanoparticle doses even if the biopersistence of TiC NPs in the lung is 2 times lower than SiC NPs [6]. This level of micronuclei production is quite lower than those produced by WC-Co. In addition, it is clear that the production of micronuclei is dose dependent as evidenced by the low and high dose of SiC NPs.

As previously reported, pulmonary toxicity of WC-Co is related to the capacity of induce both, reactive oxygen species and transient acute inflammation [12]. Here, after 72 h exposure to WC-Co particles no cellular infiltration was observed in BAL, however micronuclei formation in type II epithelial cells was present. Meanwhile, SiC induced a pronounced inflammatory response in the lung and no signs of genotoxicity were observed. From this differences, we propose that SiC produces a substantial lung toxicity through the damage of resident cells in the alveoli, as demonstrated by the significant increase of LDH concentration in BAL. When cellular integrity is compromised, several molecules are liberated from cytosol, among them the damage-associated molecular-pattern (DAMP) molecules (such as $Ca^{++}$, ATP, RNA, DNA, etc.) could trigger inflammation and promote recruitment of innate inflammatory cells such as macrophages and neutrophils towards the alveoli, even in the absence of pathogens [13].

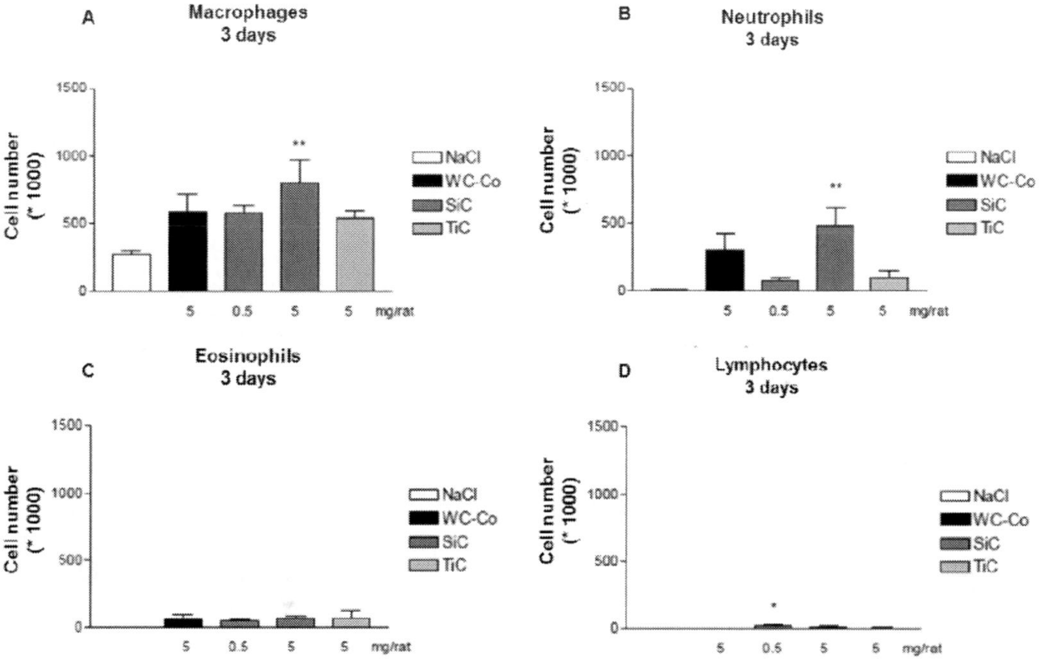


**Figure 2**. Number of (A) macrophages, (B) neutrophils, (C) eosinophils, and (D) lymphocytes after exposure. Statistical significance, between negative control (NaCl) and exposed sample, evaluated by Student-Newman-Keuls test with * for p<0.5 and ** for p<0.01.

Mechanistically, production of reactive oxygen species (ROS) is considered as the major player in NP induced toxicity [14]. It is possibly that SiC, but not TiC, induce free radicals that arise from redox reactions between their highly oxidized surface [6] and the biological milieu. For SiC, the aforementioned process is more relevant because their increased surface/volume ratio make it more capable to produce in ROS in higher amounts compared to TiC and WC-Co. Therefore, given the relatively similar hydrodynamic diameter of both SiC and TiC NPs, this is a NP surface-chemical mediated effect.

Depending on the severity, ROS could elicit different cellular responses: at low levels, an increase in proliferation and antioxidant defense is promoted, in mild levels DNA damage and release of inflammation mediators, whereas at high concentration, ROS provoke excessive organelle and DNA damaged leading to cell death [15]. Therefore, one possibility by which SiC did not shown DNA damage to type II cells is due to an overwhelming ROS production and subsequent cell death, since micronuclei formation could only be detected in cells without compromised viability, in which mild DNA damage allowed them progress to cell division.

The impact of SiC NPs was studied recently in a Whole Body Exposure (WBE) model [4], evaluating the impact of a SiC nanoaerosol when the whole respiratory pathway is taken into account. The toxicological analysis was found very similar to this study, where a limited inflammatory response was evidenced by a elevated presence of macrophages and neutrophils 24 h after an increment in the LDH activity.

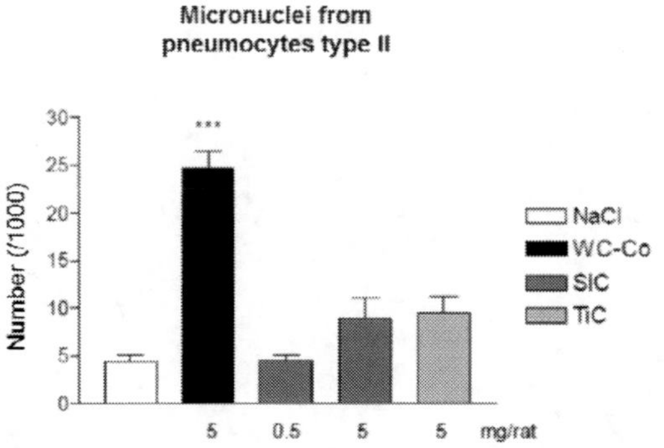


**Figure 3**. Number of micronuclei from pneumocytes type II after exposure. Statistical significance, between negative control (NaCl) and exposed sample, evaluated by Student-Newman-Keuls test with *** for $p<0.01$.

## 4. Summary

There is a dose dependent production of micronuclei with both SiC and TiC NPs exposed by intratracheal instillation. This genotoxicity, however seems not related to the prominent inflammatory response and more specially the relation with the neutrophils presence. The high dose, 5 mg, of SiC NPs produced the most important inflammatory response. TiC NPs, while not induced a inflammatory response, produced a similar level of micronuclei like SiC NPs.

**Acknowledgements.** This work was supported by the Service Public de Wallonie (SPW) – Direction générale opérationnelle – Economie, Emploi et Recherche (DGO6), Départément des Programmes de Recherche (Nanotoxico Project, SPW/FUNDP research convention No. 516252). Thanks to Dr. Georgina B. Vega for her insightful dialogues about the inflammatory response.

## References

1. Oberdörster, G., V. Stone, and K. Donaldson, *Toxicology of nanoparticles: A historical perspective.* Nanotoxicology, 2007. **1**(1): p. 2-25.
2. SCP. *Silicon Carbide Products.* [cited 2014 January 8, 2014]; Available from: http://www.scprobond.com/.
3. Ferro-TiC. *Titanium Carbide Applications.* 2013 [cited 2013 December 12, 2013]; Available from: http://www.ferro-tic.com/.
4. Laloy, J., et al., *Acute inflammatory response in rats after exposure to silicon carbide nanoaerosol in a whole-body exposure model.* J Nanopar Res. Submitted, 2014.
5. Lozano, O., et al., *What is the global and local biopersistence dynamics of silicon carbide nanoparticles in rat lungs after chronic inhalation exposure? A general biopersistence descriptive model.* In preparation, 2015.

6. Lozano, O., et al., *Development of a PIXE analysis method for the determination of the biopersistence of SiC and TiC nanoparticles in rat lungs.* Nanotoxicology, 2012. **6** (3): p. 263-271 (doi:10.3109/17435390.2011.572301).
7. Lozano, O., et al., *Effects of SiC nanoparticles orally administered in a rat model: biodistribution, toxicity and elemental composition changes in feces and organs.* Toxicol Appl Pharmacol, 2012. **264**(2): p. 232-245. doi: 10.1016/j.taap.2012.08.004
8. Laloy, J., et al., *Can TiC nanoparticles produce toxicity in oral administrationto rats?* Toxicology Reports, 2014. **1**: p. 172-187.
9. Mejia, J., et al., *Are stirring and sonication pre-dispersion methods equivalent for in vitro toxicology evaluation of SiC and TiC?* J Nanopart Res, 2012. **14**: p. 815-832.
10. M, D.B., et al., *In vivo genotoxicity of hard metal dust: induction of micronuclei in rat type II epithelial lung cells.* Carcinogenesis, 2003. **24**(11): p. 1793-1800.
11. J, M., et al., *Respiratory toxicity of multi-wall carbon nanotubes.* Toxicol Appl Pharmacol, 2005. **207**(3): p. 221-31.
12. De Boeck, M., et al., *In vivo genotoxicity of hard metal dust: induction of micronuclei in rat type II epithelial lung cells.* Carcinogenesis, 2003. **24**(11): p. 1793-1800.
13. Wynn, T.A., *Integrating mechanisms of pulmonary fibrosis.* The Journal of Experimental Medicine, 2011. **208**(7): p. 1339-1350.
14. Nel, A., et al., *Toxic potential of materials at the nanolevel.* Science (New York, N.Y.), 2006. **311**(5761): p. 622-7.
15. Xia, T., et al., *Comparison of the Abilities of Ambient and Manufactured Nanoparticles To Induce Cellular Toxicity According to an Oxidative Stress Paradigm.* Nano Lett, 2006. **6**: p. 1794 - 1807.

Journal of Physics: Conference Series **617** (2015) 012018 doi:10.1088/1742-6596/617/1/012018

# Organ weight changes in mice after long-term inhalation exposure to manganese oxides nanoparticles

**T Zeman[1], M Buchtová[2], B Dočekal[3], I Míšek[2], J Navrátil[1], P Mikuška[3], O Šerý[2] and Z Večeřa[3]**

[1] Department of Population Protection, University of Defence, Kounicova 65, 662 10 Brno, Czech Republic

[2] Institute of Animal Physiology and Genetics, v.v.i., Academy of Sciences of the Czech Republic, Veveří 97, 602 00 Brno, Czech Republic

[3] Institute of Analytical Chemistry, v.v.i., Academy of Sciences of the Czech Republic, Veveří 97, 602 00 Brno, Czech Republic

E-mail: tomas.zeman2@unob.cz

**Abstract**. Recently, it has been proven that manganese from inhaled particles of manganese compounds can accumulate in the internal organs of laboratory animals. Nevertheless, there were only a few researches dealing with changes in body morphology induced by inhalation of these particles, even though results of some studies indicate existence of such changes. The aim of our research was to assess the effect of inhaled manganese oxides nanoparticles on weight of internal organs. For this purpose a long-term inhalation experiment on laboratory mice was performed, during which the mice were exposed to $MnO.Mn_2O_3$ nanoparticles in concentration $2 \times 10^6$ particles/cm$^3$ for 17 weeks, 24 hours a day, 7 days a week. Manganese oxides nanoparticles were synthesized continuously via aerosol route in a hot wall tube flow reactor using thermal decomposition of metal organic precursor manganese(II)acetylacetonate in the flow tube reactor at temperature 750 °C in the presence of 30 vol% of oxygen. It was proven that inhaled nanoparticles can influence the weight of internal organs of mice. Moreover, it was discovered that the resulting change in weight of selected organs is disproportional. The mice from the experimental group had statistically significantly lighter kidneys, liver and spleen and heavier pancreas compared to the mice from the control group.

## 1. Introduction

Toxic effects of manganese were probably firstly described by prof. John Couper of the University of Glasgow in 1837 [1]. He observed gradual development of paraplegia in five workers who crushed manganese oxides in a factory producing bleaching powder. In two of them, who were exposed to manganese for the longest time, these symptoms did not disappear even after discontinuation of exposure and subsequent treatment. Since then a number of similar cases of manganese neurotoxicity have been described. Based on these cases a condition known as manganism (Mn-induced Parkinson-like disease) was described. Manganism is an irreversible progressive brain disease with similar symptoms like idiopathic Parkinson's disease, connected with accumulation of manganese in the brain tissue. The first stage of the disease is usually manifested by impaired fine motor coordination, fatigue, headache, loss of appetite, apathy, insomnia and diminished libido [2, 3]. In the next stage follow:

psychotic behaviour, slurred speech, slow and clumsy gait, rigidity of facial muscles, speech disorder, tremor, dystonia, balance disorders and severe muscle stiffness, which can even lead to immobility. Neurotoxicity of manganese is probably associated with damage of basal ganglia, especially in the areas of globus pallidus and corpus striatum [4]. Neither the exact mechanism of manganese action nor the reason for damage of the basal ganglia is known. However, it is assumed that this could be a consequence of oxidative stress or inflammatory processes [2, 5, 6, 7]. It seems probable that high doses of manganese also induce changes in the levels of the neurotransmitters dopamine, glutamate (glutamic acid) and $\gamma$-aminobutyric acid, which may serve as another explanation for neurotoxicity of manganese [8].

Idiopathic Parkinson's disease and manganism are considered different diseases with different symptoms and aetiologies [9]. Unlike patients with Parkinson's disease patients suffering from manganism do not respond to treatment by dopamine substitution, exhibit abnormalities in the areas of globus pallidus and corpus striatum, where manganese selectively accumulates, and also have some different symptoms [2, 9, 10, 11]. Recently, however, it has been shown that exposure to manganese may also contribute to occurrence of idiopathic Parkinson's disease itself, or at least it can accelerate its progression [12]. Racetech et al. [13] came up with the hypothesis that high doses of manganese appear as manganism, while long-term excessive intake of lower doses of manganese can be one of the previously unknown causes triggering idiopathic Parkinson's disease. This statement was supported by the research during which they found out that the symptoms of welders affected by parkinsonism and idiopathic Parkinson's disease patients did not differ significantly. In welders, however, these symptoms occurred on average 17 years earlier than in patients with idiopathic Parkinson's disease. Lucchini et al. [14] also came to a similar result when they demonstrated dependence of parkinsonism incidence in the province of Brescia in northern Italy on the amount of manganese in settled dust collected in different villages, as well as an increased incidence of parkinsonism in people who lived near the ferroalloy plants. There are also indications that long-term chronic exposure to manganese can negatively affect intellectual and motor skills of children [15, 16, 17].

Manganese gets into the bloodstream primarily by ingestion and inhalation [18] provided that the ingestion is usually considered less dangerous way of exposure [9]. Inhaled particles of manganese compounds are partially trapped in the mucus of the lower respiratory tract, partly penetrate the alveoli and from there they get into the blood [19]. The possibility of the passage of manganese compounds nanoparticles from the nasal cavity directly to the brain through the olfactory nerve is also often discussed [9]. In the past, it had been proven that manganese from inhaled particles of manganese compounds can accumulate in the internal organs of laboratory animals, especially in the lungs, liver, pancreas and gall bladder [10, 20, 21, 22, 23]. Results of some studies [3, 23, 24] additionally suggest that inhaled manganese compounds particles may also influence the constitution of animals, but conclusive evidence has not been found yet. Most toxic effects were observed after the inhalation exposure to higher concentrations of manganese [9]. In the past neurotoxicity due to inhalation of manganese compounds particles was observed in some specific occupations, particularly in miners in manganese ore mines, workers in the steel mills and foundries, especially those where ferromanganese and steel are melted, and factories producing "dry cells". Also welders belong among highly exposed occupation groups [19]. Welding electrodes usually contain manganese which then forms a part of the welding fume [25]. In addition, it provably contains ultrafine particles of size <100 nm, which easily penetrate the respiratory system [21]. The welding fume contains manganese primarily in the form of $Mn^{2+}$ and $Mn^{3+}$ [22, 23, 25].

The aim of our research was to extend current knowledge on the effects of inhaled nanoparticles of $MnO.Mn_2O_3$ on the constitution of mice. We conducted a long-term inhalation experiment, during which laboratory mice were exposed to nanoparticles of $MnO.Mn_2O_3$ for 17 weeks, 24 hours a day, 7 days a week. On previously chosen days, mice were collected from the cages and autopsied. During the autopsy selected internal organs were weighed.

Journal of Physics: Conference Series **617** (2015) 012018 doi:10.1088/1742-6596/617/1/012018

## 2. Materials and Methods

### *2.1. Experimental design*

All mice (adult males, ICR line, weighing about 24 g at the beginning of the experiment) were allowed to acclimate to laboratory conditions for at least 1 week before the experiment began. The mice were provided with a commercial diet and water ad libitum. The experimental work was performed in accordance with the ethical approval of the Institute of Animal Physiology and Genetics (no. 081/2010). The experiment started with 80 laboratory mice which were randomly divided into 2 equally sized groups: a fresh-air control and the experimental group exposed to nanoparticles of MnO.Mn2O3. Both groups were placed in identical inhalation chambers and received identical feeding. During the experiment 2 mice from the control group had to be euthanized for health reasons and those were not further analyzed. At chosen time intervals (figure 1) mice were taken out of the cage, euthanized and autopsied one by one. During the autopsy the brain, lungs, heart, liver, kidneys, spleen, pancreas, testes and thymus were weighed.

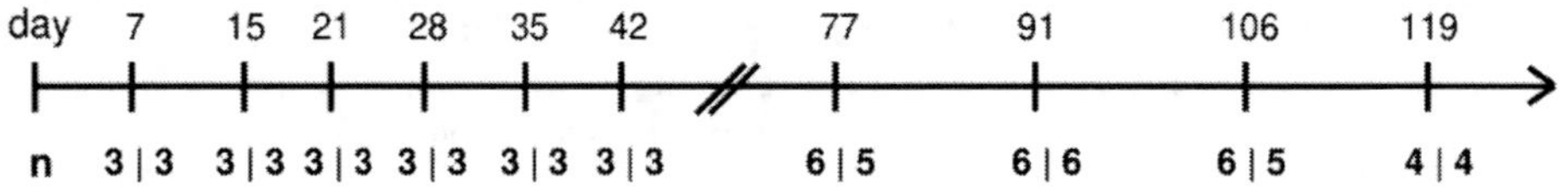


**Figure 1.** Overview of the days when mice were collected from cages during the inhalation experiment, n marks the number of mice taken from experimental group | control group.

### *2.2. Nanoparticles generation and measurement*

Manganese oxides ($MnO.Mn_2O_3$) nanoparticles (MnONPs) were synthesized continuously via aerosol route in a hot wall tube flow reactor using a thermal decomposition of metal organic precursor manganese(II)acetylacetonate in the flow tube reactor (in vertical position) at temperature 750 °C in the presence of 30 vol% of oxygen. The vapours of manganese(II)acetylacetonate were generated from solid form of acetylacetonate in saturator at temperature 160 °C and released vapours were transported by nitrogen (purity 5.5, flow rate 0.5 l/min) into the flow reactor. The total flow rate of nitrogen/oxygen mixture through the flow reactor was 2 l/min. At the output from the reactor, MnONPs transported in nitrogen flow were mixed with air (1 l/min). The concentration of produced MnONPs at the reactor output was in the range 1-3 $\times$ $10^7$ particles/cm$^3$ and the size of generated nanoparticles $MnO.Mn_2O_3$ was in the range 7-50 nm. Before entering the inhalation chamber MnONPs in the mixture of $N_2$, $O_2$ and air (at total flow rate of 3 LPM) were further diluted using filtrated, humidified air (20 L/min) at temperature 21 °C resulting in the MnONPs concentration 2 $\times$ $10^6$ particles/cm$^3$. Concentration of nanoparticles and particle size distribution in the inhalation chamber were continuously measured by SMPS (model 3936L72, TSI).

### *2.3. Statistical analysis*

Differences between the group exposed to MnONPs and the control group in organ weight were statistically evaluated. Statistical testing was carried out in two successive steps. First, the exact p-value was computed separately for each day when the mice were taken out and autopsied using one-tailed Wilcoxon Rank Sum test. In case that the exposed and the control group do not differ in organ weight, these p-values computed for different collection days should be uniformly distributed over the interval [0, 1] with median being equal to 0.5 [26]. In the second step, this null hypothesis was tested at the 0.05 significance level using two-tailed Wilcoxon Sign Rank test. Statistically significant results indicate that organ weight in both groups differs at least in one collection day. Software R [27] was used for all statistical computations.

## 3. Results

Statistically significant differences in organ weight of the liver, kidneys, spleen and pancreas were found (table 1). The liver, kidneys and spleen were considerably lighter in the mice from the group exposed to MnONPs compared to the mice from the control group (figure 2). Pancreas was also affected by MnONPs inhalation, but with the opposite effect. It was proven that the mice from the exposed group had heavier pancreas with higher manganese weight than the mice from the control group.

**Table 2.** Comparison of organ weight in mice from the exposed group and the control group for assessed internal organs, n.s. - statistically not significant result, C - control group, E - exposed group.

| organ | result[a] | p-value[b] |
|---|---|---|
| brain | n.s. | - |
| lungs | n.s. | - |
| heart | n.s. | - |
| liver | C > E | < 0.01 |
| kidneys | C > E | < 0.05 |
| spleen | C > E | < 0.05 |
| pancreas | E > C | < 0.05 |
| testes | n.s. | - |
| thymus | n.s. | - |

[a] indicates which of the groups had higher organ weight
[b] computed using two-tailed Wilcoxon Sign Rank test

## 4. Discussion

Only few researches dealing with the influence of manganese on body weight have been carried out so far [7, 24, 28]. During the experiment performed by Santos et al. [7], two groups of rats were treated intraperitoneally with 25 mg of $MnCl_2.4H_2O$/kg every 48 hours. The first group received a total of 4 doses, the second group of 8 doses. The authors tested whether the exposed groups differed from the control groups in body weight, but did not find any statistically significant differences. However, it was not an inhalation experiment, nor nanoparticles were used, so this comparison has only a limited information value.

Comparable long-term inhalation experiment was conducted by Dorman et al. [28] who studied changes in organ weight in macaques due to exposure to inhaled $MnSO_4$. Even they did not find any statistically significant differences in the weights of these organs between the exposed groups and the control group in the monkeys which were euthanized immediately after the end of the experiment. They found only a relative reduction of about 17% in heart weight in the monkeys which were left alive for 90 days after the end of exposure. Nevertheless, the monkeys from the control group were euthanized immediately after the end of the inhalation experiment and were thus of different age. As noted by the authors themselves, the monkeys were still growing during the experiment, so their result cannot be considered decisive.

Torrente et al. [24] gave rats 275 and 550 mg of $MnCl_2$/kg/day in drinking water for 19 weeks, during which the rats were weighed regularly. In the rats from the exposed groups a lower body weight and a lower weight gain together with less food and water intake were discovered in comparison with the control group. Regarding the cause of lower body weight of exposed rats, Torrente et al. suggested that this could be a consequence of aphagia due to lesions in the mesolimbic dopaminergic system induced by excessive doses of manganese. Under these conditions, reduced food intake would undoubtedly influence the weight of internal organs, but it is not clear whether these changes would be as disproportionate as we observed in our sample.

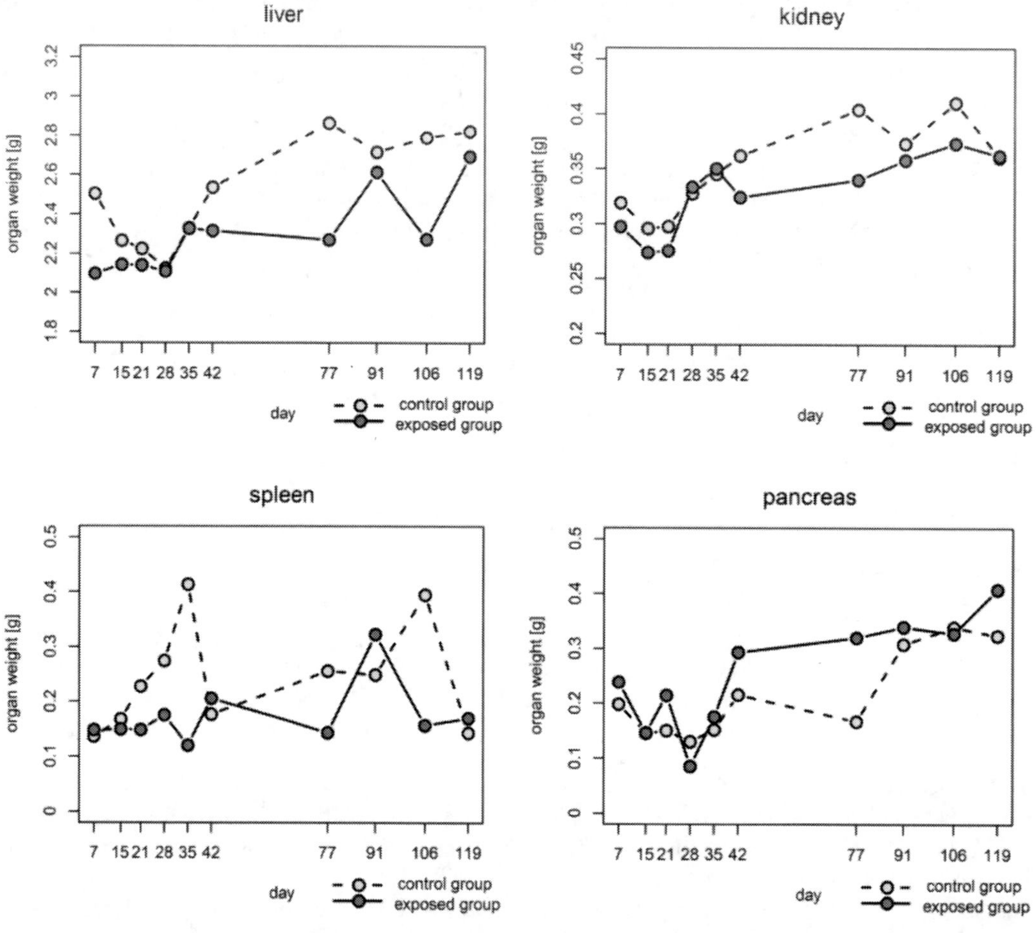


**Figure 2.** Development of median weight of the liver, kidneys, spleen and pancreas in mice during the 119 days of the experiment; green point marks the median of the control group; red point marks the median of the group exposed to MnONPs; dashed line connects adjacent medians of the control group; solid line connects adjacent medians of the exposed group.

## 5. Conclusion

A long-term inhalation experiment was carried out. During the experiment laboratory mice were exposed to $MnO.Mn_2O_3$ nanoparticles in concentration $2 \times 10^6$ particles/cm$^3$ for 17 weeks, 24 hours a day, 7 days a week. It was proven that the manganese oxides nanoparticles cause changes in the weights of some organs. The observed disproportionality of these changes is a very interesting phenomenon. While the weight of the liver, kidney and spleen in the exposed group was lower than in the control group, in the pancreas it was the contrary. However, clarification of the cause of this effect is not possible based on our experiment. For this purpose, it would be appropriate to carry out a new experiment, during which some other variables would be monitored, in particular the amount of consumed food and the level of physical activity of mice.

**Acknowledgements**
This study was co-supported by the Grant Agency of the Czech Republic (P503/12/G147, P503/11/2315) and the Ministry of Education, Youth and Sports of the Czech Republic (SV14-FEM-K106-08-ZEM).

**References**

[1] Couper J 1837 *Am. J. Med. Sci.* **21** 522–3
[2] Tuschl K, Mills P B and Clayton P T 2013 *Int. Rev. Neurobiol.* **110** 277–312
[3] Williams M, Todd G D, Roney N, Crawford J, Coles C, McClure P R, Garey J D, Zaccaria K and Citra M 2012 *Toxicological Profile for Manganese* (Atlanta: U.S. Department of Health and Human Services, Public Health Service, Agency for Toxic Substances and Disease Registry)
[4] Cercosimo M G, Koller W C 2006 *NeuroToxicology* **27** 340–6
[5] Martinez-Finley E J, Gavin C E, Aschner M and Gunter T E 2013 *Free Radic. Biol. Med.* **62** 65–75
[6] Milatovic D, Zaja-Milatovic S, Gupta R C, Yu Y and Aschner M 2009 *Toxicol. Appl. Pharm.* **240** 219–25
[7] Santos D, Milatovic D, Andrade V, Batoreu M C, Aschner M and Marreilha dos Santos A P 2012 *Toxicology* **292** 90–8
[8] Fitsanakis V A, Au C, Erikson K M and Aschner M 2006 *Neurochem. Int.* **48** 426–33
[9] Michalke B and Fernsebner K 2014 *J. Trace Elem. Med. Biol.* **28** 106–16
[10] Brenneman K A, Wong B A, Buccellato M A, Costa E R, Gross E A and Dorman D C 2000 *Toxicol. Appl. Pharm.* **169** 238–48
[11] Zatta P, Lucchini R, van Rensburg S J and Taylor A 2003 *Brain Res. Bull.* **62** 15–28
[12] Martin C J 2006 *NeuroToxicology* **27** 347–9
[13] Racette B A, McGee-Minnich L, Moerlein S M, Mink J W, Videen T O and Perlmutter J S 2001 *Neurology* **56** 8–13
[14] Lucchini R G, Albini E, Benedetti L, Borghesi S, Coccaglio R, Malara E C, Parrinello G, Garattini S, Resola S and Alessio L 2007 *Am. J. Ind. Med.* **50** 788–800
[15] Hernández-Bonilla D, Schilmann A, Montes S, Rodríguez-Agudelo Y, Rodríguez-Dozal S, Solís-Vivanco R, Ríos C and Riojas-Rodríguez H 2011 *NeuroToxicology* **32** 615–21
[16] Torres-Agustín R, Rodríguez-Agudelo Y, Schilmann A, Solís-Vivanco R, Montes S, Riojas-Rodríguez H, Cortez-Lugo M and Ríos C 2013 *Environ. Res.* **121** 39–44
[17] Bouchard M, Laforest F, Vandelac L, Bellinger D and Mergler D 2007 *Environ. Health Perspect.* **115** 122–7
[18] Fechter L D, Johnson D L and Lynch R A 2002 *NeuroToxicology* **23** 177–83
[19] Antonini J M, Santamaria A B, Jenkins N T, Albini E and Lucchini R 2006 *NeuroToxicology* **27** 304–10
[20] Elder A, Gelein R, Silva V, Feikert T, Opanashuk L, Carter J, Potter R, Maynard A, Ito Y, Finkelstein J and Oberdörster G 2006 *Environ. Health Perspect.* **114** 1172–1178
[21] Zimmer A T and Biswas P 2001 *J. Aerosol Sci.* **32** 993–1008
[22] Minni E, Gustafsson T E, Koponen M and Kalliomaki P-L 1984 *J. Aerosol Sci.* **15** 57–68
[23] Gonser M J, Lippold J C, Dickinson D W, Sowards J W and Ramirez A J 2010 *Welding Journal* **89** 25–33
[24] Torrente M, Colomina M T, Domingo J L 2005 *Toxicology* **211** 59-69
[25] Jenkins N T and Eagar T W 2005 *Welding Journal* **84** 87–93
[26] Murdoch D J, Tsai Y-L and Adcock J 2008 *Am. Stat.* **62** 242–5
[27] R Core Team 2013 *R: A Language and Environment for Statistical Computing* (Vienna: R Foundation for Statistical Computing)
[28] Dorman D C, Struve M F, Marshall M W, Parkinson C U, James R A and Wong B A 2006 *Toxicol. Sci.* **92** 201–10

Journal of Physics: Conference Series **617** (2015) 012019 doi:10.1088/1742-6596/617/1/012019

# Interlab study on nanotoxicology of representative graphene oxide

**Nelson Durán[1,2,*], Diego S.T. Martinez[2,3], Giselle Z. Justo[4], Renata de Lima[5], Vera Lúcia de Castro[6], Gisela A. Umbuzeiro[7], Edison Barbieri[8], Marcela Durán[1,11], Patricia S. Melo[9], Oswaldo L. Alves[2,10], Wagner J. Fávaro[2,11]**

[1]Chem. Inst (IQ). Universidade Estadual de Campinas (UNICAMP), SP, Brazil.
[2]NanoBioss, IQ-UNICAMP, SP, Brazil.
[3]Braz. Nanotechnol. Nat. Lab. (LNNano), SP. Brazil.
[4] Fed. Univer. São Paulo (UNIFESP), SP, Brazil.
[5]Univer. Sorocaba (UNISO)/Biotechnol, SP, Brazil.
[6]EMBRAPA/Lab. Ecotoxicol. Biosaf. SP, Brazil.
[7]Fac. Tech.. Ecotox. Environ. Microbiol. Lab., UNICAMP,SP, SP, Brazil.
[8]Fisheries Inst. São Paulo, Ecotox. Lab., SP, Brazil.
[9]Biochem-METROCAMP, SP, Brazil.
[10]Chem. Inst., Solid Chem. Lab. IQ-UNICAMP, SP, Brazil.
[11]Depart. Struct. Funct. Biol. IB-UNICAMP, SP, Brazil.

duran@iqm.unicamp.br

**Abstract.** The graphene sample GO:Single-layer graphene oxide, purity 99%, thickness 0.7-1.2 nm (AFM); ~300-800nm X&Y dimensions is the standard size <450 nm & 1-20 μm lateral dimensions. Cheap Tubes Inc., Bratleboro, USA was selected for our study. Exhaustive characterization of GO was afforded. It exhibited thermal stability over 60°C and it was suspended in deionized water after ultrasonication (1 mg/mL) (stable 10 days). All the biological fluids used in the different assays were used as control of the colloidal suspension stability. Then, all the studies were carried out within that stability period. The cytotoxicity assays were carried out by the Alamar Blue (Resazurin) reduction, MTT and flow cytometry assays in mouse embryonic fibroblast cells (3T3), human keratinocytes (HaCaT), colorectal cancer cells (Caco-2/HCT 116), Lewis lung cancer cells (3LL), acute myeloid leukemia cells (KG-1, Jurkat, Kasumi-1) and chronic myeloid leukemia cells (K562, Lucena) and no significant toxicity was found after exposition to 0.1-100 μg/mL for 24 and 48 h. Breast cancer cells, MCF-7, showed a 20% reduction on cell viability at 24 and 48 h. No cytotoxicity were found in lymphocytes, Chinese hamster ovary cells (CHO) and human macrophage cell line (U937) at 0.1-50 μg/mL, but 30-50% survival inhibition was observed at 100 μg/mL. A dose-dependent increase in apoptosis was observed in some cells (Kasumi-1, Jurkat and K562 cells). In the case of CHO and 3T3 cells, greater levels of necrosis with increasing concentrations of GO (>50 μg/mL) were observed. Genotoxic study using the Comet assay showed slight DNA damage in lymphocytes at all concentrations tested, while more significant effects was observed in CHO cells. Econanotoxicity was carried out by lethality assays in the nematode *Caenorhabditis elegans,*d in the freshwater coelenterate *Hydra, Daphania* amd in Shrimp with no signs of toxicity at concentrations varying from 0.1-100 μg/mL of GO. However, death and disintegration of *Hydra* was observed after exposition to 100 μg/mL for 72 h. In *in vivo* studies, no changes in biochemical parameters of Fischer 344 rats were observed after the i.p. administration of GO. Some black agglomerates were found in the intraperitoneal cavity of rats injected with GO. However, in Fisher 344 rats-bearing prostate tumors, treatment with GO (up to 100 μg/mL) negatively affected the hepatic parameters, whilst in the renal ones, an improvement was observed. Studies are in progress to understand the mechanisms involved in the uptake of GO by RES. GO appears as a potential non-toxic *in vitro* and *in vivo* assays at the concentrations used in this interlab experiments.

## 1. INTRODUCTION

Graphene oxide (GO) is a material with potential in many different applications and with thermal stability [1]. GO is not considered till now as representative compound in the actual studies in the EU (e.g. REACH, OECD). Recently, European Commission's JRC Institutes IRMM (Belgium) and IHCP (Ispra, Italy) introduced the term 'representative test material' (RTM), and provided a frame for its use [2]. The definition of RTM is a material from a single batch, homogeneous and stable with respect to one or more specified properties (ISO/AWI TS 16195). The ISO Tech. Comm. Nanotechnol. (TC 229) has recently published a new standard: ISO/TR 16197:2014 – Nanotechnologies – Compilation and description of toxicological screening methods for manufactured nanomaterials. This standard is a Technical Report related to *in vitro* and *in vivo* methods that can be useful for the toxicological and ecotoxicological measurement, complementing the ISO/TS 27687:2008 and ISO/TS 80004-1. In view of these documents, our Brazilian Network on Nanotoxicology-GIGENANOTOX (MCTI/CNPq) decided to follow and evaluate these new rules for a selected representative compound, GO, by different nanotoxicological assays.

## 2. MATERIAL AND METHODS

### *2.1. Sample*

The graphene sample GO:Single-layer graphene oxide, purity 99%, thickness 0.7-1.2 nm (AFM); ~300-800nm X&Y dimensions is the standard size <450 nm & 1-20 μm lateral dimensions from Cheap Tubes Inc., Bratleboro, USA was selected for our study. Exhaustive characterization of GO was afforded by spectroscopic and microscopies techniques..

### *2.2. Debris determination in GO*

To isolate oxidation debris fragments, a suspension of sample raw-GO (0.5 mg $mL^{-1}$ ) was produced through sonication for 30 min in an ultrasound bath. The suspension was brought to reflux in a 0.1 mol $L^{-1}$ NaOH solution for 90 min, but resulting in a light supernatant and a black pelletized sediment. The black pellet was separated from the supernatant by filtration, reprotonated with a 1.0 mol $L^{-1}$ HCl solution, dialyzed, and finally dried by lyophilization thus yielding a black powder (sample df-GO: debris-free graphene oxide)[3].

### *2.3. Colloidal stability of GO in different biological media used;*

The standard ASTM E2524-08 (Standard test method for analysis of haemolytic properties of nanoparticles) was followed. It was added 1.5 mL of an stock GO solution (1 mg/mL) to 13.6 mL of each culture medium or water from different sources in Falcon tube (15 mL). It was transferred 1 mL of these solution to Epperdorf tubes, in order to visualize the experiment. The time of observation was 0-120 h. An aliquot of 200 μL of each samples collected from these experiments were transferred to plate with 96 wells and the absorption was measured at 450 nm in the ELISA protocol. At the same time it was photographed all the eppendorf tubes at the time interval of 6 and 144 h.

### *2.4. Hemolysis*

The hemolytic assay was carried out following the standard procedure from ASTM E2524-08 (Standard test method for analysis of hemolytic properties of nanoparticles.

Hemolytic activity assay was performed according to our earlier report [4]. Briefly, fresh human blood (volunteer at ethical conditions) was collected in a centrifuge tube containing anticoagulant, trisodium citrate (3.2%), and was centrifuged at 3000 rpm for 10 min. The supernatant was discarded and only the erythrocytes were collected. Erythrocytes were further washed three times with PBS (pH = 7.4). A

10% (v/v) suspension of erythrocytes in PBS was prepared and 1.9 mL of this erythrocyte solution was placed in a 2 mL centrifuge tube and 0.1 mL of GO)Stock solution 1 mg/mL) (up to 100 μg/mL) in PBS ( pH 7.4) was added to it. The tubes were then incubated for 1h at 37°C. For comparison, Triton X-100 (0.2%) and PBS were taken as the positive and negative controls, respectively. After incubation the tubes were centrifuged at 3000 rpm for 10 min, and absorbance of the supernatant was taken at 570 nm in UV-visible spectrophotometer.

### *2.5. Cytotoxicity*

The cytotoxicity assays were carried out by the Resazurin reduction, MTT and flow cytometry assays in mouse embryonic fibroblast cells (3T3-ATCC CCL-163), human keratinocytes (HaCaT-Amsterdam Medical Center, The Netherland)), colorectal cancer cells (Caco-2-ATCC HTB-30)/HCT 116-ATCC CCL-247), Lewis lung cancer cells (3LL-ATCC CRL-1642), acute myeloid leukemia cells (KG-1-ATCC CCL-246, Jurkat-ATCC TIB-152, Kasumi-1-ATCC CRL-2724) and chronic myeloid leukemia cells (K562-TCC CCL-243, Lucena-Inst. Medical Biochemistry-UFRJ, Brazil), breast cancer cells (MCF-7-ATCC HTB-22), Chinese hamster ovary cells (CHO-K1-ATCC CCL-61), breast cancer cells (SKBR3-ATCC HTB-30) and human macrophage cell line (U937-ATCC CRL-1593.2). All the cell were checked for mycotoxin by PCR and Flouorescence assays.

### *2.6. Genotoxicity*

The genotoxicity of GO was studied using 3T3, CHO-K1 and Jurkat cells. In brief: The lymphocytes, CHO-K1 and 3T3 cells were exposed to the GO suspension (stock solution 1 mg/mL) for a period of 1 hour. Negative and positive controls employed phosphate buffered saline (PBS) and H2O2 (0.1 mM), respectively. The comet assay was performed as described by Azqueta et al. [5]. Lymphocytes: Each treatment used 10 μL of lymphocytes in 110 μL of low melting point agarose (0.8%), and the mixture was placed onto microscope slides that had been pre-coated with normal melting point agarose (1.5%). After polymerization, the cover slips were removed and the slides were treated with an ice-cold lysis solution for 20-30 minutes. All treatments were then incubated in an electrophoresis buffer for 20 minutes, followed by electrophoresis for 20 minutes at 1.3 V/cm. After electrophoresis, the slides were stained and analyzed by optical microscopy. Scores of between 0 and 4 were assigned according to the quantity of DNA in the tail and the length of the tail: Score 0 corresponded to intact cells, with no damage caused by the exposure; Score 1 corresponded to cells with minimal damage; Score 2 to average damage; Score 3 to severe damage; and Score 4 to cells with maximum damage. In this visual method, the number of cells found for each score was multiplied by the value of the score and the values were summed at the end of the analysis of each slide. Since the score depended on the number of cells observed, an index of tail damage (TD) was created by dividing the score given for the slide by the number of cells analyzed on the slide [5, 6-8].

### *2.7. Tali analysis*

Cell viability was measured using a Tali™ Apoptosis Kit consisting of Annexin V Alexa Fluor® 488 and Propidium Iodide (Invitrogen), and a Tali™ image-based cytometer, which enabled the numbers of viable, apoptotic, and dead cells to be counted. The cells that had been treated with SH-MNPs (both MSA- or DMSA-coated MNPs) and SNO-MNPs at nanoparticle concentrations (0.01; 0.05 and 0.1 mg/mL), nanoparticles were centrifuged and concentrated to 1 x $10^6$ cells/mL. 100 μL aliquots of sample were prepared according to the specifications of the kit, and the tests were performed in triplicate [9].

### *2.8. Salmonella/microsome assay testing*

*S. typhimurium* TA98 *his*D3052, Δ(*uvr*B, *bio*), *ffa*, $Ap^r$; frameshift mutation, pKM101 was used for these experiments, following the OECD 471 standard from OECD. It is based upon scoring bacterial growth (colonies) on selective agar plates after exposure of the bacterial cells to a test chemical, either by incorporation of the test chemical into the agar plates, or by pre-incubation prior to plating out [10].

### *2.9. Econanotoxicity*
These experiments were carried out by lethality assays in the nematode *Caenorhabditis elegans* and in the freshwater coelenterate *Hydra.*
***2.9.1. C. elegans**:*
***2.9.1.1. Maintance of the C. elegans cutures*** : Strain N2 was purchased from *Caenorhabditis* Genetic Center (Brazil) following the standards procedures[11]. The nematodes were cultivated at 20°C in Petri plates in nematodes medium (NGM) seeding with *E. coli* strain OP50 for feeding. To obtain synchronized cultures, pregnant hermaphrodites nematodes were lysed in an alkaline hypochlorite solution. All the assays with the nematodes were carried out in a liquid medium (S-Medium or M9) using plates with of 24 or 96 wells.
***2.9.1.2. Lethality assay***: This measure was evaluated by alive nematodes. After exposure of to GO for 72 h at 20°C, the nematodes were observed in an inverted microscope and classified as dead or alive (n = 120 nematodes). Each assay consisted in seven different concentrations of GO (0.1-100 µg/mL) and one as control (triplicated assays).
***2.9.1.3. Growth*** : The growth was evaluated by the nematode body length. He body length was determined usinf the plane surface of the nematodes using a software Axiovision ZEISS (n = 50 nematodes)
***2.9.1.4. Statistic analises estatística*** : For this analysis was used the GraphPad Prism 5.0 program (La Jolla, CA, USA). To verify the occurrence of significant differences between the study groups was used in the experiments analysis of variance (ANOVA). Where significant differences were found, the Tukey test was used to test existing minor differences between all groups. To assess significant differences between two groups, Student's t test was used unpaired. In all analyses, α was 5% ($p <0.05$).

***2.9. 2. Hydra:*** When exposed to toxicants, hydras can display a broad range of morphological changes indicative of either sub lethal or lethal effects. Increasing degrees of intoxication departing from normalcy are characterized by stages showing clubbed tentacles, shortened tentacles and the ''tulip phase'', which leads to the death of the organism.
*Hydra attenuata (Cnidaria)* was cultured in hydra medium (TES). It was fed 3 times per week with *Artemia nauplii.* Cultures were maintained at controlled conditions of temperature and light. Toxicity tests were carried out in microplates to which the GO will be added. Hydras were exposed to samples for 96 h. Three nonbudding hydra will be selected to be exposed to each concentrations tested. Seven test concentrations (0.1 µg/mL; 1.0 µg/mL; 5.0 µg/mL; 10 µg/mL; 25 µg/mL; 50 µg/mL and 100 µg/mL) and a control were used. The progressive effects of the GO concentrations were recorded daily by rating the morphological status of the individual polyp and assigning a score from 10 if normal to 0 if disintegrated [12, 13]. Median scores of progressive responses were analyzed at each exposure time for each GO tested concentration.

### *2.9.3. Daphnia acute toxicity test (48h).*
*Daphnia similis* Straus 1820 (Crustacea, Cladocera) was used following the OECD guideline [14]. Young daphnids, aged less than 24 h at the start of the test, are exposed to the GO at a range of concentrations for a period of 48 hours. Immobilization is recorded at 24 h and 48 h and compared with control values. The results were analyzed in order to calculate the EC50 at 48h.

### *2.9.4.* **Acute ecotoxicity on freshwater Shrimp**
On this experiment 60 shrimps (*Palaemon pandaliformis*) were used. The shrimp specimens were acclimatized in the salinity of 30%to 20°C with constant aeration, cleaning and daily renewal of the water for a period of 5 days prior the experiments. Before starting the experiment, the animals were

kept individually in respirometric tubular chambers with continuous water circulation for 60 min in order to minimize the stress caused by handling. At the end of the acclimation period GO was separately introduced in an amount determined in respirometers with a precision pipette in order to achieve the desired final concentration. The difference between the concentrations determined at the beginning and end of the confinement (after 30 minutes) were used to calculate the oxygen consumption (mL/g/h) and ammonia excretion (µl/mL/h) taking into account the respirometer volume and the animal's wet weight. The dissolved oxygen was determined through the Winkler method and the ammoniacal nitrogen based upon the phenol hypochlorite method. Average oxygen specific consumption and ammonia excretion by shrimp were evaluated through variance analysis [15].

***2.9.5. The Ames Salmonella/microsome mutragenicity assay***
*Salmonella typhimurium* TA98 *his*D3052, Δ(*uvr*B, *bio*), *ffa*, $Ap^r$; frameshift mutation, pKM101 was used for this mutagenic study. It was followed the OECD 471 Bacterial Reverse Mutation Test, OECD, Paris, France (1997) standard procedure [10].

***2.10. In vivo rats toxicity study***
***2.10.1. Genotoxicity: Micronucleous assay* [16, 17]**.
A total of 25 rats (Fischer 344) were divide in 5 groups (n = 5 animals): negative control group (NCT): 0.3 mL of physiological solution (0.9% intraperitoneal (i.p.) administration; Positive control group (PCT): dose of 1.5 mg/mL of *N-metil-N-nitrosouréia* (MNU - *Sigma, St. Louis, MO, EUA*) by i.p. administration; GO group-1 (GO 1): dose of 0.5 mg/mL of GO by i.p. administration; GO group-2 (GO 2): dose of 5 mg/mL of GO by i.p. administration; GO group-3 (GO 3): dose of 25 mg/mL of GO by i.p. administration.
Animals were euthanized 24 hours after the administration of chemical agents. Femurs were removed and cleaned; and the epiphysis was sectioned to expose the spinal canal. The bone marrow was aspirated with 1 mL fetal bovine serum (FBS) and placed in 2 ml Falcon tube containing FBS. The bone marrow was suspended several times to form a suspension with FBS. Subsequently, the samples were centrifuged at 1000 rpm for 5 minutes. The supernatant was removed and the precipitate was carefully aspirated with a pipette. Next, 100 µL of the precipitate of each sample was applied to histological slide and smear done with the aid of the cover slip. The smears were dried for 24 hours at room temperature. After this period, samples were fixed in absolute methanol for 15 min and stained with Giemsa (1:10 dilution in sodium phosphate buffer pH 6.8).
Samples were analyzed under a light microscope with 100x oil immersion objective and the number of micronuclei was counted in 3000 polychromatic erythrocytes (PCEs) per animal. The identification of micronuclei followed the criteria described by Fenech [18].
***2.10.2. Toxicological and biochemical analyses*:** It was used 20 rat (Fischer 344) and divided in four groups (n = 5 animals/group). The treatment will be exactly the same as above for four weeks. After 4 weeks of treatment, the animals were euthanized and blood samples were collected by cardiac puncture (left ventricle). The serum was separated by centrifugation at 6000 rpm, 5 min. and used to investigate the hepatotoxic effects, nephrotoxic, cardiotoxic of different concentrations of GO. This was measured through the analysis of alanine aminotransferase (ALT) that is for specific marker activity for hepatic damage; activity of aspartate aminotransferase (AST), indicative marker of liver and heart damage and circulating levels of urea and creatinine to check kidney function. Readings were performed by microplate reader Multi-Mode Microplate Reader Synergy H1M (Bio-Tek Instruments, USA) with temperature control and analyzed by Kcjúnior software (Bio-Tech Instruments, Inc., Winooski, VT, USA). For the analysis of systemic toxicity of different GO concentrations fragments of different organs were collected: pancreas, spleen, lungs, heart and kidneys; of all animals in each experimental group. Different organs were fixed in Bouin solution for twelve hours. After fixation, the fragments were washed in 70% ethanol, with subsequent dehydration in an ascending concentration of ethanol.

Subsequently, the fragments were diaphanized in xylene for 2 h and included in a polymers (Paraplast Plus, ST. Louis, MO, USA). Then the materials were sectioned in a Leica RM 2165 rotary microtome (Leica, Munich, Germany) with a thickness of 5 μm and stained with hematoxylin-eosin and photographed on Zeiss Axiophot light microscope (Zeiss, Munich, Germany).

***2.10.3. Histopathology***: Histopathology of various organs will be evaluated and correlated with the toxicity levels of inflammation. The degree of inflammation will be assessed by a semi-quantitative scale: 0 = no inflammation, 1 = minimal inflammation (less than five lymphocytes in an area of 0.25 $mm^2$), 2 = moderate inflammation (scattered throughout the tissue mononuclear inflammatory cells but stroma still was visible), 3 = intense inflammation (mononuclear inflammatory cells infiltrating tissues deeply [19].

## 3. RESULTS AND DISCUSSION

The graphene oxide sample GO from Cheap Tubes Inc. (USA) was selected for our study. Exhaustive characterization of GO was afforded. Characterization through XRD, XPS, NMR, Raman, FTIR, UV-Vis, TG, DSC, TEM, EDS, DLS and NTA, showed that these analyses corresponded to a single-layer graphene oxide, purity 99%, without Debris and it was stable in water suspension (1 mg/mL) for 10 days. From these analyses was clear that the representative GO from a commercial source it is single layer and a good sized distribution and easily suspended in water. From this analysis it was clear that the representative GO from a commercial source it is single layer and a good sized distribution.

It exhibited thermal stability over 60°C and it was suspended in deionized water after ultrasonication (1 mg/mL) (stable 10 days). All the biological fluids used in the different assays were used as control of the colloidal suspension stability (Fig.1).

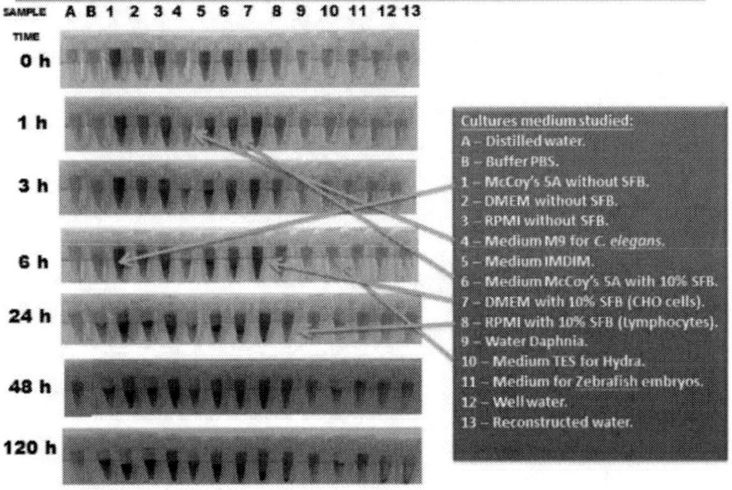


Figure 1. Colloidal stability of GO (1 mg/mL) in different biological media used in the different toxicological assays.

The Fig. 1 shows that in different media the GO showed a short period of stability. In general it was suggested for all the team that used the suspension freshly prepared during the 1-3 h period. The same results were corroborated by the UV-Vis spectroscopy. Then, all the studies were carried out within that stability period.

The hemolytic assay following the procedure from ASTM E2524-08 showed completely absence of hemolysis by GO up to 100 ug/mL,

The cytotoxicity assays were carried out by the Alamar Blue (Resazurin) reduction, MTT and flow cytometry assays in mouse embryonic fibroblast cells (3T3), human keratinocytes (HaCaT), colorectal cancer cells (Caco-2/HCT 116), Lewis lung cancer cells (3LL), acute myeloid leukemia cells (KG-1, Jurkat, Kasumi-1) and chronic myeloid leukemia cells (K562, Lucena) and no significant toxicity was found after exposition to 0.1-100 μg/mL for 24 and 48 h. Breast cancer cells, MCF-7, showed a 20% reduction on cell viability at 24 and 48 h. No cytotoxicity were found in lymphocytes, Chinese hamster ovary cells (CHO) and human macrophage cell line (U937) at 0.1-50 μg/mL, but 30-50% survival inhibition was observed at 100 μg/mL (Table 1).

Table 1. Viability of cell cultures on graphene oxide at 24h.

| Cell Culture | Characteristics | MTT assay | Comments | Resazurin | Comments |
|---|---|---|---|---|---|
| 3T3 | Mouse embryonic fibroblast cells | Up to 50 ug/mL no toxicity | >100 ug/mL 20% inhibition | Up to > 100 ug/mL no toxicity | |
| HaCat | Human keratinocytes | | | Up to 50 ug/mL no toxicity | |
| Caco-2 | Colorectal adenocarcinoma cells | | | Up to > 100 ug/mL no toxicity | |
| HCT 116 | Colorectal carcinoma cells | | | Up to > 100 ug/mL no toxicity | |
| 3LL | Lewis lung cancer cells | | | Up to > 100 ug/mL no toxicity | |
| KG-1 | Acute myeloid leukemia cells | | | Up to > 20 ug/mL no toxicity | >80 ug/mL 15% inhibition |
| Jurkat | Acute myeloid leukemia cells | Up to > 100 ug/mL no toxicity | | Up to > 100 ug/mL no toxicity | |
| Kasumi-1 | Acute myeloid leukemia cells | Up to 50 ug/mL no toxicity | >100 ug/mL 20% inhibition | Up to 50 ug/mL no toxicity | >80 ug/mL 20% inhibition |
| K562 | Chronic myeloid leukemia cells | Up to 50 ug/mL no toxicity | >100 ug/mL 50% inhibition | Up to > 100 ug/mL no toxicity | |
| Lucena | Chronic myeloid leukemia cells | | | Up to > 60 ug/mL no toxicity | >80 ug/mL 10% inhibition |
| MCF-7 | Breast cancer cells | | | Up to 50 ug/mL no toxicity | >80 ug/mL 10% inhibition |
| SKBR3 | Breast cancer cells | | | Up to 20 ug/mL no toxicity | >50 ug/mL 40% inhibition |
| CHO | Chinese hamster ovary cells | Up to 50 ug/mL no toxicity | >100 ug/mL 70% inhibition | | |
| U937 | Human leukemic monocyte lymphoma cell | | | Up to 30 ug/mL no toxicity | >40 ug/mL 50% inhibition |
| Lympho-cytes | Peripheral blood | >25 ug/mL 50% inhibition | | | |

Note: All he cells were checked out for mycoplasma by the PCR and fluorescence assay. All the cell cultures were from ATCC. Similar results at 48 h.

In some cases there are differences in the cell sensibilities when is used Alamar Blue or MTT assays. But in general, at different laboratory the results of cytotoxicity is within the experimental errors when it was used the same cytotoxicity assay. For example, in 3T3 and K562 the MTT assay was more sensible than Resazurin assay. However, almost identical results in the Jurkat and Kasumi cell cultures in the MTT and Resazurin assays were observed.

A dose-dependent increase in apoptosis was observed in some cells (Kasumi-1, Jurkat and K562 cells). In the case of CHO and 3T3 cells, greater levels of necrosis with increasing concentrations of GO (>50 µg/mL) were observed (Table 2).

**Table 2. Graphene oxide (GO) necrosis and apoptosis**

| Cell Culture | Characteristics | MTT assay | Comments | Cell Death |
|---|---|---|---|---|
| Lymphocytes | Peripheral blood | >25 ug/mL 50% inhibition | | Apoptosis |
| Kasumi-1 | Acute *myeloid leukemia* cells | Up to 50 ug/mL no toxicity | >100 ug/mL 20% inhibition | Apoptosis |
| K562 | Chronic myeloid leukemia cells | Up to 50 ug/mL no toxicity | >100 ug/mL 50% inhibition | Apoptosis |
| 3T3 | Mouse embryonic fibroblast cells | Up to 50 ug/mL no toxicity | >100 ug/mL 20% inhibition | Necrosis |
| CHO | Chinese hamster ovary cells | Up to 50 ug/mL no toxicity | >100 ug/mL 70% inhibition | Necrosis |

Genotoxic study using the Comet assay showed slight DNA damage in lymphocytes at all concentrations tested, while more significant effects were observed in CHO cells.

Econanotoxicity was carried out by lethality assays in the nematode *Caenorhabditis elegans*, in the freshwater coelenterate *Hydra attenuate, Daphnia similis* and on *Shrimp.*
Chronic GO exposure on survival of *Caenorhabditis elegans* (N2 strain; 120 nematodes) at L4 stage expressed as the percentage of control viability (100%) showed no toxicity up to 100 ug/mL. The study on GO on survival (120 nematodes) and body length (50 nematodes) of *Caenorhabditis elegans* (N2 strain) at L1 stage after 72 h of exposure at 20°C showed also no toxic effect in this stage. No effect up to 100 µg/mL of GO acting on *Daphnia similis* was found. Lethality assays in the freshwater coelenterate *Hydra* with showed no signs of toxicity at concentrations varying from 0.1-50 µg/mL of GO. At 50 mg/mL of GO, the tentacles was shortened after 72 h of exposure. However, death and disintegration of *Hydra* was observed after exposition to 100 µg/mL for 96 h.

In the Salmonella/microsome mutagenicity assay using 10-250 µg/plate no effect was observed.

No acute ecotoxicity was verified up to 3.0 µg/mL (Exposure time: 24h) to shrimp. Absence of effects on oxygen consumption and ammonia excretion up to 1.0 µg/mL (Exposure time: 24h)

In *in vivo* studies, no changes in biochemical parameters of Fischer 344 rats were observed after the i.p. administration of GO. Some black agglomerates were found in the intraperitoneal cavity of rats injected with GO. This was also observed previously in Balb/c mice receiving GO i.p. [20]. This probably did not affect the final absorption of GO via intraperitoneal administration, since in another experiment with this GO (100 µg/mL) on Fisher 344 rats-bearing prostate tumors, treatment with GO negatively affected the hepatic parameters, whilst in the renal ones, an improvement was observed. Studies are in progress to understand the mechanisms involved in the uptake of GO by RES.

## 4. CONCLUSIONS

Finally, through all of these experimental assays, such as cytotoxicity, genotoxicity, mutagenicity, ecotoxicity and *in vivo* toxicity assays, the GO appeared as a potential new drug (~50-100 µg/mL) for several diseases *in vivo* treatment with very low limitations.

**Acknowledgement:** Support from the Brazilian Network of Nanotoxicology (CIGENANOTOX) (MCTI/CNPq), INOMAT (MCTI/CNPq), NanoBioss (MCTI) and FAPESP are acknowledged.

## References


[1] Zhou and Bongiorno A 2013 *Sci. Rep.* **3** 2484

[2] Roebben, G, Rasmussen K, Kestens V, Linsinger T P J, Rauscher H,Emons H and Stamm H 2013 *J. Nanopart. Res.* **15** 1455

[3] Faria A F, Martinez D S T, Moraes A C M, Costa M E H M, Barros E B, Souza Filho A G, Paula A J and Alves O L 2012 *Chem. Mater* **24** 4080

[4] Durán N, Makita Y and Innocentini L H 1979 *Biochem. Biophys. Res. Commun.* **88** 642

[5] Azqueta A, Gutzkow K B, Brunborg G and Collins A R 2011 *Mutat. Res.* **724** 41.

[6] Dusinská M, Collins A, Kazimírová A, Barancoková M, Harrington V, Volkovová K, Staruchová M, Horská A, Wsólová L, Kocan A, Petrík J, Machata M, Ratcliffe B and Kyrtopoulos S 2004 *Mutat. Res.* **553** 91

[7] Dusinska M and Collins A R 2008 *Mutagenesis* **23** 191

[8] Collins A R, Oscoz A A, Brunborg G, Gaivão I, Giovannelli L, Kruszewski M, Smith C C and Stetina R 2008 *Mutagenesis* **23** 143

[9] Lima R, Feitosa L O, Ballottin D, Marcato P D, Tasic L and Durán N 2013 *J. Phys. Conf. Ser.* **429** 012020

[10] OECD 1997 Guideline 471 Ames bacterial reverse mutation screening assay (Non-GLP), Paris, France

[11] Brenner S 1973 *Brit. Med. Bull.* **29** 269

[12] Wilby O K 1988 *Proc. Workshop Ass. Franc. Teratologie* 108

[13] Wilby O K and Tesh J M 1990 *Toxicol. in Vitro* **4** 582

[14] OECD 2004 Guideline for Testing of Chemicals: Daphnia sp. Acute Immobilization Test. OECD Guideline No 202

[15] Barbieri E, Bondioli, A C V, De Melo C B and Henriques, M B 2014 *Marine Freshwater Behaviour Physiol.* **47** 1

[16] Schmid W 1975 *Mutat. Res.* **31** 95

[17] Mersch J, Beauvais M M and Nagel P 1996 *Mutat. Res.* **371** 47

[18] Fenech M 2000 *Mutat. Res.* **455**, 81

[19] Offersen B V, Knap M M, Marcussen N, Horsman M R, Hamilton-Dutoit S and Overgaard J. 2002 *Br. J. Cancer* **87** 1422

[20] Yang K, Gong H, Shi X, Wan J, Zhang Y and Liu Z 2013 *Biomaterials* **34**, 2787

Journal of Physics: Conference Series **617** (2015) 012020 doi:10.1088/1742-6596/617/1/012020

# Nitric oxide-releasing polymeric nanoparticles against *Trypanosoma cruzi*

**A B Seabra[1*], N A Kitice[1], M T Pelegrino[1], C A C Lancheros[2], L M Yamauchi[2], P Pinge-Filho[3] and S F Yamada-Ogatta[2]**

[1]Exact and Earth Sciences Departament, Universidade Federal de São Paulo, Rua São Nicolau 210, CEP 09913030, Diadema, SP, Brazil,
[2]Department of Microbiology, Universidade Estadual de Londrina, Campus Universitário, CEP 86055-990, Londrina, PR, Brazil,
[3]Department of Phatological Sciences, Laboratory of Immunology, Universidade Estadual de Londrina, Campus Universitário, CEP 86055-990, Londrina, PR, Brazil

*Email: amedea.seabra@gmail.com, amedea.seabra@unifesp.br

**Abstract**. Chagas disease, also known as American trypanosomiasis, is a potentially life-threatening illness caused by the protozoan parasite, *Trypanosoma cruzi* (*T. cruzi*), and the disease remains a major health problem in many Latin American countries. Several papers report that the killing of the parasite is dependent on the production of nitric oxide (NO). The endogenous free radical NO is an important cellular signalling molecule that plays a key role in the defense against pathogens, including *T. cruzi*. As *T. cruzi* is able to compromise host macrophages decreasing endogenous NO production, the administration of exogenous NO donors represents an interesting strategy to combat Chagas disease. Thus, the aims of this study were to prepare and evaluate the antimicrobial activity of NO-releasing polymeric nanoparticles against *T. cruzi*. Biocompatible polymeric nanoparticles composed of chitosan/sodium tripolyphosphate (TPP) were prepared and used to encapsulate mercaptosuccinic acid (MSA), which is a thiol-containing molecule. Nitrosation of free thiols (SH) groups of MSA were performed by the addition of equimolar amount of sodium nitrite ($NaNO_2$), leading to the formation of S-nitroso-MSA-containing nanoparticles. These polymeric nanoparticles act as spontaneous NO donors, with free NO release. The results show the formation of nanoparticles with average hydrodynamic diameter ranging from 270 to 500 nm, average of polydispersity index of 0.35, and encapsulation efficiency in the range of 99%. The NO release kinetics from the S-nitroso-MSA-containing nanoparticles showed sustained and controlled NO release over several hours. The microbicidal activity of S-nitroso-MSA-containing nanoparticles was evaluated by incubating NO-releasing nanoparticles (200 - 600 μg/mL) with replicative and non-infective epimastigote, and non-replicative and infective trypomastigote forms of *T. cruzi*. In addition, a significant decrease in the percentage of macrophage-infected (with amastigotes) and NO-releasing nanoparticle-treated cells was observed. Taken together, our results reveal a potent toxic effect of NO-releasing polymeric nanoparticles against different life cycle forms of *T. cruzi*, indicating that the encapsulation of the NO donor S-nitroso-MSA represents an interesting approach to combat and to prevent Chagas disease.

## 1. Introduction

Chagas disease, caused by the heteroxenic trypanosomatid *Trypanosoma cruzi* (*T. cruzi*) [1] is endemic in Latin America with an estimated 7-8 million people infected [2]. However, the intense migration of people from endemic areas for non-endemic countries has contributed for the globalization of this disease [3,4].

The life cycle of *T. cruzi* involves a vertebrate and invertebrate hosts, alternating between replicating (epimastigote and amastigote) and infectious forms (trypomastigote). The natural transmission of the protozoan to mammalian host occurs through the excreta containing metacyclic trypomastigotes of an infected hematophagus reduviid bug of the subfamily Triatominae. Within vertebrate host, the parasite can invade a wide range of nucleated cell. After intracellular multiplication as amastigotes, they turn into trypomastigotes, which are released and can invade adjacent cells or spread hematogenously to distant sites. Once in the bloodstream, the trypomastigotes may be ingested by the invertebrate host, turn into epimastigotes which multiply actively and transform in metacyclic trypomastigotes in its gut [5]. Alternatively, the bloodstream trypomastigotes may be transmitted by blood transfusion [6].

In Brazil, benznidazole is currently available for the specific treatment of Chagas disease. The effectiveness of this drug is dependent on various factors such as the disease stage, dose, age, geographical origin of the patient and parasite drug susceptibility. In addition, this substance can produce serious side effects, which may lead to discontinuation of the treatment [7]. Thus, there is an urgent need for development of new strategies for the control of Chagas disease.

The endogenous found molecule nitric oxide (NO) is not only a physiological mediator of important process such as the dilation of blood vessels [8,9], but also plays a key role in the immunological system against several pathogens [10-12]. Upon pathogen infection, activated macrophages and other immune cells inhibit pathogen replication by releasing several effector molecules, including NO [13]. Therefore, NO is an important defense molecule against infectious organisms, including *T. cruzi* [13-15]. As a free radical, NO is unstable and it can be readily inactivated in the biological system. In order to increase the life-time of NO, several works describe the preparation of small molecule weight S-nitrosothiols (RSNOs), an important class of NO donor, used in several biomedical applications [11,16-18]. However, RSNOs are thermally unstable in aqueous solutions, leading to free NO release [19]. In this context, RSNOs have been successfully incorporated in biocompatible polymeric nanoparticles to promote a sustained NO release over several hours [12,20-23].

In this context, the present work describes the preparation of polymeric biocompatible and biodegradable nanoparticles composed by chitosan/ tripolyphosphate (TPP) containing S-nitroso-mercaptosuccinic acid (S-nitroso-MSA), an RSNO, which act as spontaneous NO donor. The antimicrobial efficacy of NO-releasing polymeric nanoparticles was evaluated against different forms of *T. cruzi*. The results indicate the potential of NO-releasing nanoparticles in the treatment and combat of Chagas` diseases.

## 2. Methods

### *2.1. Synthesis of MSA-containing chitosan/TPP nanoparticles*

Chitosan nanoparticles containing MSA were prepared through ionotropic gelation process using chitosan and sodium tripolyphosphate (TPP), as previous described [12]. Briefly, 0.1 g of chitosan was dissolved in 100 mL of 1% acetic acid. Through magnetic stirring at room temperature, 5 mL of the TPP (0.6 mg/mL) solution was added dropwise into 15 mL of a 1 mg/mL chitosan solution. The final mixture was stirred to prepare an aqueous solution with chitosan/TPP particles. MSA-containing chitosan nanoparticles were prepared by adding MSA into the chitosan solution. The final concentration of MSA in the chitosan/TPP nanoparticles was 400 mmol/L, which corresponds to 60 mg/mL.

*2.2. Synthesis of S-nitroso-MSA-containing chitosan/TPP nanoparticles*
Free thiol groups of MSA-containing chitosan/TPP nanoparticles were nitrosated by adding equimolar amount of sodium nitrite ($NaNO_2$) directly into the polymeric aqueous dispersion (pH 4.0). The final solution was homogenized with magnetic stirring for 30 min and protected from light. S-nitroso-MSA-chitosan/TPP nanoparticle formation was characterized by the S-NO group absorption bands at either 336 nm ($\varepsilon = 922$ mol $L^{-1}cm^{-1}$) or 545 nm ($\varepsilon = 18$ mol $L^{-1}cm^{-1}$) using a UV–Visible spectrophotometer (Agilent, model 8453, Palo Alto), as previous described [24].

*2.3. The average size and size distribution for polymeric particles in aqueous medium*
The average size for MSA-containing chitosan/TPP nanoparticles were measured using photon correlation spectroscopy (PCS) (Nano ZS Zetasizer, Malvern Instruments Co.) at 25 °C in polystyrene cuvettes with a 10 mm path length.

*2.4. MSA encapsulation efficiency in chitosan/TPP nanoparticles*
The encapsulation efficiency of MSA in chitosan/TPP nanoparticles were measured by the UV–vis method, as already described [12]. Briefly, free MSA was separated from polymeric nanoparticles by ultracentrifugation, by using a Microcon centrifugal filter device containing ultrafiltration membranes (MWCO 10,000, Millipore). The amount of free MSA in the ultrafiltrates was measured by titration with the thiol-reacting 5,50-dithiobis-(2-nitrobenzoic acid) (DTNB), based on the absorbance band at 412 nm ($\varepsilon = 14.15$ mmol $L^{-1}cm^{-1}$) of the 2-nitro-5-thiobenzoate anion, which is generated in the reaction of MSA with DTNB. The percentage of MSA encapsulation was determined by the described equation:

$$(\%) = (\text{mass of MSA encapsulated/mass of MSA total}) \times 100 \qquad \text{(Eq. 1)}$$

*2.5. Kinetics free NO release from S-nitroso-MSA-chitosan/TPP nanoparticles*
The kinetics for NO release from S-nitroso-MSA-chitosan/TPP nanoparticles were monitored by following the spectral changes at 336 nm and 545 nm, which are solely associated with S-N bond cleavage and free NO release [12,19] by using an Uv-Visible spectrophotometer (Agilent 8453). The kinetics curves were monitored at 28 and 37 °C for 24 h. The amount of NO released was calculated from the amount of S-nitroso-MSA decomposed, as previous described [25,26]. The initial rates of NO release through S-nitroso-MSA decomposition were determined through linear regression of the curve slopes, as previous described [27].

*2.6. Parasite and mammalian cell culture*
Epimastigote forms of *T. cruzi* Y strain [28] were obtained from four-day incubation culture in liver infusion tryptose (LIT) medium [29] supplemented with 10% (v/v) heat-inactivated fetal bovine serum (FBS), at 28 °C. Trypomastigote forms were obtained from the supernatant of infected LLC-MK2 cell line after 120 h post-infection at 37 °C. LLC-MK2 cells were cultured in RPMI 1640 medium supplemented with 10% (v/v) of FBS, 2 mmol/L *L*-glutamine, 100 IU/mL penicillin, 100 µg/mL streptomycin and 2.5 µg/mL amphoterecin B, in 5% $CO_2$ at 37 °C. The peritoneal cells were prepared from BALB/c mice obtained from the breeding colonies of the animal facility of the *Centro de Ciências Biológicas* of *Universidade Estadual de Londrina*, Londrina, Paraná, Brazil. The experimental protocol was approved by the Ethics Animal Experimentation Committee of the Universidade Estadual de Londrina. Mice were injected intraperitoneally with 1.5 mL of 3% thioglycolate and peritoneal cells were collected four days later by injecting sterile cold 10 mL of 50 mmol/L sodium phosphate buffer pH 7.4 containing 0.15 mol/L NaCl (PBS). The obtained cells were

centrifuged at 500 g for 10 min at 4°C and resuspended in RPMI 1640 medium (Invitrogen-Gibco, USA) supplemented with 10% heat inactivated fetal bovine serum (Invitrogen-Gibco, USA), 50 μg/mL gentamicin, 100 IU/mL penicillin, 100 μg/mL streptomycin and 2 mmol/L glutamine.

*2.7. Antitrypanosomal activity*
Epimastigotes of *T. cruzi* in logarithmic growth phase (1 x $10^6$ cells/mL) were added in 24-well plates (Techno Plastic Products, Switzerland) containing LIT medium with different concentrations of S-nitroso-MSA nanoparticles (0 – 600 μg/mL). The cultures were incubated at 28°C, and cell growth was estimated by direct counting in a hemocytometer (Improved Double Neubauer) after 48 h. Trypomastigotes (1 x $10^6$ cells/mL) were added in 96-well plates (Techno Plastic Products, Switzerland) containing RPMI 1640 medium with different concentrations of S-nitroso-MSA nanoparticles (0-400 μg/mL) and the plates were incubated at 37 °C. Parasites were counted, as above, after 10, 30, 60 e 120 min. Tests containing medium alone, medium plus MSA-containing nanoparticle (without the NO donor S-nitroso-MSA), medium with sodium nitrite ($NaNO_2$) were employed as controls. The experiments were performed in triplicate in two different occasions. The results were expressed as the minimal concentration of the fraction that causes 50% ($IC_{50}$, inhibitory concentration) growth inhibition of epimastigotes, and 50% ($EC_{50}$, effective concentration) lysis of trypomastigotes.

*2.8. Interaction with macrophage cells*
Murine peritoneal macrophages were added to a 24-well culture plate (Techno Plastic Products, Switzerland), containing a round cover glass, at a density of 5 x $10^5$ cells/well, and incubated for 24 h, in 5% $CO_2$ at 37°C. Trypomastigotes (at a parasite to macrophage ratio of 5:1) were inoculated into the monolayer and the plate was incubated for 3 h. After this period, the plates were washed with sterile PBS to remove the extracellular parasites. The fresh culture medium containing S-nitroso-MSA nanoparticles (50-400 μg/mL) were added to the cell culture, and the plates were incubated as above. Additionally, the macrophages monolayer was: (i) pre-treated with S-nitroso-MSA nanoparticles and then infected with trypomastigotes; (ii) and the infection and treatment were carried out simultaneously. The cover glasses of controls (medium alone, medium plus MSA-chitosan nanoparticles, and medium plus sodium nitrite) and tests were fixed with methanol (Merck, Brazil), stained with Giemsa after 24, 48 and 72 h incubation and permanently prepared in ERV-Mount resin. The percentage of infected host cells and the mean number amastigotes/500 cells were determined by direct counting in light microscopy.

## 3. Results and Discussion

*3.1. Synthesis of MSA or S-nitroso-MSA-chitosanTPP nanoparticles*
In order to investigate the role of NO against *T. cruzi* protozoa, this unstable molecule need to be delivered to an application site in a controlled and sustained manner. Thus, biodegradable and biocompatible polymeric nanoparticles are interesting vehicles for drug delivery [21,22]. In particular, chitosan is natural polysaccharide broadly used in pharmacological applications [30]. Due to the presence of free amino groups, chitosan has a positive charge and can react with negatively charged biomolecules, such as DNA [30,31]. In addition, chitosan is reported to have antimicrobial properties [30], and thus, chitosan with the NO donor, S-nitroso-MSA, might potentiate antimicrobial activity in the nanoparticles. In this work, chitosan/TPP nanoparticles containing MSA were successfully prepared through ionic gelation. Cationic chitosan readily reacts with multivalent counter ions, such as TPP, to yield a network structure with spherical shape nanoparticles [12, 31].

In aqueous solution, the hydrodynamic diameter of the MSA-chitosan/TPP nanoparticles were found to be in the range of 270 to 550 nm with a polydispersity index of 0.35, which are in accordance with our previous results [12, 20]. The encapsulation efficiency of MSA in chitosan/TPP nanoparticle solution was found to be 99 ± 1%, indicating the high affinity of MSA to the nanoparticle.

S-nitroso-MSA-chitosan/TPP was obtained by nitrosating of free thiol groups of MSA-containing nanoparticles. An equimolar amount of sodium nitrite ($NaNO_2$) was added to the acidified MSA-containing polymeric nanoparticle solution. In acidified aqueous solution, sodium nitrite will form nitrous acid ($HNO_2$), which is the nitrosating agent, leading to the formation of S-nitroso-MSA, according to Figure 1.

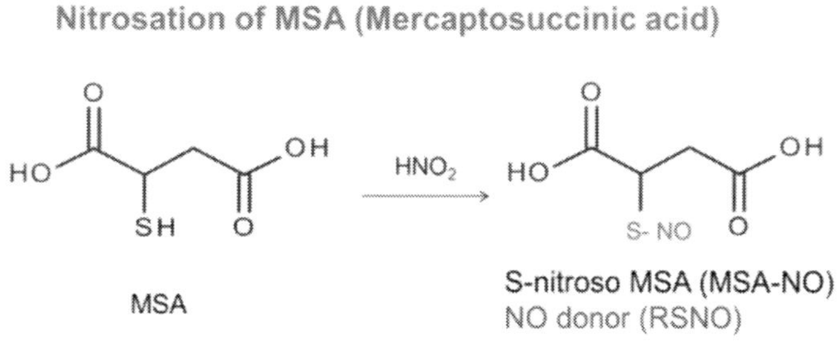


**Figure 1.** Schematic representation of the nitrosation of MSA yielding S-nitroso-MSA.

*3.2. NO release profile from S-nitroso-MSA-containing chitosan nanoparticles in aqueous solutions*

The NO release profile from S-nitroso-MSA-chitosan/TPP nanoparticles in aqueous solution were monitored by following the spectral changes at 336 and 545 nm absorption bands, associated the S-N bound cleavage with free NO release [12]. The kinetic curves were monitored for 24 h at 28 and 37°C, since these two temperatures were used in the *in vitro* experiments with *T. cruzi*. Figure 2 shows the NO release profile from S-nitroso-MSA encapsulated in chitosan.

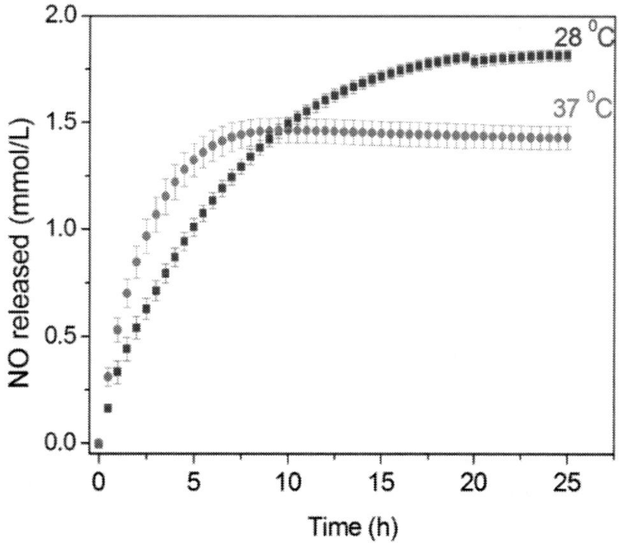


**Figure 2.** Kinects of NO release from S-nitroso-MSA-containing chitosan/TPP nanoparticles at 28 and 35 °C.

Figure 2 shows that NO is spontaneously released from S-nitroso-MSA-containing chitosan nanoparticles. The kinetic curves show an initial burst of NO release in the first 5 hours, followed by a progressively increase at lower rates, for both temperatures. A sustained NO release is observed for at least 10 h. The NO-release profile increased with the increase of the temperature, since the stability of this NO donor is dependent on the temperature, as expected [25-27]. Initial rates of NO released were calculated from the kinetic curves of Figure 2. The initial rates of NO release from S-nitroso-MSA-nanoparticles were found to be $0.13 \pm 0.01$ and $0.19 \pm 0.01$ $mmolL^{-1}h^{-1}$for 28 and 37°C, respectively. Higher rates are reported for higher temperatures [27]. The NO release from S-nitroso-MSA encapsulated in polymeric nanoparticles is reported to occur mainly through diffusion process over the pores or wall and disintegration of the hydropolymeric structure of chitosan [20].

### *3.3. Antitrypanosomal activity of NO-releasing polymeric nanoparticles*

The inhibitory effects of S-nitroso-MSA chitosan nanoparticles on insect replicating epimastigote forms of *T. cruzi* Y strain are presented in Figure 3(A). Similarly, S-nitroso-MSA chitosan nanoparticles decreased the number of motile trypomastigotes, the non-replicating and infective form, after 24 h (Figure 3(B)). In both assays, a dose-dependent antitrypanosomal effect was observed and the $IC_{50}$ and $EC_{50}$ calculated for these nanoparticles were 252 µg/mL and 91 µg/mL, respectively.

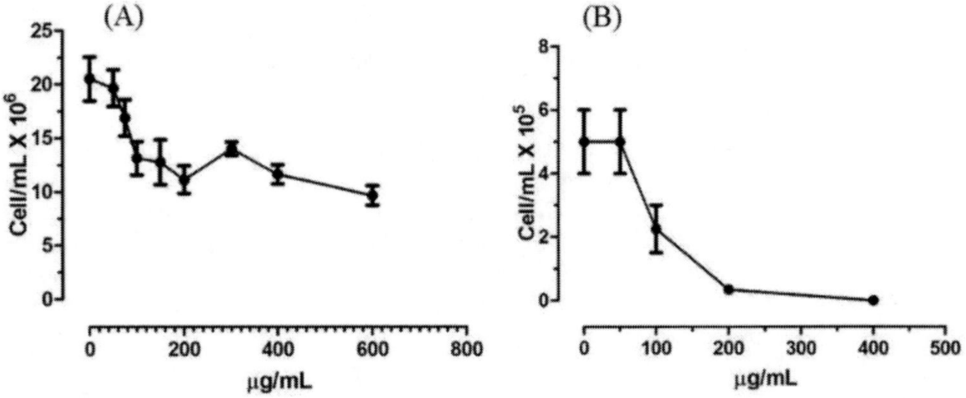


**Figure 3**. Antitrypanosomal activity of S-nitroso-MSA-containing chitosan/TPP nanoparticles in different concentrations, as indicated in the Figure, for 48 h of incubation with: (A) epimastigote form of *T. cruzi*, (B) trypomastigote form of *T. cruzi*. Temperatures of incubation: 28°C and 37°C for epimastigote and trypomastigote forms, respectively.

Figure 4 shows the antitrypanosomal activity of S-nitroso-MSA chitosan nanoparticles, at final concentration of 200 µg/mL, after 48 h of incubation with *T. cruzi* epimastigote cultures. Control groups were performed by incubating epimastigote cultures with only the medium solution, medium solution plus MSA-chitosan nanoparticles (nanoparticles without the NO donor S-nitroso-MSA), and medium with plus sodium nitrite ($NaNO_2$), the nitrosating agent of MSA to form S-nitroso-MSA. The inhibitory effect is due to the presence of S-nitroso-MSA in the chitosan nanoparticles, since MSA-nanoparticles and the sodium nitrite were not toxic (Figure 4). Therefore, the potent antitrypanosomal activity observed can be attributed to the presence of S-NO groups incorporated in the polymeric nanoparticles. S-NO groups act as spontaneous NO donor due to the homolytic bound cleavage of S-N with free NO release. The sustained NO release from S-nitroso-MSA incorporated into chitosan nanoparticles over several hours as shown in Figure 2 demonstrated potent toxic effects against *T. cruzi*. Thus, free NO release from the nanoparticles is the toxic molecule.

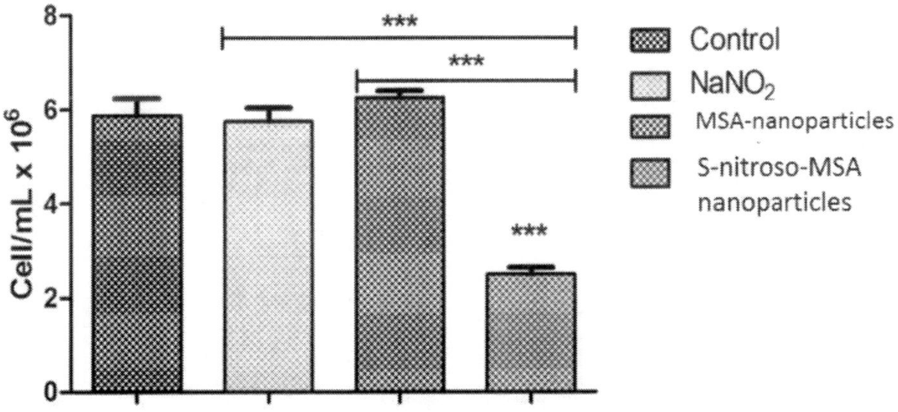


**Figure 4**. Antitrypanosomal activity of S-nitroso-MSA chitosan nanoparticles at final concentration of 200 μg/mL, after 48 h of incubation with *T. cruzi* epimastigote forms. Control groups were performed by incubating epimastigote cultures with only the medium solution, medium solution plus MSA-chitosan nanoparticles (nanoparticles without the NO donor S-nitroso-MSA), and medium plus sodium nitrite ($NaNO_2$), the nitrosating agent of MSA to form S-nitroso-MSA. Temperature of incubation: 28°C. The asterisks indicate significant difference in the antitrypanosomal activity of the group treated with NO-releasing polymeric nanoparticles in comparison with the control groups.

### *3.4. Interaction of NO-releasing polymeric nanoparticles with macrophage cells*

Because macrophages serve as effector cells in protecting against intracellular parasites, the interaction of *T. cruzi* and macrophages was studied *in vitro*. Figure 5 shows the antitrypanosomal activity of S-nitroso-MSA incorporated in chitosan nanoparticles (final concentration of 200 μg/mL) during *T. cruzi* interaction with murine peritoneal macrophages.As can be observed, S-nitroso-MSA-chitosan/TPP nanoparticles decreased the number of intracellular amastigotes, which is the replicating form found in mammalian hosts. Interestingly, NO-releasing polymeric nanoparticles were found to have a potent antitrypanosomal activity in all tested treatment conditions: before infection, during infection and after infection. Before infection means that macrophages were pre-treated with S-nitroso-MSA chitosan nanoparticles for 3 h, washed and then infected with *T. cruzi*. During infection means that macrophages were infected with *T. cruzi* and simultaneous incubated with S-nitroso-MSA nanoparticles. After infection means that macrophages were infected with *T. cruzi* and then incubated with S-nitroso-MSA nanoparticles. These results indicate that NO-releasing nanoparticles may interfere with the adhesion and/or invasion of host cells, or replication of *T. cruzi* in mammalians cells, suggesting their use in the treatment and prevention of Chagas disease.

In addition, NO is a versatile molecule in the biological system, involved in multiple defense mechanism pathways. Indeed, NO has been implicated in multiple cellular processes of the immune system and play multiple functions in infected host. The antimicrobial action of NO against pathogens is reported to occur via multiple pathways, which include DNA damage/mutation, alteration of proteins by S-nitrosation, inhibition of protein synthesis, tyrosine nitration, lipid membrane peroxidation [13]. As NO is a free radical, in the biological system, it readily reacts with superoxide ($O_2^{\bullet -}$), leading to the formation of the harmful peroxynitrite ($ONOO^-$). Tyrosine nitration by peroxynitrite is one of the major mechanism of the microbicidal action of NO [13].

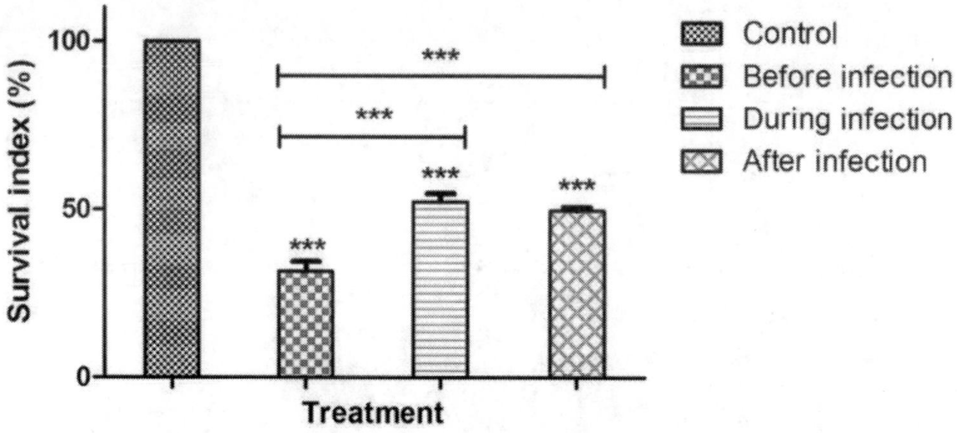


**Figure 5**. Antitrypanosomal activity of S-nitroso-MSA incorporated in chitosan nanoparticles (final concentration of 200 μg/mL) in *T. cruzi* interaction with murine peritoneal macrophages. Control: medium; Before infection: macrophages were pre-treated with S-nitroso-MSA chitosan nanoparticles for 3 h, washed and then infected with *T. cruzi*; During infection: macrophages were infected with *T. cruzi* and simultaneous incubated with S-nitroso-MSA nanoparticles; After infection: macrophages were infected with *T. cruzi* and then incubated with S-nitroso-MSA nanoparticles. Temperature of incubation: 37 °C.

## 4. Conclusions

This work describes the preparation and characterization of S-nitroso-MSA-chitosan/TPP nanoparticles and their important effect as potent antitrypanosomal agent against different life cycle forms of *T. cruzi* protozoa. The results indicate that the encapsulation of the NO donor S-nitroso-MSA represents an interesting approach to control Chagas disease.

## 5. Acknowledgements

Supported from FAPESP (Process 2012/17053-7), CNPq, CAPES, Brazilian Network in Nanotoxicology (MCT-CNPq).

## 6. References

[1] Chagas C 1909 *Mem. Inst. Oswaldo Cruz* **1** 159
[2] WHO 2014 Chagas disease (American trypanosomiasis). World Health Organization. Fact sheet no 340. http://www.who.int/mediacentre/factsheets/fs340/en/
[3] Coura J R and Vinas P A 2010 *Nature* **465,7301**: S6
[4] Bonney K M 2014 *Parasite* **21** 11
[5] Tyler K M and Engman D M 2001 *Inter. J. Parasitol.* **31** 472
[6] Rassi A J R, Rassi A and Marin-Neto J A 2010 *Lancet* **375** 1388
[7] Urbina J A 2010 *Acta Trop.* **115** 55
[8] Seabra A B, Pankotai E, Fehér M, Somlai A, Kiss L, Bíro L, Szabó C, Kollai M, de Oliveria M G and Lacza Z 2007 *Brit. J. Dermatol.* **156** 814
[9] Seabra A B, Fitzpatrick A, Paul J, de Oliveira M G and Weller R. 2004 *Brit. J. Dermatol.* **151** 977
[10] de Souza G F P, Yokoyama-Yasunaka J K U, Seabra A B, Miguel D C, de Oliveira M G and Uliana SRB 2006 *Nitric Oxide* **15** 209
[11] Seabra A B, Martins D, Simões M M S G, da Silva R, Brocchi M and de Oliveira M G 2010 *Artif. Organs* **34** E204

[12] Cardozo V F, Lancheros C A C, Narciso A M, Valereto E C S, Kobayashi R K T, Seabra A B and Nakazato G 2014 *Int. J. Pharm.* **473** 20
[13] Bañuls C, Rocha M, Rovira-Llopis S, Falcon R, Castello R, Herance J R, Polo M, Blas-Garcia A, Hernandez-Mijares A and Victor M. Victor 2014 *Curr. Pharm. Desing* **20** 4679
[14] Da Silva A, Paim F C, Santos R C V, Sangoi M B, Moresco R N, Lopes S T A, Jaques J A, Baldissarelli J, Morsch V M and Monteiro S G 2012 Exp. Parasitol. **132** 166
[15] Gutierrez F R S, Mineo T W P, Pavanelli W R, Guedes P MM and Silva J S 2009 *Mem. Inst. Oswaldo Cruz* **104** 236
[16] Amadeu T P, Seabra A B, de Oliveira M G and Costa A M A 2007 *J. Eur. Acad. Dermatol. Venereol.* **21** 629
[17] Georgii J L, Amadeu T P, Seabra A B, de Oliveira M G and Costa AMA. 2011 *J. Tissue Eng. Regen. Med.* **5** 612
[18] Seabra A B, da Silva R, de Souza G F P and de Oliveira M G 2008 *Artif. Organs* **32** 262
[19] de Oliveira M G, Shishido S M, Seabra A B and Morgon N H 2002 *J. Phys. Chem. A* **106** 8963
[20] Marcato P D, Adami L F, Barbosa R M, Melo P S, Ferreira I R, de Paula L, Duran N and Seabra A B 2013 *Curr. Nanosci.* **9** 1
[21] Seabra A B and Durán N 2010 *J. Mat. Chem.* **20** 1624
[22] Seabra A B and Duran N 2012 *Curr. Nanosci.* **8** 520
[23] Seabra A B, Marcato P D, de Paula L B and Duran N 2012 *J. Nano Res.* **20** 61
[24] Seabra A B, da Silva R and de Oliveira M G 2005 *Biomacromolecules* **6** 2512
[25] Shishido S M, Seabra A B, Loh W and de Oliveira M G 2003 *Biomaterials* **24** 3543
[26] Seabra A B and de Oliveira M G 2004 *Biomaterials* **25** 3773
[27] Seabra A B, da Rocha L L, Eberlin M N and de Oliveria M G 2005 *J. Pharm. Sci.* **94** 994
[28] Silva L H P and Nussenzweig V 1953 *Folia Clin Biol.* **20** 197
[29] Camargo E P 1964 *Rev Inst Med Trop São Paulo* **6** 93
[30] Vinsova J and Vavrikova E 2011 *Curr. Pharm. Design* **17** 3596
[31] Tripathy S, Das S, Charkraborty SP, Sahu S K, Pramanik P, and Roy S 2012 *Int. J. Pharm.* **434** 292

# Graphene oxide sheets-based platform for induced pluripotent stem cells culture: toxicity, adherence, growth and application

**Marcela Durán[1,2], Patricia F. Andrade[2,3], Nelson Durán[2,3*], Angela C.M. Luzo[4], Wagner J. Fávaro[1,2]**

[1]Depart. Struct. Funct. Biol. IB-UNICAMP, SP, Brazil.
[2]NanoBioss, IQ-UNICAMP, SP, Brazil
[3]Chem. Inst (IQ). Universidade Estadual de Campinas (UNICAMP), SP, Brazil.
[4]Mol. Biol. Lab.-Hemocenter-Med. Fac. (UNICAMP), SP, Brazil

duran@iqm.unicamp.br

**Abstract.** It was prepared the graphene oxide (GO) sheets by suspension of GO in ultrapure deionized water or in Pluronic F-68 using a ultrasonicator bath. Total characterization of GO sheets was carried out. The results on suspension of GO in water showed excellent growth and cell adhesion. GO/Pluronic F-68 platform for the growth and adhesion of adipose-derived stem cells (ASCs) that exhibits excellent properties for these processes. GO in water suspension exhibited an inhibition of the cell growth over 5 μg/mL *In vivo* study with GO suspended in water (100 μg/mL) on Fisher 344 rats via i.p. administration showed low toxicity. Despite GO particle accumulates in the intraperitoneal cavity, this fact did not interfere with the final absorption of GO. The AST (aspartate aminotransferase) and ALT (alanine aminotransferase) levels (liver function) did not differ statistically in all experimental groups. Also, creatinine and urea levels (renal function) did not differ statistically in all experimental groups. Taking together, the data suggest the great potential of graphene oxide sheets as platform to ACSs, as well as, new material for treatment several urological diseases.

## 1. INTRODUCTION

The need to find a durable material, of easy use, and which does not interfere significantly in the growth and differentiation of stem cells for the construction of a scaffold for use in urologic surgery, with the purpose of reducing infections, regeneration times and even graft rejection during reconstitution in patients with urethral structure was conducted a broad survey of information about this and came to the consensus of this project: using graphene oxide, a widely studied nanomaterials which has been presenting numerous beneficial results when in contact with the stem cells. Advanced techniques for the growth, differentiation and proliferation of mesenchymal adipose stem cells will be used, as well as the characterization of graphene oxide sheets. Detailed analysis will be performed at each stage of this project to evaluate the interference of these nanocomposites to stem cells, the mucoadhesivity and especially in vivo toxicology of these compounds involved to build the final scaffolds. There are a few *in vivo* studies with regard to the biocompatibility of graphene oxide (GO). Previously, it was developed a synthesis of thermo sensitive hydrogel system based on GO by adding an amount of Pluronic (F-126 and F-68) as a physical crosslinker, without any chemical modification of GO [1]. It was reported that the gel formation was long lasting and stable and did not show any severe chronic inflammatory response. Besides this, it was found that adipose-derived stem cells (ASCs), showed increased adhesion when grown on GO films [2]. However, there is still some controversy in terms of cell adhesion morphology when Mesenchymal stem cells (MSCs) are in contact with GO. Probably the inconsistency may be accounted for by the difference in cell types, the GO impurities, the substrates, and the manufacturing methods of GO. In view of this facts, it was used a representative GO from CheapTubes-USA (Go-single layer).

## 2. MATERIAL AND METHODS

### *2.1. Sample*

The graphene sample GO:Single-layer graphene oxide, purity 99%, thickness 0.7-1.2 nm (AFM); ~300-800nm X&Y dimensions is the standard size <450 nm & 1-20 μm lateral dimensions from Cheap Tubes Inc., Bratleboro, USA was selected for our study.

### *2.2. Graphene oxide sheets preparation*

Colloidal dispersions of individual graphene
oxide sheets in water at the concentration of 3 mg/mL were prepared with the aid of ultrasound (Fisher Scientific FS60 ultrasonic cleaning bath) in 20 ml batches. Graphene oxide paper was made by filtration of the resulting colloid through an Anodisc membrane filter (47 mm in diameter, 0.2 mm pore size; Whatman), followed by air drying and peeling from the filter [3].

### *2.3. In vivo rats toxicity study*

***2.3.1. Toxicological and biochemical analyses***: It was used 20 rat (Fischer 344) and divided in four groups (n = 5 animals/group). The treatment will be exactly the same as above for four weeks. After 4 weeks of treatment, the animals were euthanized and blood samples were collected by cardiac puncture (left ventricle). The serum was separated by centrifugation at 6000 rpm, 5 min. and used to investigate the hepatotoxic effects, nephrotoxic, cardio toxic of different concentrations of GO. This was measured through the analysis of alanine aminotransferase (ALT) that is for specific marker activity for hepatic damage; activity of aspartate aminotransferase (AST), indicative marker of liver and heart damage and circulating levels of urea and creatinine to check kidney function. Readings were performed by microplate reader Multi-Mode Microplate Reader Synergy H1M (Bio-Tek Instruments, USA) with temperature control and analyzed by Kcjúnior software (Bio-Tech Instruments, Inc., Winooski, VT, USA). For the analysis of systemic toxicity of different GO concentrations fragments of different organs were collected: pancreas, spleen, lungs, heart and kidneys; of all animals in each experimental group. Different organs were fixed in Bouin solution for twelve hours. After fixation, the fragments were washed in 70% ethanol, with subsequent dehydration in an ascending concentration of ethanol. Subsequently, the fragments were diaphanized in xylene for 2 h and included in a polymers (Paraplast Plus, ST. Louis, MO, USA). Then the materials were sectioned in a Leica RM 2165 rotary microtome (Leica, Munich, Germany) with a thickness of 5 μm and stained with hematoxylin-eosin and photographed on Zeiss Axiophot light microscope (Zeiss, Munich, Germany) [4].

***2.3.2. Histopathology***: Histopathology of various organs will be evaluated and correlated with the toxicity levels of inflammation. The degree of inflammation will be assessed by a semi-quantitative scale: 0 = no inflammation, 1 = minimal inflammation (less than five lymphocytes in an area of 0.25 $mm^2$), 2 = moderate inflammation (scattered throughout the tissue mononuclear inflammatory cells but stroma still was visible), 3 = intense inflammation (mononuclear inflammatory cells infiltrating tissues deeply [5].

### *2.4. Adipose-derived stem cells (ASCs)*

ASCs from Hemocenter (UNICAMP) were used [6]. The extraction protocol of mesenchymal cells and cellular cultures was based on Choudhery et al. [7]. Cells from adipose tissue samples were isolated by enzymatic digestion. Briefly, 2 mL of tissue slurry was placed in a 50-mL Falcon tube and washed vigorously five times with 5 mL of phosphate-buffered saline (PBS). Cells in the wash fraction were retained. The fatty tissue was treated with an equal volume of 0.2% collagenase type 1A (Sigma) bovine serum albumin (BSA) and gentamicin at 37°C for 30 min. Complete medium (DMEM-Low glucose), 20 mL, supplemented with 10% fetal bovine serum (FBS) was added in the digested tissue to neutralize collagenase, passed through a 40-μm filter and centrifuged at 150 g for 10

Journal of Physics: Conference Series **617** (2015) 012021 doi:10.1088/1742-6596/617/1/012021

min. The cells from both the wash fraction and the digested fraction were suspended in complete medium and counted using trypan blue and Turk's stains. To prevent spontaneous differentiation, cells were maintained at sub-confluent levels (70-80%) and were harvested with 0.05% trypsin-EDTA for use in subsequent experiments.

### *2.5. Growing stem cells in the presence of GO*

Inocula were prepared standards mesenchymal stem cells obtained from our Laboratory of Human Adipose Tissue (Medicine School-UNICAMP) in DMEM (Dulbeco's Modified Eagle Medium), and the cells were incubated for 72 hours to achieve maximum log growth for the time of incubation in an incubator at 37°C in 5% CO2 with 10% calf serum [8]. Thus, inoculum from the previous one was prepared a solution of stem cells through detachment of the adherent cells in the disposable petri dish (diametro150 mm) using 1 mL of trypsin (0.005%) for 3 minutes at 37°C and washing with 0.1 M of PBS (phosphate buffered saline). Cell viability by Trypan Blue staining technique and counting in a Neubauer chamber was carried out. This cell solution was used to inoculate the experiments (stock: $5x10^3$ cells / mL). Stem cells were inoculated at the concentration $10^5$ cells / mL in 2 mL DMEM culture medium in petri dishes along with graphene oxide solution at different concentrations (1- 50 µg/mL) and incubated at 37°C. After 24 hours, the wells were washed with the PBS (twice) and stained with Live Dead® kit and fluorescence microscopy images obtained. Other wells were leading at the same growth conditions and after 72 h were repeated the same procedure described above with ASCs.

## 3. RESULTS AND DISCUSSION

The graphene oxide sample GO from Cheap Tubes Inc. (USA) was selected for our study. Exhaustive characterization of GO was afforded. Characterization through XRD, XPS, NMR, Raman, FTIR, UV-Vis, TG, DSC, TEM, EDS, DLS and NTA, showed that these analyses corresponded to a single-layer graphene oxide, purity 99%, without Debris and it was stable in water suspension (1 mg/mL) for 10 days. From these analyses was clear that the representative GO from a commercial source it is single layer and a good sized distribution and easily suspended in water. Previously, it was shown that *in vitro* and *in vivo* did not exert any nanotoxic effect up to 100 µg/mL and besides these no hemolysis was also found [9].

AST and ALT levels (liver function) and creatinine and urea levels (renal function) did not differ statistically in all experimental groups after inoculation of Fischer 344 rats with GO (Table 1).

**Tabela 1:** Biochemical parameters of hepatic and renal functions after 24 h of i.p. administration of GO 1 (10 µg/mL) and GO 2 (100 µg/mL).

| Groups | AST (U/L) | ALT (U/L) | Urea (mg/dL) | Creatinine (mg/dL) | *P value* |
|---|---|---|---|---|---|
| Control | 51.7±2.1 a | 28.7±2.1 a | 29.7±1.5 a | 0.27±0.06 a | $P < 0.01$ |
| GO 1 (10 µg/mL) | 53.5±2.1 a | 31.0±2.8 a | 31.5±0.7 a | 0.31±0.06 a | $P < 0.01$ |
| GO 2 (100 µg/mL) | 53.0±2.8 a | 32.5±3.5 a | 27.0±2.8 a | 0.26±0.03 a | $P < 0.01$ |

Data expressed as average ± standard deviation (*n= 03*). Two measurements followed by the same minuscule letter indicate no differences between them by the Turkey test.

*In vivo* study with GO suspended in water on Fischer 344 rats via intraperitoneal administration showed low toxicity, since part of them they accumulates in the intraperitoneal cavity. Similar results with GO on bald/c mice were reported previously [10] (Fig.1).

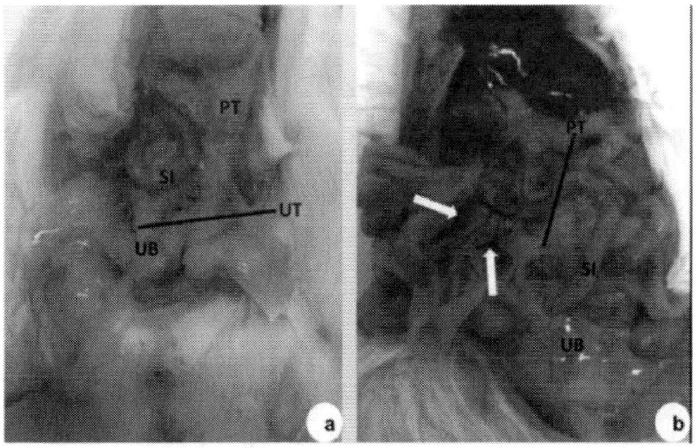


**Figure 1**: Exposure of abdominopelvic cavity in the animals from Control (a) and GO+Pluronic F-68 (b) groups. (a) Normal features of organs and peritoneum in the abdominopelvic cavity. (b) Presence of dark agglomerates (arrows) in the peritoneum. **a – b: PT** – peritoneum; **SI** – small intestine; **UB** – urinary bladder; **UT** – uterus

This probably did not affect the final absorption of GO via intraperitoneal administration, since in another experiment with this GO (100 μg/mL) on Fisher 344 rats-bearing prostate tumors, treatment with GO negatively affected the hepatic parameters, whilst in the renal ones, an improvement was observed. A significant tumor regression at those conditions was afforded [9].

*In vitro* cell cultures of a GO in water suspension exhibited no inhibition of the cell growth up to 5 μg/mL. A very effective cell growth occurred at 1 μg/mL (Fig. 2).

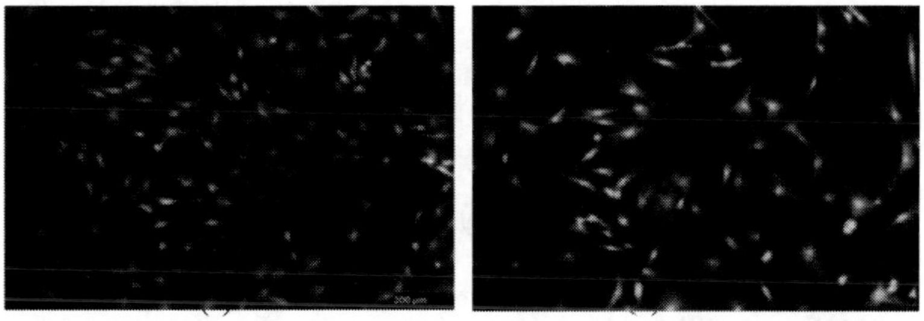

**Figure 2**. Fluorescence micrographs of steam cells. a).Control and b) After 24 h (GO - 1 μg/mL) (ASCs growth and adhesion).

In order to have a scaffold for ASCs it was prepared GO sheets by suspension of GO in ultrapure deionized water or in Pluronic F-68 using a ultrasonicator bath (Fig.3).

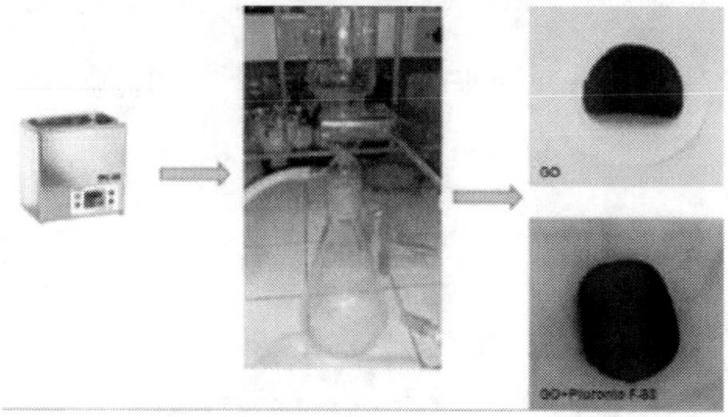

**Figure 3.** Preparation of GO and GOP68 sheets.

In a DMEM solution the GO shits were completely stable for several hours, in comparison that GO powder that was immediately suspended in the same medium. Preliminary studies indicate that these GO sheets led to ASCs growth without any inhibition.

## 4. CONCLUSIONS

GO in water suspension exhibited an inhibition of the cell growth over 5 µg/mL. GO/Pluronic F-68 sheets platform for the growth and adhesion of ASCs that exhibits excellent properties for these processes was prepared. *In vivo* study with GO suspended in water on Fisher 344 rats via i.p. administration showed no toxicity up to 100 µg/mL, in spite of part of these GOs accumulated in the intraperitoneal cavity at the first hours. Then, taking together, the data suggest the great potential of graphene oxide sheets as platform to ACSs, as well as, new material for treatment several urological diseases.

**Acknowledgement:** Support from the Brazilian Network of Nanotoxicology (CIGENANOTOX) (MCTI/CNPq), INOMAT (MCTI/CNPq), NanoBioss (MCTI) and FAPESP are acknowledged.

### References

[1] Sahu A, Choi W I and Tae G 2012 *Chem. Commun.* **48** 5820
[2] Kim J, Choi K S, Kim Y, Lim K-T, Seonwoo H, Park Y, Kim D-H, Choung P-H, Cho C-S, Kim S Y, Choung Y-H and Chung J H 2013 *J. Biomed. Mater. Res. A* **101**, 3520
[3] Dikin D A, Stankovich S, Zimney E J, Richard D. Piner R D, Geoffrey H. B. Dommett G H B, Evmenenko G, Nguyen S B T and Ruoff R S 2007 *Nature* **448** 457
[4] Fávaro W J, Nunes O S, Fabio RF Seiva F R F, Nunes I S, Woolhiser L K, Durán N and Lenaerts A J 2012 *Infect. Agents Cancer* **7** 14
[5] Offersen B V, Knap M M, Marcussen N, Horsman M R, Hamilton-Dutoit S and Overgaard J. 2002 *Br. J. Cancer* **87** 1422
[6] Rego P B, Ribeiro T B, Durate A S S, Luzo A C M, Vidal B C, Costa F F and Saad S T O 2010 *Adv. Life Sci. EMBO Meeting* **2010** 163
[7] Choudhery MS, Badowski M, Muise A and Harris D T 2013 *Cytotherapy* **15** 330
[8] Akhavan O, Ghaderi E and Akhavan A. 2012 *Biomaterials* **33** 8017
[9] Durán N, Martinez D S T, Justo G Z, de Lima R, de Castro V L, Umbuzeiro G A, Barbieri E, Durán M, Melo P S, Alves O L and Fávaro W J 2014 NanoSafe 2014, Grenoble, France.2014, Abstr. O4b-1.
[10] Yang K, Gong H, Shi X, Wan J, Zhang Y and Liu Z 2013 *Biomaterials* **34**, 2787

# Synthesis, Characterization and Cytotoxicity Evaluation of Nitric Oxide-Iron Oxide magnetic Nanoparticles

**P S Haddad[1*], T N Britos[1], M C Santos[1], A B Seabra[1], M V Palladino[2] and G Z Justo[2]**

[1]Exact and Earth Sciences Departament, Universidade Federal de São Paulo, CEP 09913030, Diadema, SP, Brazil,

[2]Department of Biological Sciences (Campus Diadema) and Department of Biochemistry (Campus São Paulo), Universidade Federal de São Paulo, CEP04044-020, São Paulo, SP, Brazil

*Email: haddadps@gmail.com, haddad.paula@unifesp.br

**Abstract**. The present work is focused on the synthesis, characterization and cytotoxic evaluation of superparamagnetic iron oxide nanoparticles (SPIONs). SPIONs have been proposed for an increasing number of biomedical applications, such as drug-delivery. To this end, toxicological studies of their potential effects in biological systems must be better evaluated. The aim of this study was to examine the *in vitro* cytotoxicity of thiolated (SH) and S-nitrosated (S-NO) SPIONs in cancer cell lines. SPIONs were prepared by the co-precipitation method using ferrous and ferric chlorides in aqueous solution. The nanoparticles ($Fe_3O_4$) were coated with thiol containing molecule cysteine (Cys) (molar ratio SPIONs:ligand = 1:20), leading to the formation of an aqueous dispersion of thiolated nanoparticles (SH-SPIONs). These particles were characterized by Fourier transform infrared spectroscopy (FTIR), X-ray diffraction (XRD), transmission electron microscopy (TEM) and vibrating sample magnetometer (VSM). The results obtnained showed that Cys-SPIONs have a mean diameter of 14 nm at solid state and present superparamagnetic behavior at room temperature. Thiol groups on the surface of the nanoparticles were nitrosated through the addition of sodium nitrite leading to the formation of S-NOCys-SPIONs (S-nitrosated-Cys-SPIONs, which act as spontaneous nitric oxide (NO) donor). The cytotoxicity of thiolated and S-nitrosated nanoparticles was evaluated in acute T cell leukemia (Jurkat cell line) and Lewis lung carcinoma (3LL) cells. The results showed that at low concentrations thiolated (Cys) and S-nitrosated (S-NOCyst) SPIONs display low cytotoxicity in both cell types. However, at higher concentrations, Cys-SPIONs exhibited cytotoxic effects, whereas S-NOCys-SPIONs protected them, and also promoted cell proliferation.

## 1. Introduction

Nanoparticles represent an attractive class of materials, which due to their small size exhibit novel properties that differ from bulk materials [1-2]. Of all the nanomaterials, superparamagnetic iron oxide nanoparticles (SPIONs) are of enormous interest due to their exclusive properties such as superparamagnetism, and biodegradability [3]. SPIONs have found applications in targeted drug delivery, magnetic resonance image (MRI) contrast enhancement, cancer diagnosis, hyperthermic treatment of tumors, magnetically mediated separation of biomolecules and waste water treatment [4-6]. In this context, iron oxide particles, such as magnetite ($Fe_3O_4$), are the most common investigated in biomedical applications [7-8]. In general, *in vitro* studies of SPIONs demonstrated little cytoxicity indicating the biocompatibility of this nanostructured systems [9-11]. Functionalization permits the adsorption of molecules of interest, such as biomolecules, giving a stable aqueous dispersion. Further, conjugation of SPIONs with biomolecules, in many cases, would provide special properties to the nanoparticles, such as drug delivery [12]. In this context, molecules containing thiol groups such as cysteine (Cys) can be nitrosated, leading to the formation of S-nitroso-Cys-SPIONs. The formation of S-nitroso-groups (S-nitrosothiols) on the surface of SPIONs leads to nitric oxide (NO)-releasing iron oxide magnetic nanoparticles, as previously reported by our research group [7-11]. NO is involved in several physiological and pathophysiological processes, such as the control of vascular tone, the inhibition of platelet aggregation, and the immune response against microbes [13]. Thus, there is a great interest in the investigation of NO-releasing vehicles that are able to stabilize and release NO locally direct to the target site, in diverse biomedical applications. This work reports studies *in vitro* of the cytotoxicity of thiolated (SH) and S-nitrosated (S-NO) SPIONs in cancer cell lines. SPIONs were prepared by the co-precipitation method using ferrous and ferric chlorides in aqueous solution. The nanoparticles were coated with thiol containing molecule Cys (molar ratio SPIONs:ligand = 1:20), giving thiolated nanoparticles (Cys-SPIONs). Further Cys-SPIONs were nitrosated with an excess of an aqueous solution of sodium nitrite ($NaNO_2$) at pH= 6. These particles were characterized by Fourier transform infrared spectroscopy (FTIR), X-ray diffraction (XRD), transmission electron microscopy (TEM) and vibrating sample magnetometer (VSM). The cytotoxicity of thiolated and S-nitrosated nanoparticles was evaluated in acute T cell leukemia (Jurkat cell line) and Lewis lung carcinoma (3LL) cells.

## 2. Methods

### *2.1. Synthesis of Cys-SPIONs*

SPIONs were synthesized by using a co-precipitation method, as previously reported [10]. In brief, 4.0 mL of $FeCl_3{\cdot}6H_2O$ and 1.0 mL of $FeCl_2{\cdot}4H_2O$ (molar ratio 2:1), prepared in 1.0 mol/L HCl, were mixed and stirred, while a volume of 50 mL of $NH_4OH$ (0.7 mol/L) was added as precipitator. At this stage, the black suspension formed was decanted magnetically followed by the addition of Cys in molar ratio $Fe_3O_4$: ligand 1:20. This mixture was then stirred for 1 hour. The dispersion was centrifuged and the new precipitate was washed several times with ethanol leading to a SPIONs covered with Cys.

### *2.1. Determination of the amount of free thiol groups on the surface of SPIONs*

The amount of free thiol groups (–SH) on the surface of the nanoparticles was measured by the 5,5'-dithiobis-(2-nitrobenzoic acid) (DTNB) reaction, based on the absorbance at 412 nm ($\varepsilon$ = 14,150 $molL^{-1}cm^{-1}$) of the 2-nitro-5-thiobenzoate anion ($TNB^{2-}$) generated in the reaction of –SH groups with DTNB [15]. Cys-SPIONs were added to 3.0 mL of DTNB (0.01 $molL^{-1}$) in PBS buffer (pH 7.4) with 1 $mmolL^{-1}$ of ethylenediaminetetraacetic acid. After 5 min of incubation, the suspensions were filtered by centrifugal ultrafiltration using a Microcon centrifugal filter device containing ultrafiltration membranes (MWCO 10-kDa molar mass cut-off filter, Millipore, Billerica, MA, USA). The supernatant was placed into a quartz cuvette, and the intensity of the absorption band at 412 nm was measured in an Uv-Vis Spectrophotometer (Agilent, model 8553, Palo Alto, CA, USA). The experiments were carried out in triplicates.

### *2.2. Preparation of S-nitrosated SPIONs*
Cysteine nanoparticles (Cys-SPIONs) were nitrosated leading to the formation S-nitroso-Cys-SPIONs by adding excess of $NaNO_2$ at slight acid pH [14]. Excess of unreacted nitrite was removed from SPIONs suspension by centrifugal ultrafiltration.

### *2.3. In vitro experiments*
Cancer cell lines (K562, Lucena, Jukart and 3LL cells) were treated with different concentrations of Cys-SPIONs and S-nitroso-Cys-SPIONs for 24 h. Cytotoxicity assays were performed by tetrazolin reduction MTT (3-(4-5-dimethylthiazolyl-2)-2,5-diphenyltetrazolium bromide).

## 3. Results and Discussion

The amounts of free thiol (-SH) groups on the surface of Cys-SPIONs were determined by the reaction of the thiolated nanoparticles with a thiol-specific reagent, DTNB [15]. The quantity of 12 ± 8 μmol SH per gram of Cys-SPIONs was found, indicating a high extend of functionalization of SPIONs with Cys. Moreover, Cys-SPIONs were investigated by our research group by several physicochemical techniques, such as, FTIR, X-ray diffractometric, TEM and the hysteresis curves (data no shown). Therefore, the combined results obtained by all these techniques gave support to deduce that the covering of cysteine onto nanoparticles surface produced crystalline cysteine magnetite (Cys-$Fe_3O_4$) with spherical morphologies and narrow sizes distribution. Superparamagnetic behavior at room temperature was also observed.

Figure 1 shows the schematic representation the nitrosation of free thiol groups (SH) adsorbed on $Fe_3O_4$ surface (Cys-$Fe_3O_4$) by the addition of $NaNO_2$ in aqueous solution at pH slight acid, leading to the formation of S-nitroso-Cys-$Fe_3O_4$.

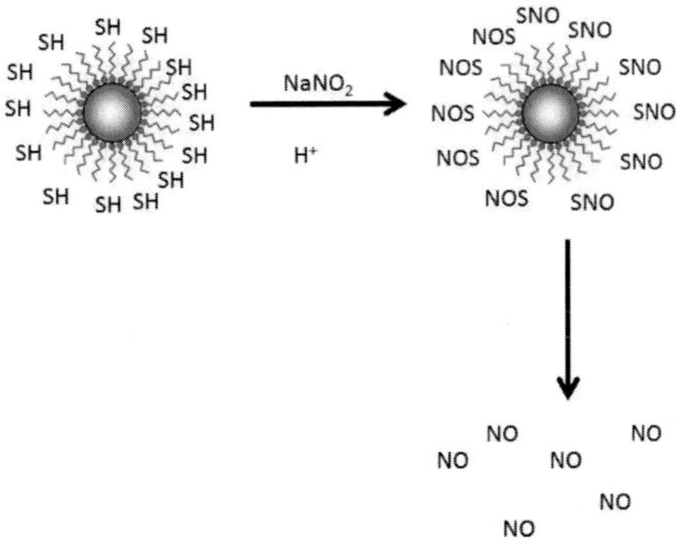


**Figure 1.** Schematic representation of the functionalization of SH-SPIONs with NO molecules.

S-nitroso groups (S-NO) act as spontaneous NO donor due to the homolytic S-N cleavage with free NO release (Fig. 1). Free NO released from the surface of SPIONs may have several effects in biological system. In order to propose the use of SPIONs in biomedical application, it is important to evaluate the cytotoxicity of the NPs. In this context, the *in vitro* toxicities of thiolated (Cys-SPIONs) and S-nitrosated (S-nitroso-Cys-SPIONs) nanoparticles were evaluated in acute T cell leukemia (Jurkat cell line) and Lewis lung carcinoma (3LL) cells.

Fig. 2 shows the cell viability of K562 cells after incubation with Cys-$Fe_3O_4$ and S-nitroso-Cys-$Fe_3O_4$ nanoparticles. It can be observed that both Cys-$Fe_3O_4$ and S-nitroso-Cys-$Fe_3O_4$ nanoparticles did not significantly decrease the cell viability at concentrations from 0.0001 to 0.01 mg/mL, when compared to the control cells. However, at nanoparticle concentrations of 0.25 and 0.5 mg/mL, thiolated nanoparticles (Cys-SPIONs) slight decrease cell viability. In contrast, S-nitrosated nanoparticles did no cause toxic effects, even at higher concentrations.

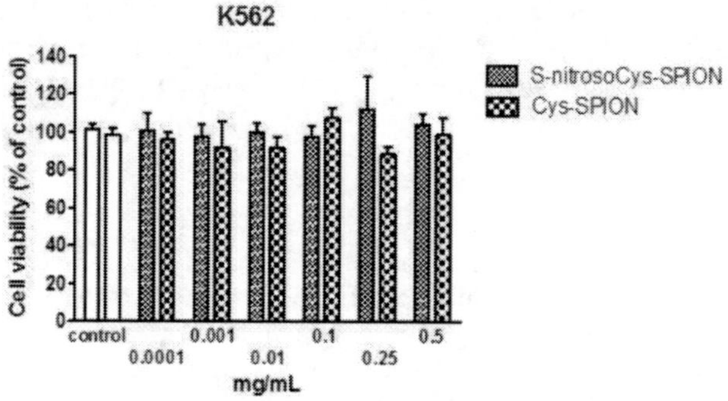


**Figure 2.** Cytotoxicity of Cys-SPIONs and S-nitroso-Cys-SPIONs in K562 cells. Chronic myeloid leukemia (K562) cells (1x105/mL) were treated with different concentrations of Cys-SPIONs and S-nitroso-Cys-SPIONs for 24 h, and afterwards, cell viability was analyzed by the resazurin reduction assay. The results represent means ± SD of one experiment run in sextuplicate. Data were expressed relative to control cell viability (100%).

Fig. 3 represents the viability cell of Lucena cells. Similarly, it was observed that nanoparticles did not reduce the cells viability at the lowest tested concentrations. By increasing the nanoparticles concentrations (0.1 up to 0.5 mg/mL), a discrete increase in cell viability upon cell incubation with S-nitroso-Cys-SPIONs could be noted, while a slight decrease in cell viability upon cell incubation with Cys-SPIONs could be observed.

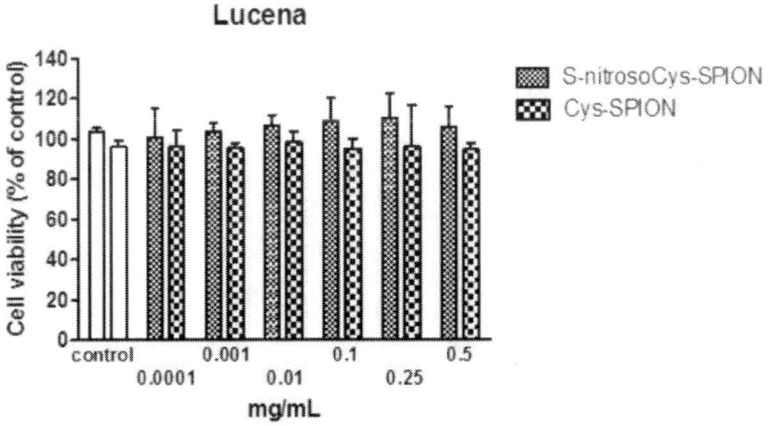


**Figure 3.** Cytotoxicity of Cys-SPIONs and S-nitroso-Cys-SPIONs in Lucena cells. Multidrug resistant chronic myeloid leukemia (Lucena) cells ($1x10^5$/mL) were treated with different concentrations of Cys-SPIONs and S-nitroso-Cys-SPIONs for 24 h, and afterwards, cell viability was analyzed by the resazurin reduction assay. The results represent means ± SD of one experiment run in sextuplicate. Data were expressed relative to control cell viability (100%).

Interesting results are shown for the essays using Jurkat and 3LL cells (Figs. 4 and 5, respectively). These cancer cells are more resistant, and a more evident cytotoxic effects of Cys-SPIONs could be observed, whereas S-nitroso-Cys-SPIONs protected the cells, and also promoted cell proliferation.

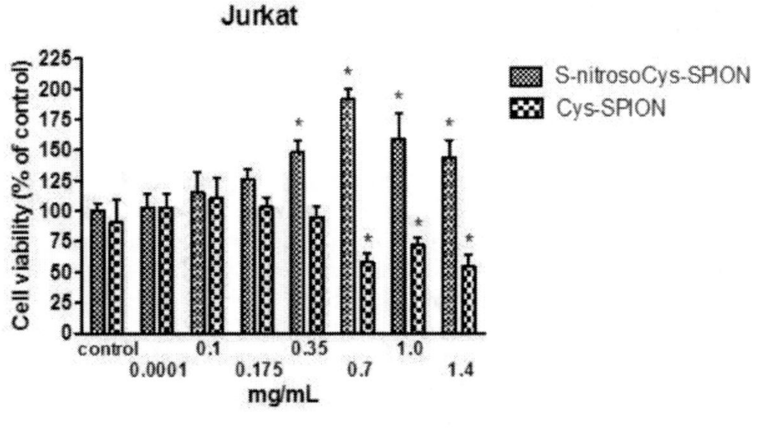


**Figure 4.** Cytotoxicity of Cys-SPIONs and S-nitroso-Cys-SPIONs in Jurkat cells. Jurkat cells ($1x10^5$/mL) were treated with different concentrations of Cys-SPIONs and S-nitroso-Cys-SPIONs for 24 h, and afterwards, cell viability was analyzed by the resazurin reduction assay. The results represent means ± SD of one experiment run in sextuplicate. Data were expressed relative to control cell viability (100%). *$p<0.05$ compared with control cells.

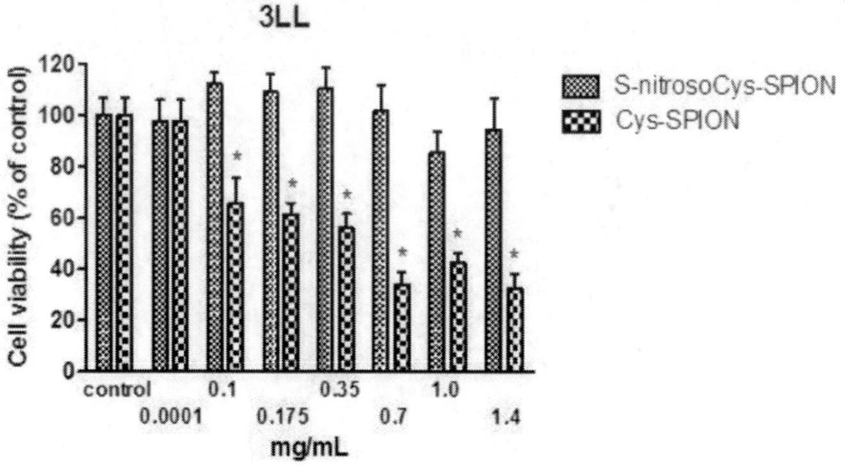


**Figure 5.** Cytotoxicity of Cys-SPIONs and S-nitroso-Cys-SPIONs in 3LL cells. 3LL cells ($1x10^4$/well) were treated with different concentrations of Cys-SPIONs and S-nitroso-Cys-SPIONs for 24 h, and afterwards, cell viability was analyzed by the resazurin reduction assay. The results represent means ± SD of one experiment run in sextuplicate. Data were expressed relative to control cell viability (100%). *$p<0.05$ compared with control.

Taken together, the preliminary results indicate the toxicity of Cys-SPIONs and S-nitroso-Cys is dependent on the cell type. In the case of K562 and Lucena cell lines, a significant decrease in cell viability was not observed for both nanoparticles at all tested concentrations. However, for Jurkat and 3LL cell lines, thiolated SPIONs were found to be toxic at higher tested concentrations, whereas S-nitrosated-SPIONs rescue cell toxicity. This interesting effect can be assigned to the complex biochemistry of NO, which is involved in several mechanism pathways in the biological system. Moreover, NO is reported to have a dichotomous effect on cancer cell, acting either to promote cell proliferation or to kill cancer cells. Further studies are required to better understand the mechanism involved.

## 4. Conclusions

SPIONs functionalized with Cys were successfully synthesized. Free thiol groups on the surface of the nanoparticles were nitrosated through the addition of sodium nitrite leading to the formation of S-nitroso-Cys-SPIONs. The cytotoxicity results showed that at low concentrations Cyst-SPIONs and S-nitroso-Cys-SPIONs exhibited low cytotoxicity in all cell types investigated. However, at higher concentrations, Cys-SPIONs exhibited cytotoxic effects, whereas S-nitroso-Cys-SPIONs protected them, and also promoted cell proliferation, in the case of Jurkat and 3LL cell lines. Further studies are undertaken to better investigate this effect.

## 5. Acknowledgements

Supported from FAPESP (Process 2014/13913-07, 2012/17053-7, CNPq, CAPES and Brazilian Network in Nanotoxicology (MCT-CNPq).

## 6. References


[1] Murray B C, Kagan C R, Bawendi M G 2000 *Annu. Rev. Mater. Sci.* **30** 545

[2] Jenders J J M, Altan C L, Bomans P H H, Arakaki A, Bucak S, de With G and Sommerdijk N A J M 2014 *Cryst. Growth Des.* dx.doi.org/10.1021/cg500816z

[3] Sangeetha J and Philip J 2013 *RSC Advances* **3** 8047

[4] Majewski P and Thierry B 2007 *Crit. Rev. Solid State Mat. Sci.* **32** 203

[5] Laurent S, Forge D, Port M, Roch A, Robic C, Elst L V and Muller R N 2008 *Chem.Rev.* **108**, 2064

[6] Agasti S S, Rana S, Park M H, Kim C K, You C C and Rotello V M 2010 *Adv. Drug Delivery Rev.* **62** 316

[7] Haddad, P S and Seabra A B 2012 Biomedical Applications of Magnetic Nanoparticles *Iron Oxides: Structure, Properties and Applications* Nova Science Publishers vol. 1, pp 165–188

[8] Haddad P S, Duarte E L, Baptista M S, Goya G F, Leite, C A P and Itri R 2004 *Progr.Colloid Polym.Sci.* **128** 232

[9] de Lima R, Oliveira J L, Ludescher A, Molina M M, Itri R, Seabra A B and Haddad P S 2013 *J. Phys. Conf. Ser.* **429** 012034

[10] de Lima R, Oliveira J L, Murakami P S K, Molina M M, Itri R, Haddad P S and Seabra A B 2013 *J. Phys. Conf. Ser.* **429** 012021

[11] Seabra AB, Pasquoto T, Ferrarini ACF, Santos MD, Haddad PS and de Lima R. *Chem Res Toxicol* 2014 **27** 1207

[12] Sangeetha J and Philip J, 2012 *Colloids Surf. A* **406** 52

[13] Seabra A B and Duran N 2010 *J. Mat. Chem.* **20** 1624

[14] Kemp M, Go Y M and Jones D P 2008 *Free Radic. Biol. Med.* **44** 921

[15] Seabra A B, Martins D, Simões M M S G, da Silva R, Brocchi M and de Oliveira M G 2010 *Artif. Organs* **34** E204

Journal of Physics: Conference Series **617** (2015) 012023 doi:10.1088/1742-6596/617/1/012023

# Hsp70 as an indicator of stress in the cells after contact with nanoparticles

**Šárka Hradilová[1], Markéta Havrdová[2], Aleš Panáček[1], Libor Kvítek[1], Radek Zbořil[1]**

[1] Regional Centre of Advanced Technologies and Materials, Department of Physical Chemistry, Faculty of Science, Palacky University, 17. listopadu 12, 771 46 Olomouc, Czech Republic
[2] Regional Centre of Advanced Technologies and Materials, Department of Experimental Physics, Faculty of Science, Palacky University, 17. listopadu 12, 771 46 Olomouc, Czech Republic

E-mail: sarka.hradilova@upol.cz

**Abstract.** In recent years, production of nanoparticles is increased and thus grows our contact with them too. Question of safety is closely related to the issue of use nanoparticles. There are a number of tests that monitor the viability, ROS production, the effect on the DNA and cell cycle, however, rarely encountered studies on stress in the cells after contact with nanoparticles. Heat shock proteins (HSP) are among the substances that can be used for monitoring stress in cells. HSP are structures with a chaperone activity. They are evolutionarily very old, conservative and they are found with a high degree of homology in prokaryotes and eukaryotes including humans. They exist at low concentrations under physiological conditions, while in the denaturing conditions e.g. high or low temperature, radiation, exposure to chemicals, heavy metals, or nanoparticles their expression is changed. HSPs are involved in maintaining homeostasis in the cell that the denatured protein conformations allow recovery to the original stage. One of the most common proteins from HSP family is Hsp70 - protein with a molecular weight of 70 kDa. The level of Hsp70 in a cell after exposure to the stress changes depending on the stress level to which the cell is exposed to and a time period during which lasted stressful conditions. Our research monitors stress levels of cells manifesting by Hsp70 production after contact with silver nanoparticles. Nanoparticles show different toxicity towards different types of target cells, which is reflected in the values of $IC_{50}$ – concentration that kills 50 % tested cells. Concentration of test substance toxic to one cell type may be innocuous to cells of another type. $IC_{50}$ obtained from the MTT assay provides a suitable default data and if multiples of $IC_{50}$ values are used, we can compare and generalize. Studies can be used to compare stress levels in cells that show different sensitivity to the tested nanoparticles compared with cells under optimal growth conditions. The study was done on two types of mouse fibroblasts NIH-3T3 and L929. While NIH-3T3 cells exhibit stress response proportional to the concentration of silver nanoparticles, for L929 cells this was not observed.

## 1. Introduction

Nanoparticles are widely used because they give us the advantages that are associated with their nano dimension. Higher utilization of nanomaterials, however, brings risks as well. Nanoparticles are toxicologically tested to determine the $IC_{50}$, ROS, membrane potential, it is monitored DNA damage, cell cycle and many other parameters. With the exception of a few studies [1-3] is not paid attention to the stress induced by contact with nanoparticles. Stress at the cellular level is manifested production of proteins known as stress proteins. The best known include heat shock proteins – Hsp. The heat shock proteins are group of proteins that provide the stress response of cells when exposed to high temperature, extreme pH, heavy metals or nanoparticles. Heat shock proteins are evolutionarily ancient structures, with a high degree of homology. In addition to the repair of damaged or improperly formed proteins participates in a number of immunological processes [4]. Hsp occur both in prokaryotes and in eukaryotes (including man). Hsp70 - protein with a molecular weight of 70 kDa is expressed in stressful situations and appears to be a suitable biomarker [5-9].

If the stress upon contact with nanoparticles is studied, *Drosophyla melanogaster* [1], *Japanese medaka* [10] or other aquatic organisms [11] are tested. Less frequent testing is performed on isolated cell lines [3, 12].

Silver nanoparticles are among the most widely used nanoparticles for their antibacterial properties. They are used whenever is possible to use their abilities to kill microbes. The widespread use of silver nanoparticles leads to intense contact with this kind of nanoparticles, so it is necessary to carefully consider their toxicity, even at the cellular level. The novelty of our study is that have not been tested the same concentration of the nanoparticles, but their multiples derived from $IC_{50}$ values. Our study explores the stress caused by contact with silver nanoparticles on two very similar lines of mouse fibroblasts, but they have a different toxicological response - their $IC_{50}$ is slightly different.

## 2. Materials and methods

*Silver nanoparticles*

Silver nanoparticles (NPs) prepared by modified Tollens method [13]. As the reducing sugar maltose was used. The average size of the particles was 26±2.5 nm.

*Buffer*

Lysis buffer - 3 ml 5M NaCl, 10 ml 10% Triton X, 5 ml Tris pH 8, 82 ml water, prior to use 1 tablet od protease inhibitors (cOmplete Protease Inhibitor Cocktail Tablets, Roche) per 10 mL of buffer was added. All other used chemicals were from Sigma and were used without further purification.

*Viability*

Viability of two cellular types (NIH-3T3 and L929) was evaluated using the MTT assay. Cells were seeded at a density $10^4$ cells per well in 96-well plates. After spreading (4 hours), cells were treated by increasing concentration (0.25 – 30 μg/mL) of the silver nanoparticles and incubated 24 hours. Subsequently, 20 μl of MTT solution (5 mg/mL) was added in the each well and kept in a $CO_2$ incubator (NIH-3T3 cells for 4 hours, L929 cells for 1.5 hours). Finally, the blue formazan crystals were dissolved by 100 μL of dimethyl sulfoxide (DMSO). The absorbance was measured at 570 nm using a multiplate reader Infinite 200 PRO (Tecan). The dose-response curves were obtained and the inhibitory concentration ($IC_{50}$) determined. Values of $IC_{50}$ were used in following tests.

*Cell cultures for nanoparticles and heat shock testing*

Cell (NIH-3T3 and L929) were seeded in 6 well plate at a density 2.5 x $10^5$ cells per well and cultivated in DMEM (Dulbecco's Modified Eagle medium, ), at 37°C in a moisturized 5% $CO_2$ incubator. After 24 hours, the silver nanoparticles were added and incubate 24 hours. Following this period, cells were washed 3 times with PBS, trypsinised and centrifuged 10 min at 4°C in speed 6 000 g. Supernatant was removed and pellet resuspended in 0.5 ml chilled lysis buffer, subsequently sonicated for 30 s. Each sonication was repeated 5 times and samples were kept on ice between each step. Supernatant containing proteins was obtained by centrifugation 20 min at 4°C in speed 10 000 g. Positive controls were prepared by the same protocol – cells exposed to a temperature 43°C for 1, 2

Journal of Physics: Conference Series **617** (2015) 012023 doi:10.1088/1742-6596/617/1/012023

and 3 hours. After 24 hours of recovery and spreading, cells were lysed as described above.

*Total protein content*

Total protein content was determined by Bradford method using BSA as a standard. The absorbance was measured at 595 nm using a multiplate reader Infinite 200 PRO (Tecan).

*SDS-Page and Western blotting*

Supernatants were diluted 1:1 with SDS sample buffer (2x) and loaded on 10% polyacrylamide gel. Then proteins were transferred on a PVDF membrane and incubated with anti-Hsp70 (1:1000, Sigma H5147) overnight. The Hsp70 complex was detected using HRP-conjugated secondary antibody (Sigma A9917) and ECL (Amersham). Quantification of Hsp70 was performed by densitometry analysis using free software ImageJ.

*Characterization techniques*

Transmission electron microscopy (TEM) images were obtained using a transmission electron microscope (JEOL JEM-2010) operating at 160 kV with a point-to-point resolution of 1.9 Å. The samples dispersed in deionized water were sonicated for $\approx$ 1 min. Then, a drop of very diluted dispersion was placed on a holey carbon-coated copper grid and allowed to dry at ambient temperature. Hydrodynamic size of silver nanoparticles was performed on Malvern Instrument Zetasizer Nano-ZS instrument

*Statistical analysis*

All measurements were performed at least three times and the results are given in the form: average value $\pm$ standard deviation.

## 3. Results

Silver nanoparticles used in our experiments were characterised by TEM (see figure 1) and dynamic light scattering (DLS). The particle sizes obtained from the TEM and DLS were well correlated; the average particle size was 26 $\pm$ 2.5 nm.

MTT assay provides $IC_{50}$ - the concentration of test compound that kills 50% tested cells (see figure 2). $IC_{50}$ was 12 μg/mL for NIH-3T3 cells and 15 μg/mL for L929 cells. It means that silver nanoparticles are more toxic for NIH-3T3. To test the stress response caused by silver nanoparticles were chosen two concentrations - one equals $IC_{50}$ and second half of $IC_{50}$ (6 and 7.5 μg/mL).

First positive controls were prepared by incubating the cells in temperature higher than its optimum (figure 3). Total protein was measured by Bradford method, Hsp70 quantification was done by densitometry analysis. Subsequently Hsp70 content in control and test samples were calculated. Content of Hsp70 under physiological conditions was defined as 100%; the other values were related to this value. Figure 4 shows relative changes in Hsp70 level after 24 hours contact with silver nanoparticles. As is evident from Figure 3 and 4, both cell lines respond differently to heat shock, and on contact with the nanoparticles. The reaction of NIH-3T3 cells to a higher temperature is rapid. After one hour incubation at 43°C is seen increase the level of Hsp70 to 300%. A maximum (385%) is reached during the three hour incubation. In contrast, the reaction of L929 cells is gradual, the maximum (419%) is reached at two hours incubation and then decreases to 253% after three hours.

Cells react differently on contact with nanoparticles too. Here it is possible to observe an increase of the Hsp70 content, although not as high as that induced by heat stress. For NIH-3T3 cells is evident increase to 196 and then to 318%, while for L929 cells is observed to rise to 128% at a dose of 1/2 $IC_{50}$ and then drop to 113%. In the case of NIH-3T3 cells, both tested doses were significantly different from the controls and there is noticeable concentration dependence. L929 cells respond differently, there is no concentration dependence. A dose equivalent to half the $IC_{50}$ is significantly different from controls, while dose equal to the $IC_{50}$ is no different.

Figure 1. TEM image of silver nanoparticles

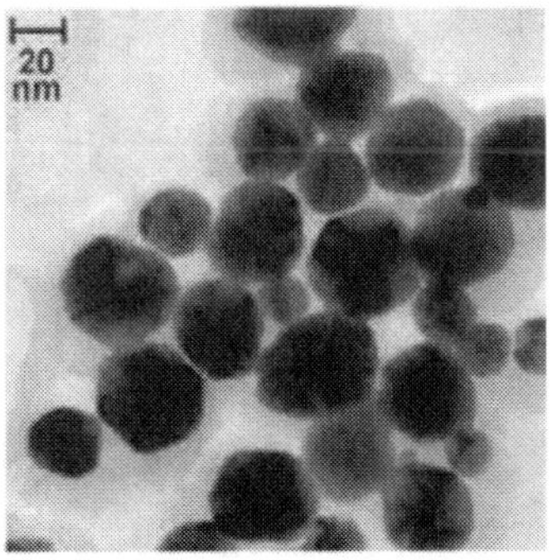


Figure 2. MTT tests

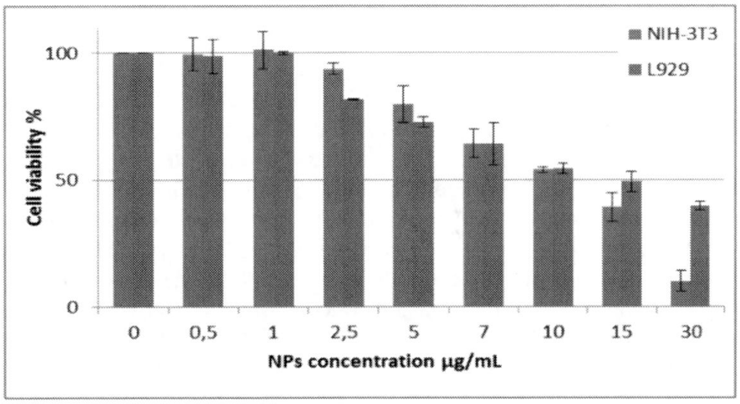


Figure 3. Relative Hsp70 changes – heat shock

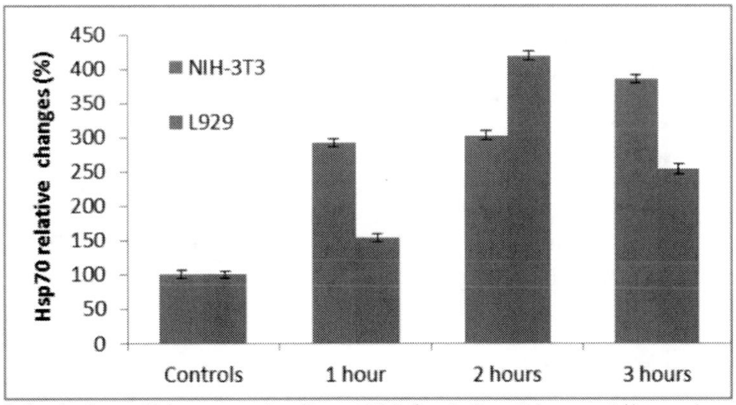


Controls – physiological condition, tested cells – 43°C for 1, 2 or 3 hours. All tested samples are significantly different from controls, ($P<0.05$).

Figure 4. Relative Hsp70 changes – contact with silver nanoparticles

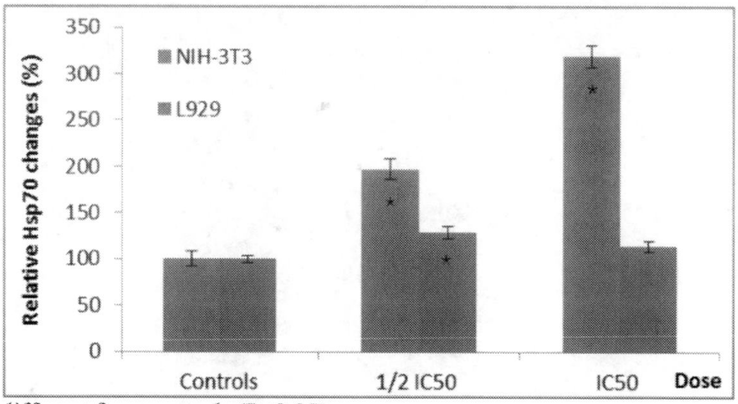


* Significantly different from controls ($P<0.05$).

### 4. Discussion

As already mentioned, increased use of nanoparticles also entails some risks. Commonly performed tests monitor viability, ROS production and other parameters. Negligible number of studies follows the stress at the cellular level. The importance of stress monitoring just mentions in his work Mukhopadhyay et al. [9] especially with regard to the environment.

When whole organisms or cell cultures are exposed to elevated temperatures, they respond by synthesizing a small number of highly conserved proteins called heat shock proteins, because first observation was connected with heat temperature induction [14, 15]. This response is universal. Hsp are involved in a wide range of processes in cells. Maintain homeostasis, participating in de novo folding of nascent proteins and repair of damaged ones [16], play role in a number of events associated with the immune system [4]. One of the most frequently studied Hsp is Hsp70. Hsp70 occur across species [15] with high degree of homology. It is often described as a suitable biomarker of stress [5]. This stress can be caused by high or low temperature [17], by treatment with radiation, certain drugs, heavy metals [9] and nanoparticles [1, 5, 10]. The stress response to silver nanoparticles was chosen for our research.

The first part of our research was to prepare the positive controls. Positive control (cells exposed to a higher temperature) served as loading controls for SDS-PAGE and Western blots. When comparing the stress response of both cell lines to thermal shock, it is obvious that the reaction is different (see figure 3). For NIH-3T3 cells it is seen the increase of Hsp70 after one hour incubation. Longer incubation at elevated temperature leads to an increase of Hsp70. Increased production of protein indicates a higher level of damage. L929 cells respond to thermal shock by slower increase in Hsp70 after one hour exposure. This may mean that L929 cells have better tolerance to temperature rise, if it not takes too long. If cell is exposed to higher temperatures for a longer period (2 hours), it is observed a dramatic increase in Hsp70. Then there is a decrease, probably due to cell death. Two hours exposure at 43°C is the limit, after which the cell is unable to repair denatured proteins and other stay at this temperature is lethal. Ability to cope with fluctuations in temperature allowed survival of the species, an organism with better tolerance survived. If the temperature rise is slow, isolated cells or whole organism have time to prepare (stress proteins induction). These stress proteins act protectively towards different stressors than the original was [18]. The next step of our experiment was to investigate the stress response after contact with silver nanoparticles. The silver nanoparticles prepared by a modified Tollens process are toxic for both NIH-3T3 and L929 cells; $IC_{50}$ is 12 μg/mL and 15 μg/mL respectively. Contact with these nanoparticles induces stress which can be monitored via Hsp70 in both cell lines. The response of cell lines upon contact with silver nanoparticles was different. NIH-3T3 cells responded production of Hsp70 proportional to the concentration of nanoparticles, L929 cells generate maximum at a dose equal to half the $IC_{50}$, and a higher dose of nanoparticles did not lead to a further increase. Conversely, a decrease from 128% to 113% relative to controls was observed. Kaur and Tikka [3] investigated the effect of surface charge on the cytotoxicity of silver nanoparticles using Hsp70 as just one of biomarkers. They performed tests on cancer line A431 (human epithelial carcinoma), A549 (human lung carcinoma) and murine RAW264.7 macrophages. They used relatively high doses of silver nanoparticles (up to 100 µg /mL). Maximum levels of rise Hsp70 they observed in A431 (11-fold increase), in the remaining two lines increase was about 3-fold. This finding is consistent with what we observed. Like us on L929 cells, they also observed a higher production of Hsp70 on the lower dose and explain to the fact that higher concentrations are less stressful for cells. We believe that the explanation is rather that high doses are so toxic that it exceeds cell´s reparative capabilities and cells die. Only the live cell can produce proteins. Low numbers of cells produce low levels of Hsp70. Lim et al [12] tested a low dose of 5 nm silver nanoparticles on human macrophage cell line (U937) and they observed increase in the range of 1.2 to 2.3-fold.

## 5. Conclusion

The stress response caused by heat shock and exposure to silver nanoparticles was tested. Murine fibroblasts exposed to a temperature higher than their optimum react increased production of Hsp70. Compared to control cells cultured in ideal conditions, level of Hsp70 is increased until 4-fold. Our results show that the silver nanoparticles (size 26 nm, modified Tollens method) show signs of toxicity to murine fibroblasts L929 and NIH-3T3. $IC_{50}$ values obtained by MTT assay was 15 and 12 µg/mL. Denaturation of proteins due to the contact with silver nanoparticles was found. Presence of denatured proteins is outwardly manifested by expression of Hsp70, which is involved in the repair in order to restore physiological state of the cell. NIH-3T3 cells exhibited a clear concentration dependency, while L929 responded markedly to lower concentrations of silver nanoparticles. It is possible to say that Hsp70 seems suitable biomarker for stress testing in cells upon contact with silver nanoparticles. The question is whether it can be used for other types of nanoparticles. This will be the subject of our further research.

**Acknowledgment.** The authors gratefully acknowledge the support by the project LO1305 of the Ministry of Education, Youth and Sports of the Czech Republic, by the Operational Program Education for Competitiveness - European Social Fund (project CZ.1.07/2.3.00/20.0056 of the Ministry of Education, Youth and Sports of the Czech Republic), and by Internal Grant of Palacky University in Olomouc (PrF_2013_031).

## 6. References

[1] Ahamed M, Posgai R, Gorey TJ, Nielsen M, Hussain SM, Rowe JJ. Silver nanoparticles induced heat shock protein 70, oxidative stress and apoptosis in Drosophila melanogaster. Toxicology and Applied Pharmacology 2010;242:263-9.

[2] Ho CC, Huang LJ, Yang RC. Silver nanoparticles induces heat shock response and provides an anti-inflammatory effect in Clone 9 cells. Faseb J 2013;27.

[3] Kaur J, Tikoo K. Evaluating cell specific cytotoxicity of differentially charged silver nanoparticles. Food and Chemical Toxicology 2013;51:1-14.

[4] Weigl E, Kopecek P, Raska M, Hradilova S. Heat shock proteins in immune reactions. Folia Microbiol 1999;44:561-6.

[5] Ait-Aissa S, Pandard P, Magaud H, Arrigo AP, Thybaud E, Porcher JM. Evaluation of an in vitro hsp70 induction test for toxicity assessment of complex mixtures: comparison with chemical analyses and ecotoxicity tests. Ecotox Environ Safe 2003;54:92-104.

[6] Ait-Aissa S, Porcher JM, Arrigo AP, Lambre C. Activation of the hsp70 promoter by environmental inorganic and organic chemicals: relationships with cytotoxicity and lipophilicity. Toxicology 2000;145:147-57.
[7] Gibney E, Gault J, Williams J. The use of stress proteins as a biomarker of sub-lethal toxicity: induction of heat shock protein 70 by 2-isobutyl piperidine and transition metals at sub-lethal concentrations. Biomarkers 2001;6:204-17.
[8] Gupta SC, Sharma A, Mishra M, Mishra RK, Chowdhuri DK. Heat shock proteins in toxicology: How close and how far? Life Sciences 2010;86:377-84.
[9] Mukhopadhyay I, Nazir A, Saxena DK, Chowdhuri DK. Heat shock response: hsp70 in environmental monitoring. Journal of Biochemical and Molecular Toxicology 2003;17:249-54.
[10] Chae YJ, Pham CH, Lee J, Bae E, Yi J, Gu MB. Evaluation of the toxic impact of silver nanoparticles on Japanese medaka (Oryzias latipes). Aquatic Toxicology 2009;94:320-7.
[11] Wong SWY, Leung PTY, Djurisic AB, Leung KMY. Toxicities of nano zinc oxide to five marine organisms: influences of aggregate size and ion solubility. Anal Bioanal Chem 2010;396:609-18.
[12] Lim DH, Jang J, Kim S, Kang T, Lee K, Choi IH. The effects of sub-lethal concentrations of silver nanoparticles on inflammatory and stress genes in human macrophages using cDNA microarray analysis. Biomaterials 2012;33:4690-9.
[13] Panacek A, Kvitek L, Prucek R, Kolar M, Vecerova R, Pizurova N, et al. Silver colloid nanoparticles: Synthesis, characterization, and their antibacterial activity. Journal of Physical Chemistry B 2006;110:16248-53.
[14] Ritossa F. New puffing pattern induced by temperature shock and dnp in drosophila. Experientia 1962;18:571-&.
[15] Srivastava PK, Menoret A, Basu S, Binder RJ, McQuade KL. Heat shock proteins come of age: Primitive functions acquire new roles in an adaptive world. Immunity 1998;8:657-65.
[16] Richter K, Haslbeck M, Buchner J. The Heat Shock Response: Life on the Verge of Death. Molecular Cell 2010;40:253-66.
[17] Krebs RA, Feder ME. Tissue-specific variation in Hsp70 expression and thermal damage in Drosophila melanogaster larvae. Journal of Experimental Biology 1997;200:2007-15.

Journal of Physics: Conference Series **617** (2015) 012024 doi:10.1088/1742-6596/617/1/012024

# *In vivo* nanotoxicology of hybrid systems based on copolymer/silica/anticancer drug

**C P Silveira[1*], A J Paula[2], L M Apolinário[3], W J Fávaro[3,4] and N Durán[1,4,5]**

[1]Chemistry Institute, Universidade Estadual de Campinas (UNICAMP), Campus Universitário Zeferino Vaz, s/n, Cidade Universitária, CEP 13083-870, Campinas, SP, Brazil,
[2]Department of Physics, Universidade Federal do Ceará (UFC), Avenida da Universidade, 2853, Benfica, CEP 60020-181, Fortaleza, CE, Brazil,
[3]Department of Environmental Engineering, Universidade Estadual Paulista (UNESP), Avenida Três de Março, 511, CEP 18087-180, Sorocaba, SP, Brazil,
[4]Farmabrasilis R&D Division, Brazil,
[5]Nanobioss Laboratory, Chemistry Institute, Universidade Estadual de Campinas (UNICAMP), Campus Universitário Zeferino Vaz, s/n, Cidade Universitária, CEP 13083-870, Campinas, SP, Brazil

*Email: camila.silveira@iqm.unicamp.br,

**Abstract**. One of the major problems in cancer therapies is the high occurrence of side effects intrinsic of anticancer drugs. Doxorrubicin is a conventional anticancer molecule used to treat a wide range of cancer, such as breast, ovarian and prostate. However, its use is associated with a number of side effects like multidrug resistance and cardiotoxicity. The association with nanomaterials has been considered in the past decade to overcome the high toxicity of these drugs. In this context, mesoporous silica nanoparticles are great candidates to be used as carriers once they are very biocompatible. Taking into account the combination of nanoparticles and doxorrubicin, we treated rats with chemically induced prostate cancer with systems based on mesoporous silica nanoparticles and a thermoreversible block copolymer (Pluronic F-127) containing doxorrubicin. Preliminary results show a possible improvement in tumor conditions proportional to the concentration of the nanoparticles, opening a perspective to use mesoporous silica nanoparticles as carrier for doxorrubicin in prostate cancer treatment.

## 1. Introduction

It is known that one of the main problems related to cancer treatments is the high incidence of side effects caused by anticancer drugs. Doxorrubicin (DOX), for instance, is an antibiotic and anticancer that is commonly used to treat many types of cancers, such as breast, ovarian and prostate [1,2]. However, its efficacy is yet very limited for the occurrence of aggressive side effects, including mielosuppression, alopecia, multidrug resistance and cardiotoxicity [2,3].

In the last years, many alternatives to minimize this issue have been studied. Among them, the association with nanoparticles is promissing. Their use as drug carriers is advantageous once they can provide a sustained release and increase the effectiveness of drug delivery [4,5]. Among nanomaterials, mesoporous silica nanoparticles (MSNs) have a good perspective to be used as drug carriers because of their biocompatibility and safety profile [4]. These particles have high superficial area and great pore volume, which is very suitable to carry actives [4].

Another strategy to overcome the intrinsic toxicity of anticancer drugs is the association with materials that can provide a long-term release of the drug. This ensures a better drug distribution and avoids a premature degradation of the drug [6]. In this context, Pluronic F-127 (PF-127) is a biocompatible block copolymer, approved by the FDA, that has a transition sol-gel temperature near room temperature [7]. That is, it is liquid at low temperatures (e.g. refrigerator temperatures) and it gelates at room temperature, which makes it very suitable to reach *in situ* gelation [6].

So, to obtain preliminary results about the antitumor potential of the association MSNs + DOX + PF-127 and the role of the nanoparticles in this process, we developed three systems containing these components and tested them in rats with chemically induced prostate cancer. The MSNs were used as carriers for DOX, to reach a sustained release of the drug. The polymer, in turn, was used as the matrix of the hydrogel, to achieve *in situ* gelation (i.e. inside the animal body after inoculation) and thus reach a prolonged release of the components.

## 2. Methods

*Synthesis and characterization of mesoporous silica nanoparticles (MSNs)*

MSNs were synthesized following the protocol developed by Paula et al. (2012) [8], which consists of a sol-gel synthesis based on the methods of Stöber [9] and Bein [10]. This approach allow to produce spherical and monodisperse nanoparticles with high colloidal stability in aqueous medium. In this context, 0.75 g of cetyltrimethylammonium bromide (CTAB) were added to a 0.050 mol $L^{-1}$ ammonium hydroxide solution of pH 11. Then, 3.2 mL of absolute ethanol were added and after 15 minutes under magnetic stirring, 2.5 mL of tetraethyl orthosilicate (TEOS) were added. The reaction was maintained under reflux at 60°C for two hours. After completing the reaction, a centrifugation step was carried out to separate the products (60 minutes at 18,400 rcf). To extract the CTAB, the products were resuspended in 90 mL of absolute ethanol and 10 mL of hydrochloric acid were added (1:9 HCL:Ethanol volume ratio), sonicating the mixture for 10 minutes. Finally, to obtain ethanolic suspensions of the nanoparticles, the mixture was centrifuged (60 minutes at 18,400 rcf) following two washing steps with absolute ethanol and resuspended in absolute ethanol.

The size distribution and morphology of MSNs were analyzed by transmission electron microscopy (TEM) in the bright field mode (TEM-BF, Zeiss Libra 120, operating at 80 kV). Nitrogen-sorption assays were run to obtain information on nanoparticles surface area, pore volume and pore diameter (Accelerated Surface Area and Porosimetry System ASAP 2020 micromeritics). To calculate surface area and pore diameter, the BET (Brunauer-Emmet-Teller) and BJH (Barret-Hoyner-Halenda) methods were used, respectively, both using $N_2$ adsorption branch. The pore volume was calculated from the single-point value adsorbed at $P/P0 = \sim 0.94$.

*Synthesis and characterization of hybrid systems*

To evaluate the antitumor potential of the hybrid systems as well as the role of MSNs in the association, we developed three systems. The only difference between them was the presence and concentration of the nanoparticles (Table 1). The systems were produced by firstly adding DOX to the copolymer solution in an ice bath under magnetic stirring and adding the nanoparticles lastly to avoid aggregation. The mixture was kept under these conditions overnight. All systems were produced in physiological saline solution (0.9% w/v NaCl) to keep the same osmotic pressure than the animals' cells.

The concentration of DOX was calculated based on Das et al. (2010) [2] to reach 1.5 mg of drug per kilogram of animal body weight, considering a dose application of 0.3 mL and an animal body weight of 150 g (average). The PF-127 concentration was chosen regarding its gelation temperature ($T_{gel}$): $T_{gel}$ had to fit within a temperature range that would allow easy manipulation of the systems, avoiding gelation inside the syringe and on the needles' walls and quickly gelation of the systems once applied inside the animals bodies, avoiding the diffusion of the systems to unwanted locations and premature release. So, the systems were characterized regarding their $T_{gel}$ by adding 5 mL of the liquid mixture in a beacker in an ice bath with a thermometer. The mixture was heated in a constant rate and the temperature in which the bar stopped moving was considered the gelation temperature.

In order to corroborate the potential for prolonged release of the systems, the release profile was studied *in vitro* through a membraneless dissolution method. To do so, 0.5 mL of ht hybrid systems were added to a weighed vial and incubated at 35°C (which is the body temperature of mice used in the *in vivo* assays), using a dry bath, to reach thermal equilibrium. So, the vials were weighed and 0.5 mL of the release medium (NaCl 0.9% w/v) were put carefully in order to avoid mixture. At pre-determined intervals, the release medium was removed, the vial was weighed and a new release medium was put to avoid saturation. The dissolution rate was calculated in terms of weight loss against time.

*In vivo model*

To accomplish the *in vivo* study, seventeen 7-week-old male Fischer 344 rats were used and they were obtained from the Multidisciplinary Center for Biological Investigation (CEMIB) at University of Campinas (UNICAMP). The prostate cancer induction was performed in 13 animals and followed a new protocol based on Fávaro et al. (2014) [11] in which the animals are ready to be treated in approximately four months. Firstly, the animals received a daily subcutaneous injection of 100 mg $kg^{-1}$ testosterone cypionate diluted in 0.5 mL of peanut oil for three days. Then, the animals were anesthetized with 5 mg $kg^{-1}$ of 2% xylazine hydrochloride and 60 mg $kg^{-1}$ of 10% ketamyne hydrochloride in order to perform 0.5 cm suprapubic incision and inoculation of 0.2 mL of 15 mg $kg^{-1}$ of *n*-methyl-*n*-nitrosourea (MNU) dissolved in 0.3 mL of 1 M sodium citrate (pH 6.0) and 25% PF-127, which allows *in situ* gelation of the solution. After one week, the animals received subcutaneous injections of 5 mg $kg^{-1}$ testosterone cyprionate diluted in 5 mL of peanut oil on alternate days for 120 days.

Of the total number of animals, 4 animals comprised the healthy control group (Control, n=4, received no cancer induction) and 4 comprised the cancer group (Cancer, n=4, received no treatment), both groups received physiological saline (NaCl 0.9% w/v) for treatment. The rest of the animals were divided in 3 groups (n=3): Group S1 was treated with S1 system, Group S2 was treated with S2 system and Group S3 was treated with S3 system. The treatment consisted of a weekly-dose of intraperitoneal inoculation of 0.3 mL of the respective treatment for 30 days. To prevent gelation inside the needles, the systems were kept in ice bath before each application. The animals received water and the same solid diet *ad libitum* (Nuvilab) and were allocadted in single solid-bottom boxes lined with wood shavings in a room with controlled light and temperature (12 hours light and 12 hours dark, 20-25°C). The experimental protocol followed ethical principles in animal research. After treatment, the animals were euthanized and the occurrence of macroscopic changes were observed.

## 3. Results and Discussion

*Synthesis and characterization of MSNs*

Transmission electron microscopy images (Fig. 1a) show MSNs with spherical morphology and size distribution from 45 to 75 nm. Nitrogen-sorption experiments using BET and BJH methods revealed that the superficial area was 970 $cm^2$ $g^{-1}$ and the pore volume was 1.6 $cm^3$ $g^{-1}$, with an average pore diameter of 4.8 nm. The nitrogen-sorption isotherm (Fig. 1b) shows an adsorption pattern that resembles a type IV isotherm (IUPAC) with a subtle stepwise behavior around 0.4 $P/P_0$, which is characteristic of well-ordered mesoporous materials [12].

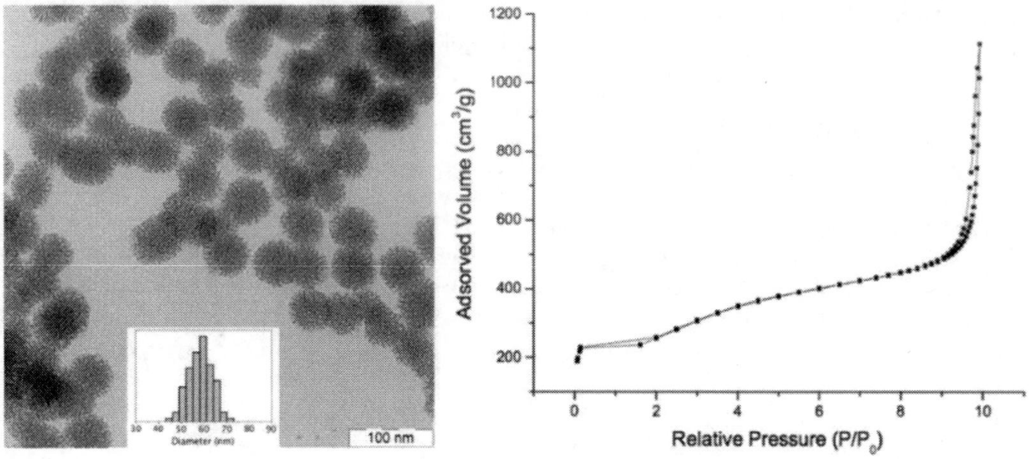


**a** **b**

**Figure 1. a)** TEM images of MSNs with size distribution histogram (measuring at least 100 nanoparticles) **b)** Nitrogen sorption isotherm of MSNs

*Synthesis and characterization of hybrid systems*

The hybrid systems were synthesized and firstly characterized regarding their $T_{gel}$ (Table 1). All systems have a $T_{gel}$ around 20°C, which allows easy manipulation of them during dose applications and a rapid gelation inside the animal body (which temperature is around 35°C).

**Table 1.** Description of the hybrid systems.

| System | Components and final concentration | Gelation temperature ($T_{gel}$, °C) |
|---|---|---|
| S1 | PF-127 (18%)<br>DOX (1.25 mg $mL^{-1}$) | 21.0 ± 0.5 |
| S2 | PF-127 (18%)<br>DOX (1.25 mg $mL^{-1}$)<br>MSNs (1.0 mg $mL^{-1}$) | 21.0 ± 0.5 |
| S3 | PF-127 (18%)<br>DOX (1.25 mg $mL^{-1}$)<br>MSNs (5.0 mg $mL^{-1}$) | 22.0 ± 0.5 |

The achievement of long-term release was confirmed during the *in vitro* dissolution experiment (Fig. 2). Taking into account the concentration gradient induced by cells permanently on the hydrogel, it is indeed expected that the hybrid system will delay the release of the components (DOX and MSNs, when it is the case) after the application.

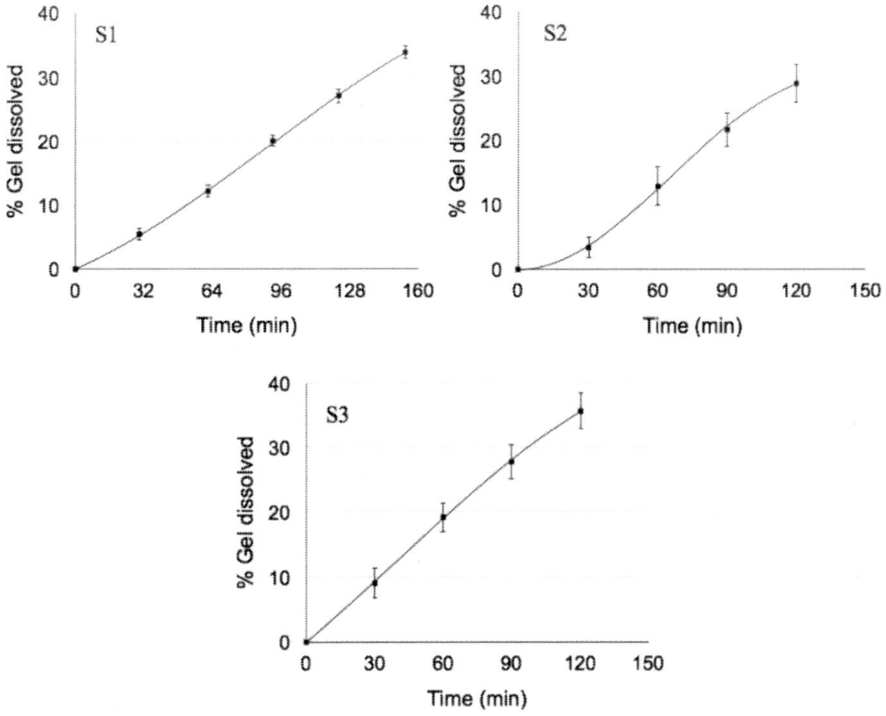


**Figure 2**. *In vitro* gel dissolution profiles of the systems.

*In vivo macroscopic analyses*

After treatment, the animals were euthanized and macroscopic changes were observed in order to get preliminary information on the antitumor potential of the systems (Table 2).

**Table 2.** Observed macroscopic changes of five experimental groups.

| Macroscopic observation | Control | Cancer | S1 | S2 | S3 |
|---|---|---|---|---|---|
| No macroscopic changes | 100% | - | 33,3% | 100% | 100% |
| Cardiomegaly | - | - | 66,6% | - | - |
| Hernia | - | - | 33,3% | - | - |

| | | | | | |
|---|---|---|---|---|---|
| Lesion in prostate base | - | 100% | 33,3% | - | - |

The results from Table 2 show a decrease in frequency of lesions that is proportional to MSNs concentration. It might be a possible improvement in tumor conditions as the MSNs are added in the formulations, calling the attention for the role of the nanoparticles in this process. Also, we can a high frequency of cardiotoxic effects in group S1 (the one treated with the system with no nanoparticles), probably caused by intrinsic toxicity of DOX, although these effects can not be seen in groups treated with systems containing nanoparticles (S2 and S3). This is particularly interesting once it suggests that the MSNs might lead to a decrease in drug toxicity.

It is known that nanoparticles are submitted to the enhanced permeability retention (EPR) effect, that allow macromolecules and particles up to 100 nm to accumulate in interstitial space of tumor cells [13]. This may increase the delivery of DOX. In turn, the prolonged and sustained release promoted by the hydrogel and the MSNs might protect the organisms from the intrinsic toxicity of the drug, increasing the efficacy of the treatment.

## 4. Conclusions

After treating rats with chemically induced prostate cancer, preliminary macroscopic analyses showed a possible improvement in tumor conditions that was proportional to MSNs concentration and a possibly cardioprotection induced by the MSNs. The results suggest that the nanoparticles might play an important role in the treatment and that they have good perspectives to minimize the side effects of DOX. It might be occurring a combined effect from DOX and MSNs, caused by EPR effect along with sustained and prolonged release of the drug. This avoids premature degradation of DOX and enables its cell internalization, achieving a better drug distribution, increasing the efficacy of the drug and possibly decreasing its toxicity.

## 5. Acknowledgements

Supported from CNPq and FAPESP.

## 6. References

[1] Singal P K, Li T, Kumar D, Danelisen I and Iliskovic N 2000 *Mol. Cell. Biochem.* **207** 77
[2] Das A, Durrant D, Mitchell C, Mayton E, Hoke N N, Salloum F N, Parl M A, Qureshi I, Lee R, Dent P and Kukreja R C 2010 *Proc. Natl. Acad. Sci. USA* **107** 1802
[3] Shen F, Chu S, Bence A K, Bailey B, Xue X, Erickson P A, Montrose M H, Beck W T and Erickson L C 2008 *J. Pharmacol. Exp. Ther.* **324** 95
[4] Mai W X and Meng H 2013 *Integr. Biol.* **5** 19
[5] Chen F, Hao H, Shi S X, Goel S, Valdovinos H F, Hernandez R, Theuer C P, Barnhart T E and Cai W B 2014 *Sci. Reports* **4** 1
[6] Heilmann S, Küchler S, Wischke C, Lendlein A, Stein C and Schäfer-Korting M 2013 *Int. J. Pharm.* **444** 96
[7] Dumortier G, Grossiord J L, Agnely F and Chaumeil J 2006 *J. Pharm. Res.* **23** 2709
[8] Paula A J, Montoro L A, Souza Filho A G and Alves O L 2012 *Chem. Commun.* **48** 591
[9] Stöber W, Fink A and Bohn E 1968 *J. Colloid Interf. Sci.* **26** 62
[10] Kecht J, Schlossbauer A and Bein T 2008 *Chem. Mater.* **20** 7207

[11] Fávaro W J, Apolinário L M, Caballero N E D, Garcia P V, Bueno C P S 2014 PIBr 10 2014 023118 8
[12] Neímark A V, Sina K S W and Thommes M 1997 Handbook of heterogeneous catalysis Volume 1, ed Ertl G, Knözinger H and Wetikamp J (Weinheim: Wiley-VHC) 721
[13] Maeda H 2003 Biomedical aspects of drug targeting, Volume 1, ed Muzykantov V P et al. (New York: Kluwer Academic Publishers) 211

Journal of Physics: Conference Series **617** (2015) 012025 doi:10.1088/1742-6596/617/1/012025

# Evaluation of the effects of nitric oxide-releasing nanoparticles on plants

**A E S Pereira[1], A M Narciso[2], A B Seabra[2*] and L F Fraceto[1,3]**

[1]Department of Biochemistry, Universidade Estadual de Campinas, Campus Universitário Zeferino Vaz, s/n, Cidade Universitária, CEP 13083-870, Campinas, SP, Brazil,
[2]Exact and Earth Sciences Departament, Universidade Federal de São Paulo, Rua São Nicolau 210, CEP 09913030, Diadema, SP, Brazil,
[3]Department of Environmental Engineering, Universidade Estadual Paulista (UNESP), Avenida Três de Março, 511, CEP 18087-180, Sorocaba, SP, Brazil

*Email: amedea.seabra@gmail.com, amedea.seabra@unifesp.br

**Abstract**. Nowadays, there are several commercially available products containing nanostructured materials. Meanwhile, despite the many benefits that can be obtained from nanotechnology, it is still necessary to understand the mechanisms in which nanomaterials interact with the environment, and to obtain information concerning their possible toxic effects. In agriculture, nanotechnology has been used in different applications, such as nanosensors to detect pathogens, nanoparticles as controlled release systems for pesticides, and biofilms to deliver nutrients to plants and to protect food products against degradation. Moreover, plants can be used as models to study the toxicity of nanoparticles. Indeed, phytotoxicity assays are required to identify possible negative effects of nanostructured systems, prior to their implementation in agriculture. Nitric oxide (NO) plays a key role in plant growth and defense, and recently, several papers described the beneficial effects due to application of exogenous NO donors in plants. The tripeptide glutathione (GSH) is an important anti-oxidant molecule and is the precursor of the NO donor, S-nitrosoglutathione (GSNO). In this context, the present work investigates the effects of different concentrations of alginate/chitosan nanoparticles, containing either GSH or GSNO, on the development of two test species (*Zea mays* and *Glycine sp.*). The results showed that the alginate/chitosan nanoparticles present a size average range from 300 to 550 nm with a polydispersity index of 0.35, and encapsulation efficiency of GSH between 45 - 56%. The NO release kinetics from the alginate/chitosan nanoparticles containing GSNO showed sustained and controlled NO release over several hours. Plant assays showed that at the concentrations tested (1, 5 and 10 mM of GSH or GSNO), polymeric nanoparticles showed no significant inhibitory effects on the development of the species *Zea mays* and *Glycine sp.*, considering the variables shoot height, root length, and dry mass. Therefore, these nanoparticles seem to have promissing uses in agriculture, and might be potencially used as controlled release systems applied by the foliar route.

## 1. Introduction

In recent years, there have been major advances in nanotechnology, which is now used in diverse areas such as biomedicine, pharmaceuticals, agriculture, amongst others [1,2]. Examples include the use of carbon nanotubes in the production of electronic devices [3], metallic nanoparticles (such as silver) as bactericidal agents [4], and a wide range of polymeric nanoparticles used as carrier systems for active agents [5,6]. Many products containing nanostructured materials are now commercially available, including paints, cosmetics, and electronic goods. Meanwhile, despite the many benefits that can be obtained from this technology, it is still necessary to understand the mechanisms in which nanomaterials interact with the environment, and to obtain information concerning their possible toxic effects.

In agriculture, nanotechnology has been used in various ways, with the aim of improving yields and the quality of products [7]. Applications include the use of nanosensors to detect pathogens, nanoparticles as controlled release systems for pesticides, and biofilms to deliver nutrients to plants and protect food products against degradation [1]. The use of nanoparticles as carrier systems for agrochemicals can improve the bioavailability and effectiveness of active agents, enabling lower dosages to be safety used and reducing damage to the environment [8].

Plants can be used as models to study the toxicity of nanoparticles and obtain information concerning possible interactions and effects of these nanosystems on processes such as germination, metabolism, and plant growth [9,10]. Phytotoxicity assays are required in order to identify any possible negative effects of nanostructured systems, prior to their implementation in agriculture [8].

The free radical nitric oxide (NO) plays important role in plant defense and growth [11, 12]. It has been reported that administration of exogenous NO donors can break seed dormancy, improve plant greening and germination, regulate iron homeostasis, and improve plant tolerance to salinity, metal toxicity, temperature and drought stress [12]. However, small molecular weight NO donors, such as S-nitrosothiols (RSNOs) are thermally and photochemically unstable [13]. Among the NO donors, S-nitrosoglutathione (GSNO) is an important RSNO, which spontaneously releases NO through the cleavage of S-N bound [14]. In order to increase the thermal and photochemical stability of NO donors, RSNOs have been incorporated into polymeric matrices, such as in alginate/chitosan nanoparticles [6]. The encapsulation of NO donors in nanoparticles is able to control the release of therapeutic amounts of NO, thus improving its beneficial effects [6].

In this context, the present work therefore investigates the effects of different concentrations of alginate/chitosan nanoparticles containing either reduced glutathione (GSH) (an important anti-oxidant molecule and the precursor of the NO donor, GSNO) or GSNO on the development of two test species (*Zea mays* and *Glycine* sp.), in order to obtain information on the toxicity of the nanoparticles, with a view to the possible use of this nanocarrier system in agricultural applications.

## 2. Methods

### *2.1. Synthesis of alginate/chitosan nanoparticles containing GSH and GSNO*

Alginate/chitosan nanoparticles (at ratio 0.75) were prepared using the ionic gelation method, as previous described [6,15]. Briefly, 0.266 g of chitosan was added to 10 mL of water containing 0.092 mL of acetic acid. After 24 h of magnetic stirring, 10 mL of water was added to the chitosan solution, and the final solution was homogenized for more 24 h. Required amounts of GSH were added to the chitosan solution and homogenized for 30 min. A volume of 1.0 mL of chitosan/GSH solution of was dropwise in 200 mL of alginate solution (50 µg/mL) at pH 4,0. This process led to the preparation of alginate/chitosan nanoparticles containing GSH in the following concentrations: 1, 5 and 10 mmol/L). In order to obtain GSNO-containing alginate/chitosan nanoparticles, equimolar amounts of sodium nitrite ($NaNO_2$) related to GSH were added to the nanoparticles. The final solution was homogenized for 8 h, protected from the light, and used immediately.

### *2.2. The average size and size distribution for polymeric particles in aqueous medium*

The average size for GSH-containing alginate/chitosan nanoparticles were measured using photon correlation spectroscopy (PCS) (Nano ZS Zetasizer, Malvern Instruments Co.) at 25°C in polystyrene cuvettes with a 10 mm path length.

*2.3. GSH encapsulation efficiency in alginate/chitosan nanoparticles*
The encapsulation efficiency of GSH in alginate/chitosan nanoparticles were measured by the UV–vis method, as already described [6]. Briefly, free GSH was separated from polymeric nanoparticles by ultracentrifugation, by using a Microcon centrifugal filter device containing ultrafiltration membranes (MWCO 10,000, Millipore). The amount of free GSH in the ultrafiltrates was measured by titration with a thiol-reacting 5,50-dithiobis-(2-nitrobenzoic acid) (DTNB), based on the absorbance band at 412 nm ($\mathcal{E}$ = 14.15 mmol $L^{-1}cm^{-1}$) of the 2-nitro-5-thiobenzoate anion, which is generated in the reaction of GSH with DTNB. The percentage of GSH encapsulation was determined by the described equation:

$$(\%) = (\text{mass of GSH encapsulated/mass of GSH total}) \times 100 \qquad \text{(Eq. 1)}$$

*2.4. Kinetics of GSNO decomposition with free NO release*
The kinetics for NO release from GSNO in alginate/chitosan nanoparticles were monitored by following the spectral changes at 336 nm, which is associated with S-N bond cleavage and free NO release [16] by using an Uv-Visible spectrophotometer (Agilent 8453). The kinetics were monitored at 35°C for 24 h. The amount of NO released was calculated from the amount of GSNO decomposed, as previous described [13]. The initial rates of NO release through GSNO decomposition were determined through linear regression of the curve slopes.

*2.5. Phytotoxicity of GSH or GSNO-containing alginate/chitosan nanoparticles*
Phytotoxicity assays were conducted using *Zea mays* and *Glycine* sp. collected in the field, in the municipality of São Miguel Arcanjo (São Paulo, Brazil). The seeds were sown in pots with upper and lower diameters of 12.5 and 9.3 cm, respectively, and heights of 9.3 cm. The pots were filled with 600 g of the Carolina Soil plant substrate. Ten seeds of the separate species were used in each pot, followed a fully randomized 7 x 3 experimental design. The pots were kept in a plant house, and at the pre-emergence stage (7 days after sowing) the nanoparticle suspensions containing GSH or GSNO (at concentrations of 1, 5, and 10 mmol/L) were sprayed. Deionized water was used as a control treatment. The plants were left for another 7 days after application of the treatments, and were then collected and analysed. Measurements were made of the heights of the aerial parts (cm) and the lengths of the roots (cm), after which the plants were placed in a drying cabinet at 60°C, for 7 days, and then weighed to determine their dry masses (g). The data obtained were presented as means and standard deviations, and statistical analysis of the differences between treatments was performed using analysis of variance (ANOVA) with the Tukey-Kramer post-hoc test

## 3. Results and Discussion

*3.1. Synthesis of GSH and GSNO-alginate/chitosan nanoparticles*
The mixture of alginate and chitosan polymers in acidified aqueous solution led to the formation of nanoparticles due to the strong electrostatic interactions of the anionic alginate chain with the cationic chitosan chain [15]. In aqueous solution, the hydrodynamic diameter of the GSH-alginate/chitosan nanoparticles were found to be in the range of 300 to 550 nm with a polydispersity index of 0.35, which are in accordance with our previous results [6,15]. The encapsulation efficiency of GSH at concentrations of 1, 5 and 10 mmol/L in alginate/chitosan nanoparticle solutions were found to be 45

± 2%, 45 ± 3% e 56 ± 5%, respectively. The nitrosation of GSH in acidified alginate/chitosan nanoparticle solution was performed by the addition of equimolar amount of sodium nitrite ($NaNO_2$). In acidified aqueous solution, sodium nitrite will form nitrous acid ($HNO_2$), which is the nitrosating agent, leading to the formation of GSNO, according to Equation 2. In this work, GSH-containing nanoparticles were nitrosated *in situ*, leading to the formation of GSNO-nanoparticles.

$$GSH + HNO_2 \rightarrow GSNO + H_2O \qquad \text{(Eq. 2)}$$

### *3.2. Kinetics of NO release from free GSNO and GSNO encapsulated in nanoparticles*

The kinetics for NO release from GSNO (1, 5 and 10 mol/L) encapsulated in alginate/chitosan nanoparticles were monitored through the spectral changes at 336 nm. The decrease in this absorption band corresponds to GSNO decomposition with free NO release. This calculation is based on the GSNO absorption band decay at 336 nm, solely associated with homolytic cleavage of the S-N bond and NO release in accordance with Equation 3 [13]. The end products of GSNO decomposition, either free or encapsulated, are NO and oxidized glutathione (GSSG) [16]:

$$2\,GSNO \rightarrow 2\,NO + GSSG \qquad \text{(Eq. 3)}$$

Figure 1 shows the NO release profile from GSNO encapsulated in alginate/chitosan nanoparticles, at different concentrations, as indicated in the Figure, at 35°C, over 24 h.

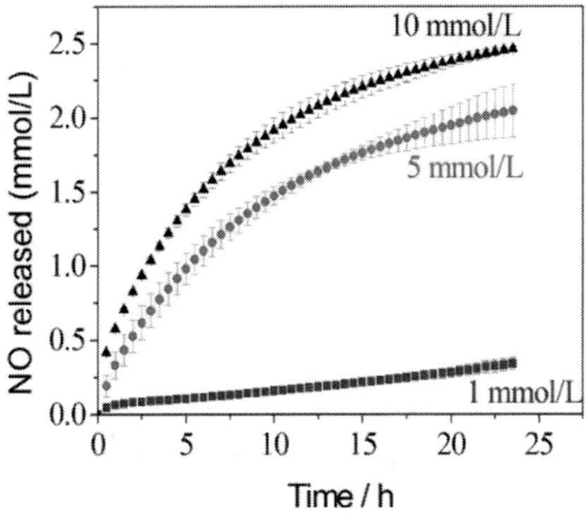


**Figure 1**. NO release profile from GSNO-containing alginate/chitosan nanoparticles at 1, 5 and 10 mmol/L, at 35°C.

It can be observed that NO is spontaneously released from GSNO-containing alginate/chitosan nanoparticles. The kinetic curves show an initial burst of NO release in the first 5 hours, followed by a progressively increase at lower rates. A sustained NO release is observed for at least 24 h. The NO-release profile increased with the increase of initial GSNO concentrations in the nanoparticles, due to the auto-catalytic effect, as previous described [13,14]. Initial rates of NO released were calculated from the kinetic curves of Figure 1. The initial rates of NO release from GSNO-nanoparticles were found to be 0.0057 ± 0.0010; 0.3088 ± 0.0066 and 0.5270 ± 0.0035 mmol/Lh for GSNO at 1, 5 and 10 mmol/L, respectively. As expected for GSNO decomposition with NO release, the initial rates increase with the initial concentration of GSNO. The NO release from GSNO-containing alginate/chitosan nanoparticles is reported to occur mainly through diffusion process over the pores or wall and disintegration of the hydropolymeric structure [15].

*3.3. Phytotoxicity of GSH or GSNO-nanoparticles*

Figure 2 shows the mean values obtained for *Zea mays* shoot lenght and root length.

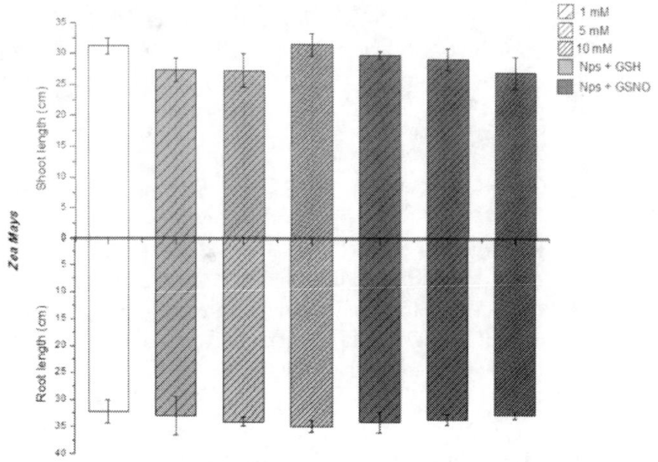


**Figure 2**. Shoot and root length for *Zea mays* after 7 days of daily treatment with GSH- or GSNO-alginate/chitosan nanoparticles (GSH or GSNO concentrations 1, 5 and 10 mM), as indicated in the Figure. The plants were left for another 7 days after application of the treatments, and were then collected and analysed.

Similarly, Figure 3 shows the mean values obtained for *Glycine* sp shoot length and root length after 7 days of treatment.

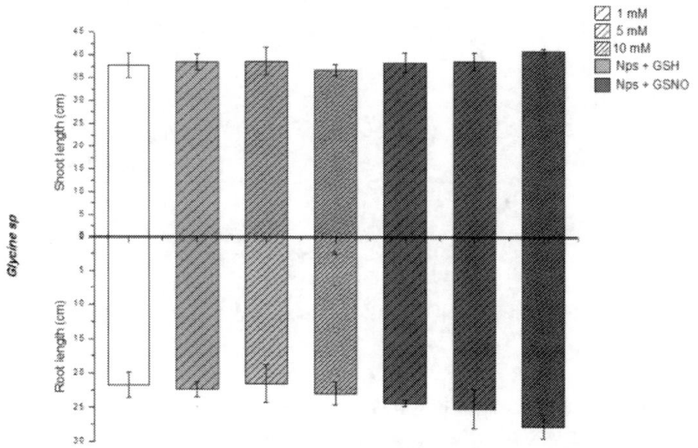


**Figure 3**. Shoot and root length for *Glycine* sp after 7 days of daily treatment with GSH- or GSNO-alginate/chitosan nanoparticles (GSH or GSNO concentrations 1, 5 and 10 mM), as indicated in the Figure. The plants were left for another 7 days after application of the treatments, and were then collected and analysed.

Figure 4 shows the values obtained for the dry masses of the specimens of *Zea mays* and *Glycine* sp after the GSH or GSNO-nanoparticles treatment.

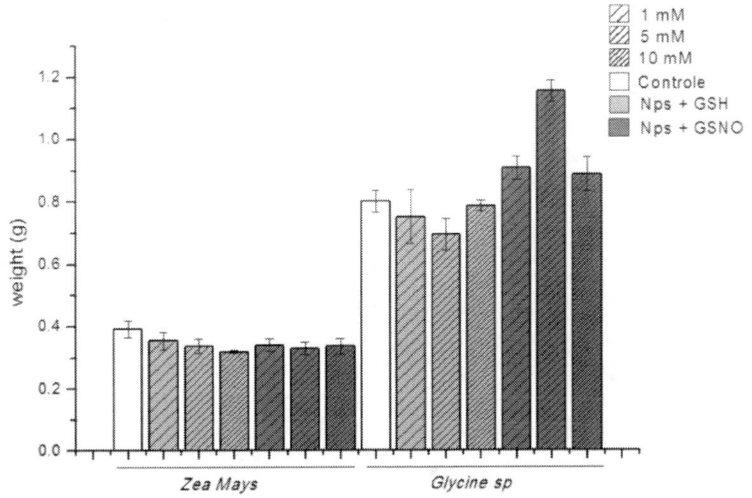


**Figure 4**. Dry masses for *Zea mays* and *Glycine* sp after 7 days of daily treatment with GSH- or GSNO-alginate/chitosan nanoparticles (GSH or GSNO concentrations 1, 5 and 10 mM), as indicated in the Figure. The plants were left for another 7 days after application of the treatments, and were then collected and analysed.

For all the concentrations tested, GSH- or GSNO-containing alginate/chitosan nanoparticles had no significant effect on the growth and development of *Zea mays* and *Glycine* sp., in terms of shoot height, root length, and dry mass. It should be noted that the increase of dry mass for *Glycine* sp upon treatment with GSNO-nanoparticles (Figure 4) were not statistically significant.

Figure 5 shows photographs of the tested plants 7 days after exposure to GSH- or GSNO-nanoparticles treatment, at the highest tested GSH or GSNO concentration 10 mmol/L, together with photographs of the corresponding controls (plants treated with deionized water).

The use of nanoparticles as carriers for active agents in agricultural applications shows considerable potential. These systems can help to improve the stability of the active agents, increase their effectiveness, and at the same time reduce the possibility of environmental contamination [1,17,18]. Despite these potential benefits, studies aimed to further investigate the effects of nanomaterials on plants remain scarce. It should be noted that characteristics of nanoparticulate systems including particle size, composition, and physicochemical properties might influence their effects during different stages of plant development [19]. In particular, alginate/chitosan are known as biocompatible and biodegradable system ideal for carrying and delivering important molecules in agriculture with no toxic effects to the environment [17,18].

On the other hand, phytotoxicity studies have shown that nanoparticles of zinc oxide and cobalt can influence the development and morphology of the species *Allium cepa* [20]. It has also been found that nanoparticles of nickel oxide can induce oxidative stress, mitochondrial dysfunction, and necrosis and apoptosis in tomato root cells [21]. Exposure of Arabidopsis thaliana to gold nanoparticles resulted in increased germination rates, greater plant development, and higher antioxidant activity [22]. Other types of nanomaterials, such as carbon nanotubes, have been found to increase germination rates and tissue development in tobacco and tomato plants [23].

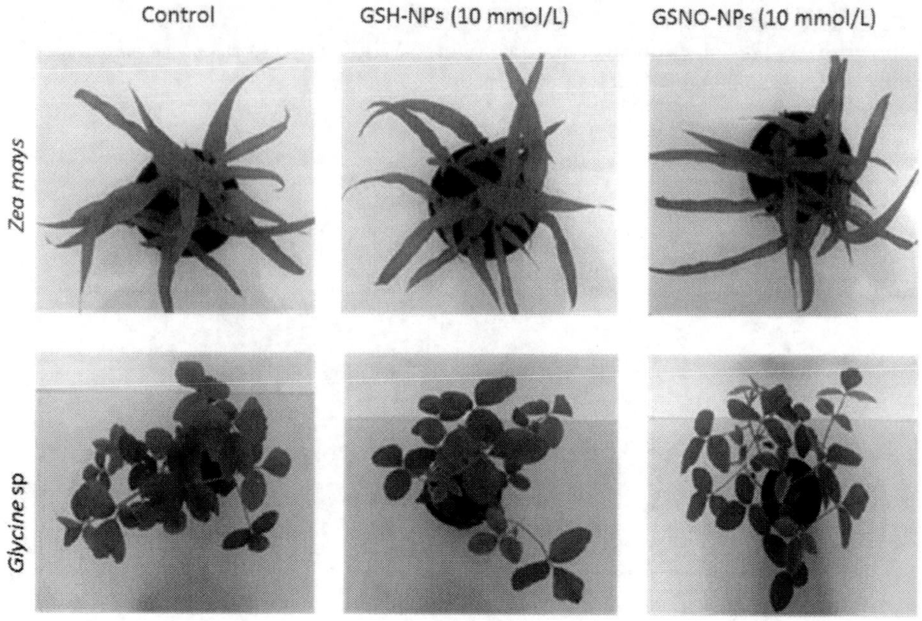


**Figure 5.** Photographs of *Zea mays* and *Glycine* sp. specimens, 7 days after application of the GSH- or GSNO-alginate/chitosan nanoparticles, at the highest test concentration (10 mmol/L of GSH or GSNO). Photographs of control specimens (treatment with deionized water) are provided for comparison.

In the present work, polymeric nanoparticles of alginate/chitosan containing GSH or GSNO (1, 5 and 10 mol/L) showed no significant effects on the development of the species *Zea mays* and *Glycine* sp, considering the variables shoot height, root length, and dry mass. These formulations therefore seem to have potential for use in agriculture as sustained release systems, although further work will be needed to investigate a wider range of concentrations and/or plant species.

## 4. Conclusions

This work describes the preparation and characterization of GSH- or GSNO-alginate/chitosan nanoparticles and their impact in plants. Nanoparticulate carrier systems have many potential applications in agriculture, where they can be used to transport a range of agrochemicals, including substances used to control pests and diseases. They can also be used in packaging to improve the quality of agricultural products. The present study showed that polymeric nanoparticles containing GSH or GSNO had no significant inhibitory effects on the development of two plant species (*Zea mays* and *Glycine* sp.), and could therefore be used commercially as controlled release systems applied using the foliar route.

## 5. Acknowledgements

Supported from FAPESP (Processes 2012/17053-7; 2013/12322-2; 2013/13096-6), CNPq, CAPES, Brazilian Network in Nanotoxicology (MCT-CNPq) and Fundunesp.

## 6. References

[1] Nair R, Varghese S H, Nair B G, Maekawa T, Yoshida Y and Kumar D S 2010 *Plant Sci.* **179** 154
[2] Wang E C and Wang A Z 2014 *Integr. Biol.* **6** 9
[3] Ghiazza M, Vietti G and Fenoglio I 2014 Carbon nanotubes: properties, applications, and toxicity *Health and Environmental Safety of Nanomaterials* Woodhead Publishing chapter 8 pp 147-174
[4] Mariselvam R, Ranjitsingh A J A A, Manthini A U R, Kalirajan K, Padmalatha C and Selvakumar P M 2014 *Spectrochim. Acta A* **129** 537
[5] Grillo R, dos Santos N Z P, Maruyama C R, Rosa A H, de Lima R and Fraceto L F 2012 *J. Hard. Mater.* **231** 1
[6] Cardozo V F, Lancheros C A C, Narciso A M, Valereto E C S, Kobayashi R K T, Seabra A B and Nakazato G 2014 *Int. J. Pharm.* **473** 20
[7] Rubilar O, Diez M C, Tortella G R, Briceno G, Marcato P D and Duran N 2014 *J. Biobased. Mater. Bio.* **8** 1
[8] Kumari A and Yadav S K 2014 *Crit. Rev. Food Sci.* **54** 957
[9] Schwabe F, Schulin R, Limbach I K, Stark W, Burge D and Nowack B 2013 *Chemosphere* **91** 512
[10] Lee W M, Kwak J I and An Y J 2012 *Chemosphere* **86** 491
[11] Shi H T, Li R J, Cai W, liu W, Fu Z W and Lu Y T 2012 *Plant Sign Behavior* **7** 438
[12] Seabra A B, Rai M and Duran N 2014 *J. Plant Biochem. Biotechnol.* **23** 1
[13] Shishido S M, Seabra A B, Loh W and de Oliveira M G 2003 *Biomaterials* **24** 3543
[14] de Oliveira M G, Shishido S M, Seabra A B and Morgon N H 2002 *J. Phys. Chem. A* **106** 8963
[15] Marcato P D, Adami L F, Barbosa R M, Melo P S, Ferreira I R, de Paula L, Duran N and Seabra A B 2013 *Curr. Nanosci.* **9** 1
[16] Seabra A B and de Oliveira M G 2004 *Biomaterials* **25** 3773
[17] Grillo R, Pereira A E S, Nishisaka C S, de Lima R, Oehlke K, Greiner R and Fraceto L F 2014 *J. Hard. Mater.* **278** 163
[18] Grillo R, Rosa A H and Fraceto L F 2014 *Int. J. Environ. Sci. Te.* **11** 1691
[19] Ma Y H, Kuang L L, He X, Bai W, Ding Y Y, Zhang Z Y, Zhao Y L and Chai Z F 2010 *Chemosphere* **78** 273
[20] Ghodake G, Seo Y D, and Lee D S 2011 *J. Hard. Mater.* **186** 952
[21] Faisal M, Saquib Q, Alatar A A, Al-Khedhairy A A, Hegazy A K and Musarrat J 2013 *J. Hard. Mater.* **250** 318
[22] Kumar V, Guleria P, Kumar V and Yadav S K 2013 *Sci. Total Environ.* **461** 462
[23] Khodakovskaya M, Dervishi E, Mahmood M, Xu Y, li Z R, Watanabe F and Biris A S 2009 *ACS Nano* **10** 3221

# Measuring Nanomaterial Release from Carbon Nanotube Composites: Review of the State of the Science

**Stacey Harper,[1] Wendel Wohlleben,[2] Maria Doa,[3] Bernd Nowack,[4] Shaun Clancy,[5] Richard Canady,[6] * and Andrew Maynard[7]**

[1] Environmental and Molecular Toxicology / Chemical, Biological and Environmental Engineering, Oregon State University, Corvallis, OR 97331

[2] BASF SE, Material Physics, 67056 Ludwigshafen, Germany

[3] US Environmental Protection Agency, Washington, DC 20460

[4] EMPA-Swiss Federal Laboratories for Material Science and Technology, Technology and Society Laboratory, CH-9014 St. Gallen, Switzerland

[5] Evonik Corporation, Parsippany, NJ 07054

[6] Center for Risk Science Innovation and Application, ILSI Research Foundation, Washington, DC 20005

[7] University of Michigan Risk Science Center, Ann Arbor, MI 48109

Email: rcanady@ilsi.org

**Abstract.** Hazard studies of "as-produced" nanomaterials are increasingly available, yet a critical gap exists in exposure science that may impede safe development of nanomaterials. The gap is that we do not understand what is actually released because nanomaterials can change when released in ways that are not understood. We also generally do not have methods capable of quantitatively measuring what is released to support dose assessment. This review presents a case study of multi-walled carbon nanotubes (MWCNTs) for the measurement challenge to bridge this gap. As the use and value of MWCNTs increases, methods to measure what is released in ways relevant to risk evaluation are critically needed if products containing these materials are to be economically, environmentally, and socially sustainable. This review draws on the input of over 50 experts engaged in a program of workshops and technical report writing to address the release of MWCNTs from nanocomposite materials across their life cycle. The expert analyses reveals that new and sophisticated methods are required to measure and assess MWCNT exposures for realistic exposure scenarios. Furthermore, method requirements vary with the materials and conditions of release across life cycle stages of products. While review shows that the likelihood of significant release of MWCNTs appears to be low for many stages of composite life cycle, measurement methods are needed so that exposures from MWCNT-composites are understood and managed. In addition, there is an immediate need to refocus attention from study of "as-produced" nanomaterials to coordinated research on actual release scenarios.

**VOCABULARY: Release** – detachment of a particle or fragment from a larger whole such as a consumer product during use. Specifically in the context of this case study, when a fragment that contains MWCNTs detaches from a MWCNT-polymer composite and is nano-sized or possesses nano-characteristics, the release is considered to be a **Nanorelease**. **Release Mechanism** – a process, such as abrasion or degradation, which results in the release of fragments and discrete MWCNTs from a MWCNT-polymer composite. **Release Measurement Method** – generation, collection, and characterization of fragments containing MWCNT or discrete MWCNTs released from a product. Method may be for detection or quantification. **Lifecycle** – a time sequence depicting the life of a MWCNT-polymer composite or product, the phases of which include MWCNT-polymer composite production, product manufacture and use, and disposal, recycling, or re-purposing. **Release Scenario** – an event or chain of events resulting in a nanorelease during a lifecycle phase of a product as described by the release mechanism, setting (e.g., workplace, dwelling, and environment), relevant media (e.g., air, water, mixed solids, and biological matrices), and form and amount of released material. **Probability of Release** – the likelihood of nanorelease, as determined by a quantitative or qualitative measurement or prediction of the total number or mass concentration of released material, for a specified release scenario and time interval. **Release Evaluation** – a total assessment of nanorelease, including the release mechanism, form of the released entity, release scenario, probability of release, and lifecycle simulation, if relevant.

## 1. Introduction

Developers of practical and beneficial products containing engineered nanomaterials lack critical information for nanomaterial use that is needed to align innovation goals with sustainable, safe product development.[1] This lack of information is causing unnecessary uncertainty regarding the safety of products using nanomaterials. To support innovative nanomaterial-based products with low life cycle risk profiles, greater information is needed on the potential hazards associated with specific exposure scenarios. Although this seems intuitive, the authors found that there are currently no standard methods to measure what is released from use of products containing nanomaterials. Furthermore, there is insufficient information to define the risk-relevant properties of nanoparticles released during stages of the life cycles of nanocomposites.[2-9] As a result, it is currently challenging to complete the exposure pathway from release to environmental transport to a receptor's exposure, and then to relate the exposure to possible health outcomes from toxicology data for a given nanomaterial in a given use.[10]

There are many claims that engineered nanomaterials are used widely and may not be well controlled.[11-13] The gap in understanding realistic nanomaterial release is sufficiently great that measuring the nanomaterial release rate from a material in a given context has rarely even been attempted[1,5,6,14] compared with a growing body of studies of the potential hazards of nanomaterials. As a result, the rate of development of hazard studies has far outpaced exposure-related studies and the situation is not improving (Figure 1). This in turn is propagating an imbalance in understanding of the risk potential of nanomaterial uses. Furthermore, although advances for some types of nanomaterials have been made for some endpoints, mechanistic evaluations of nanomaterial characteristics that predict toxicity have not led to a level of understanding that allows producers and regulators to trace a released or measured nanomaterial in the environment to a risk-relevant exposure for a receptor.[15-17] This relative lack of release information and the inability to link release data to exposure data have weakened the ability of producers, users, and regulators alike to address concerns about the use of nanotechnology.

This review addresses these issues based on the example of multi-walled carbon nanotubes (MWCNTs) and polymer composites. MWCNTs are one of the more contentious and potentially valuable nanoscale materials to have emerged in recent years. Many current and near-term applications involve integration of MWCNTs into polymer matrices to enhance thermal, mechanical, and/or electrical properties.[4] The electronics, defense, and energy sectors are leading in MWCNT-polymer applications, and MWCNT-polymers are increasingly being used in consumer sporting goods, textiles, and packaging.[19-28]

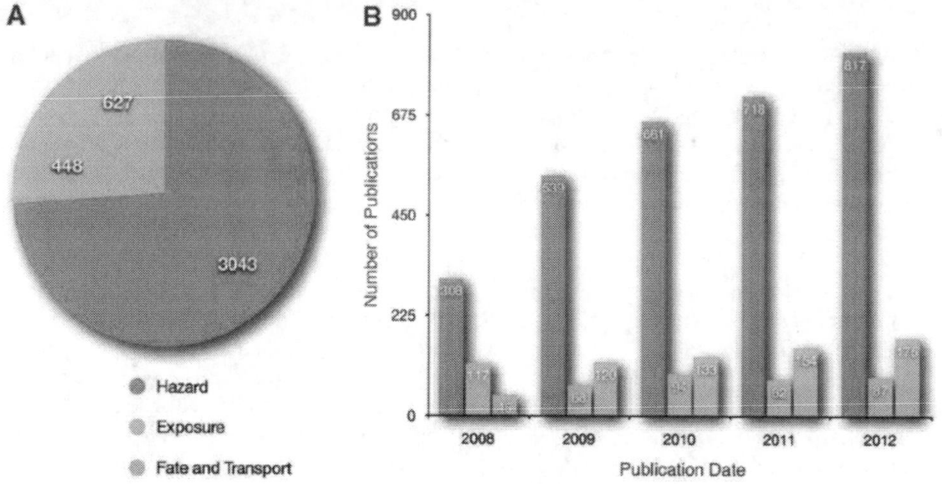


**Figure 1.** Preponderance of "hazard" environmental, health, and safety research in the literature. (A) Topic distribution analysis of the number of publications from January 2008 to December 2012. (B) Time distribution analysis of the number of papers published from 2008 to 2012. Data are from the International Council on Nanotechnology.[18]

Although MWCNTs in composites are typically tightly bound, their release through the life cycle of the nanocomposite is possible.[3] Research on hazards associated with pristine MWCNTs has been previously reviewed.[29] This information currently does not consider the quantity and form released to exposure pathways, yet this knowledge is critical to forming a quantifiable and actionable understanding of risk.[30]

Release and transport of polymer fragments to the environment occurs on a large scale, and analogous concerns exist for microplastic particles in marine and soil systems.[31,32] For example, it is known that polymeric materials can be released to the environment, degraded into small pieces (millimeters to micrometers in scale), and transported over long distances. Research has shown polymer fragments to be an important component of household dust as well as a source for human exposure to polybrominated and phosphorous-containing flame retardants.[33] Some retardants were found to be transferred to dust *via* physical processes such as abrasion from polymers.[34] It is possible that MWCNT-containing polymers in consumer products will undergo similar processes, resulting in MWCNT-containing fragments in household dust and subsequent exposure to a wide cross-section of the population. Given that global revenues for MWCNT-nanocomposites at an early stage of market penetration were on the order of $37 million in 2009,[35] and new product development is continuing, a structured analysis of our current knowledge on MWCNT release from such nanocomposites was considered necessary.

Here we provide a comprehensive review of the state of scientific understanding on the potential sources and pathways of release, pointing to situations and specific applications in which humans or the environment could encounter MWCNT releases and which methods to measure such release and exposure are needed. This review also suggests integrated, systematic approaches to increase the breadth and depth of our knowledge base on nanomaterial release from composites.

## 2. Effects of the Product Use Life Cycle on Release Likelihood and Measurement

The potential for release of MWCNTs from composites can vary throughout the life cycle of a product depending on the production and manufacturing processes, the use/misuse of the final product, and end-of-life (EOL) treatment (Figure 2). Understanding the quantitative analysis and instrumentation needs with respect to measuring release at a particular life cycle stage will depend on factors such as proximity to receptors or the nature of the media into which release occurs. For example, evaluating a life cycle stage with high potential for consumer contact could be accomplished by a simple demonstration of a lack of physical opportunity for release (*e.g.*, sealed components in a display screen), whereas specific information on the magnitude and nature of long-term release from those same components may be needed to understand contributions to downstream or downwind receptors at EOL disposal.

Both physical and chemical mechanisms of release can further affect the specific release potential for nanomaterials added to composites because they can influence whether the nanomaterial remains embedded or is transformed during the release (Figure 3).[7-9,36-55] Release mechanisms can be categorized based on the driving forces that can cause the release of MWCNTs throughout a product life cycle, such as mechanical stress, chemical processes (*e.g.*, hydrolysis and photolysis), and incineration through human and environmental processes.[3] The use of life cycle concepts combined with an understanding of the driving forces provides a framework that enables the occurrence of particular mechanisms and their potential effect on nanomaterial release to be mapped out. This type of methodology is needed if communication is to occur effectively between different actors along the value chain.

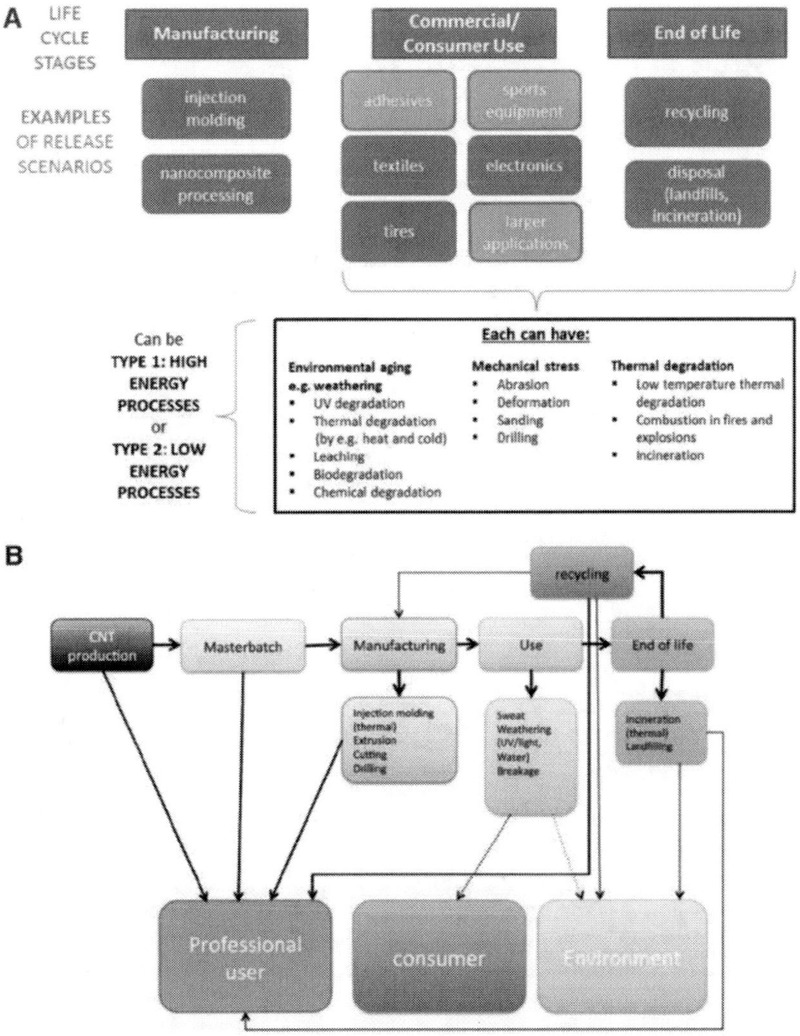


**Figure 2.** Life cycle of a nanocomposite. (A) Relevance for exposure of professionals or consumers, and environmental emission from nanofiller production to end of life/recycling. (B) Specific life cycle map of sports equipment made of MWCNT-polymer composite, detailing release processes that must be considered with the anticipated relative release probability indicated by the thickness of arrows. Reprinted (with modification) from Nowack *et al.*,[3] with permission from Elsevier.

Journal of Physics: Conference Series **617** (2015) 012026 doi:10.1088/1742-6596/617/1/012026

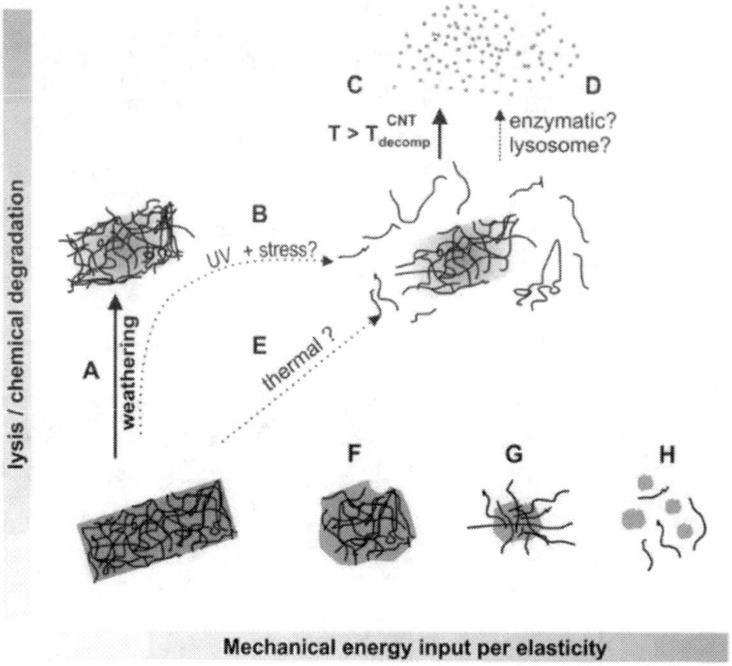


**Figure 3.** Heterogeneity of released entities for CNT composites pose measurement and risk assessment challenges unaddressed in the current literature. The chemical axis includes solvolysis, weathering, thermolysis, and incineration, whereby the matrix is degraded relatively faster and removed than the more persistent MWCNT nanofiller. The mechanical axis summarizes abrasion, normal consumer use, sanding, and shredding. Further details are available in the following publications: (release pathway A) Nguyen *et al.*,[36-38] Wohlleben *et al.*,[8] and Asmatulu *et al.*[39]; (release pathway B) Busquets-Fite *et al.*[40] and Hirth *et al.*[7]; (release pathway C) Uddin and Nyden[41]; (release pathway D) Liu *et al.*[42,43] and Kagan *et al.*[44]; (release pathway E) Ashter *et al.*,[45] Tsai *et al.*,[46] and Bello *et al.* [47]; (release pathway F) Bello *et al.*[48] and Wohlleben *et al.*[8,9] (for CNTs), and Raynor,[49] Gohler *et al.*,[50] Koponen *et al.*,[51] and Saber *et al.*[52] (for non-CNTs); (release pathway G) Cena and Peters,[53] Huang *et al.*,[54] and Hirth *et al.*[7]; and (release pathway H) Golanski *et al.*[55] (only with agglomerates).

Specific release information is particularly important for those cases in which release mechanisms may increase hazard and risk, because the knowledge of the nanomaterial content and type in a product usually decreases along the value chain of a product.[56] Consideration of the whole life cycle of nanoproducts should ensure that possible effects on humans and the environment can be systematically identified and thoroughly assessed.[57,58] For instance, combinations of polymer types and additives that affect performance at one life cycle phase may lead to instabilities in the MWCNT-polymer at another life cycle phase, complicating measurement of released MWCNTs for the life cycle of the material use. Thus, systematic evaluations of release mechanisms and potentials are critical to life cycle risk assessment that informs product development.

## 3. Specific Evaluation of Life Cycles for MWCNT-Polymer Composites

Considering the beginning of a MWCNT product life cycle, the synthesis and handling of MWCNT powders prior to formation of composites present conditions that may lead to high likelihood for release if not appropriately addressed.[59,60] However, standard industrial hygiene measures can limit exposure of professional users and the environment to a great extent. In addition, during masterbatch production in which high-concentration composites are created and during manufacturing of the final product in occupational settings, conditions for release can also be controlled and exposure of professional users to material can be minimized.[61] At increased temperatures during subsequent processing involving injection molding and extrusion, polymer can vaporize and re-condense into airborne particles (polymer fumes).

Particles released from melt processing of nanocomposites could represent a complex mixture of polymer fume particles, individual and agglomerated nanofillers, and perhaps other particles.[49,55] However, a sampling analysis of such processes did not reveal isolated MWCNTs.[47] Subsequent mechanical processes such as cutting, drilling, and sanding during manufacture or installation (*e.g.*, in construction) could additionally create particles that contain MWCNTs, as below and in Table 1.[6-9,36,41,47,48,50,53-55,61-66]

During the industrial and consumer use phases of the life cycle, environmental conditions and human activities can alter the MWCNT-polymer product. The use, characteristics, and magnitude of MWCNT-related release in response to these conditions depend strongly on the type of product and how much energy is applied to it. Again, the release potential for particles can range from zero or near-zero release due to physical barriers, to possibly significant release in applications with consistent high energy input (*e.g.*, tires).[3] Whether such particle release constitutes release of added nanomaterial in the "as added" form is subject to case-by-case evaluation. However, it is clear that for many applications currently on the market (*e.g.*, sports equipment, consumer electronics), consideration of the physical opportunity for release during use leads to the conclusion that release magnitude is anticipated to be very low.

Continuing the evaluation along the life cycle of the material, there is very little direct information to inform the behavior of MWCNT-composites, including those currently used in commerce as well as those in the disposal or recycling phase.[3,6] Established recycling schemes are available for some of the intended consumer uses of MWCNT-composites (*e.g.*, electronics waste recycling). Recycling or controlled disposal of large-scale structures, such as windmill blades or aircraft, is also likely. Release of MWCNTs from the matrix is again possible because recycling frequently involves high-energy process such as shredding, yet this is currently a largely unevaluated release scenario.

In contrast, release during waste incineration can be considered to be low given that MWCNTs should combust if incinerated at a sufficiently high temperature and provided sufficient oxygen during the process.[67] By well-ventilated incineration at temperatures above 1000°C, the carbon nanofibers[41,67] present in a composite were effectively destroyed in the flames, and the same is expected from the known decomposition temperatures of MWCNTs. If the MWCNTs should survive the incineration process, they would typically end up either in bottom ash and be landfilled, or would be carried away with the flue gas. An industrial-scale study on nanomaterials in municipal waste demonstrated that flue gas can be effectively filtered to be clear of nanoparticles[41]; therefore, remaining MWCNTs would end up in the filter ash, which is also landfilled.[68] Landfills in industrialized countries are not expected to release MWCNTs to the environment because current landfills are operated in a way that release of leachates to the environment is minimized.[69]

## 4. The Significance of MWCNT Physical Form Variation at Point of Release

The simplest and perhaps most widely held conception of MWCNT release is that free nanotubes would be released upon wear, abrasion, or degradation of a product with MWCNTs. However, a finding by many laboratories is that release of free MWCNTs is a rarity for the MWCNT-composite materials thus far evaluated.[7-9,36,40,47,48,53-55,65,70-82] The more frequently reported released entity is MWCNTs embedded in particles of the composite matrix. In fact, the nature of the MWCNT release expected for a life cycle stage can be explored by contrasting findings from studies of MWCNTs in epoxy (the most commonly studied composite) with other composites to identify emerging structure–activity relationships (SARs); in this case, "activity" refers to the probability to release fragments of different types, as highlighted in Table 1.

The systematic SAR is spanned by the chemical and mechanical axes, as shown in Figure 2. Such evaluation indicates that, unlike carbon fibers, the interfacial shear strength and length of MWCNTs predicts that they should not break under tensile stress but instead should either be pulled out from the matrix[83,84] or pull the matrix along, resulting in protrusions of MWCNTs emerging from polymer fragments. The tensile strength and bonding properties of the MWCNT would therefore tend to create a high likelihood of composite-MWCNT fragments as the dominant released moiety from MWCNT-composites studied to date. Indeed, as such qualitative SARs would predict, particles with protrusions are observed after sanding of matrices with less than 10% elongation at the break. This is true of epoxy[53,65] or cement,[8] but not the majority of thermoplastics.[8,9]

Journal of Physics: Conference Series **617** (2015) 012026 doi:10.1088/1742-6596/617/1/012026

**TABLE 1.** Areas for detection and identification before selective quantification can begin[a]

| Release Scenario (Anticipated Degradation Mechanism) | Release Pathway | Method to Simulate the Life Cycle to Induce, Detect, and Quantify Release | Morphology of Typical Released Fragments | | Key Findings and Research Needs |
|---|---|---|---|---|---|
| Outdoor use (UV degradation, hydrolysis, and weak mechanical forces) | **A** | Accelerated dry or wet weathering (ISO4892), followed by immersion to induce release into water, detected by AUC, TEM, EDX, LD, XPS | 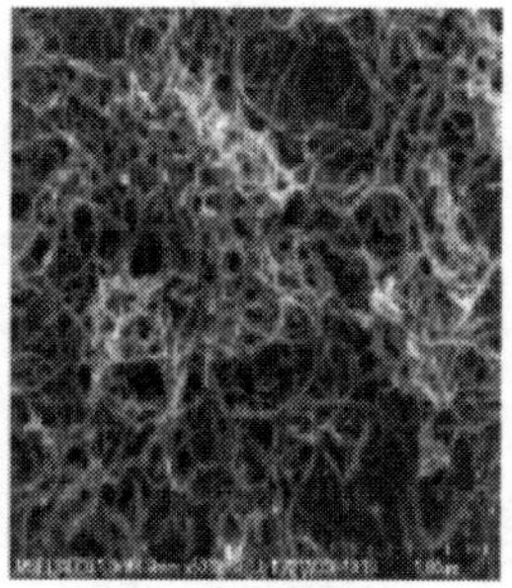 | Nguyen *et al.* found a collapsed layer of MWCNTs on MWCNT-epoxy after UV irradiation[36] | Weathering tests are highly standardized for plastics and coatings, but the MWCNT network formed on the composite essentially eliminates the release of MWCNT[36] Methods to induce a release during or after degradation are exploratory, and miss selective quantification By synergy of degradation and stress a key release scenario |
| | **B** | | 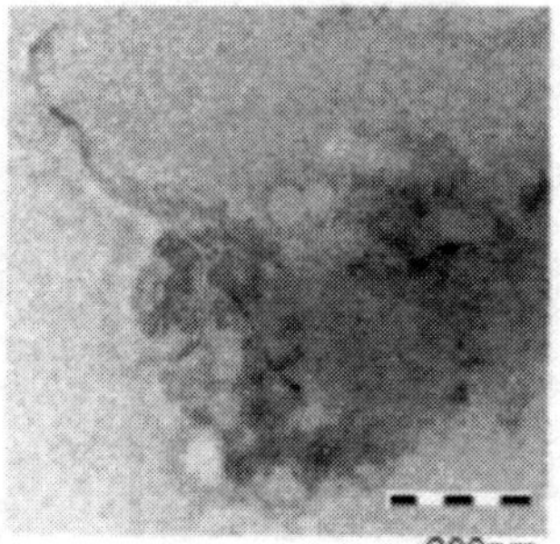 | Hirth *et al.* performed wet and dry weathering of MWCNT-polyurethane followed by immersion and shaking: release of polymer-MWCNT fragments, but free MWCNTs only by sonication[7] | |

| | | | | | |
|---|---|---|---|---|---|
| Machining by sanding, band-saw, abrasion, rotary cutting wheel, wet saw cutting, wet drilling, grinding. (moderate temperature + moderate to strong mechanical stresses: energy input per test[88] by sanding estimated is 10-fold higher than by sawing) | **F-H** | No standard simulation machines: monitoring by APS, CPC, SMPS, ESP, FMPS, LDPSA, NSAM, OPC, and sampling for characterization by SEM/TEM, EDX, AUC, LD, XPS | 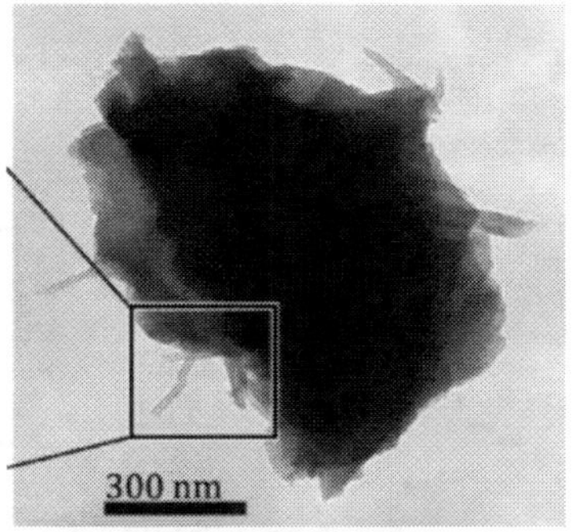 | Fragments from MWCNT-epoxy were submicrons in diameter and carried protrusions of MWCNTs that were pulled out from the matrix[48] | Many laboratory studies on different MWCNT-polymer and nanoparticle-polymer composites found no release of free nanofillers,[2-9,50,53,61-64] with exceptions linked to agglomerates in the polymer[54,55] Protrusions of MWCNTs were not observed for tough thermoplastics,[8,9] which can elongate during machining and keep MWCNTs embedded[7,65] |
| | **F-H** | |  | Fragments from MWCNT-polyoxymethylene measured microns in diameter with no protrusions[8] | |
| Machining by dry drilling or milling (moderate mechanical stress at elevated temperatures) | **E** | No standard simulation machines. Detect aerosols around machines by APS, CPC, Dust TrakTM, ESP, FMPS, TP, TSI | 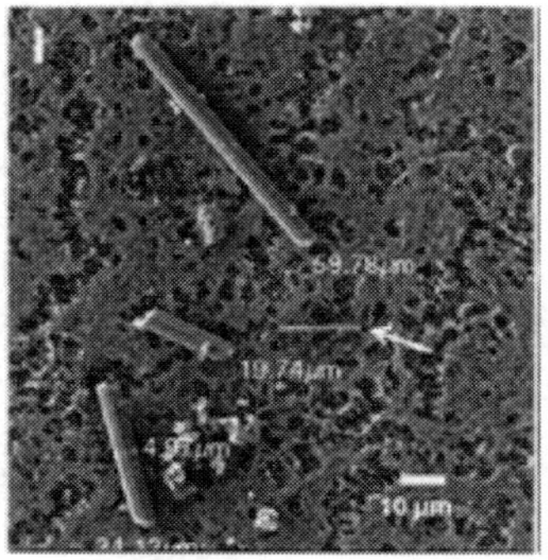 | Polymorph aerosol sampled during dry drilling of a graphite-epoxy prepreg with aligned carbon fibers and aligned MWCNTs[47] | By synergy of degradation and stress, machining with local overheating is a key release scenario |

Journal of Physics: Conference Series **617** (2015) 012026 doi:10.1088/1742-6596/617/1/012026

| | | | | | |
|---|---|---|---|---|---|
| Compounding by hot melt extrusion and Combustion (thermal degradation) | **D** | Commercial twin-screw extruders, monitored by CPC, SMPS, sampled for TEM/SEM, or TGA (ISO 11358) with off-gases monitored and sampled | 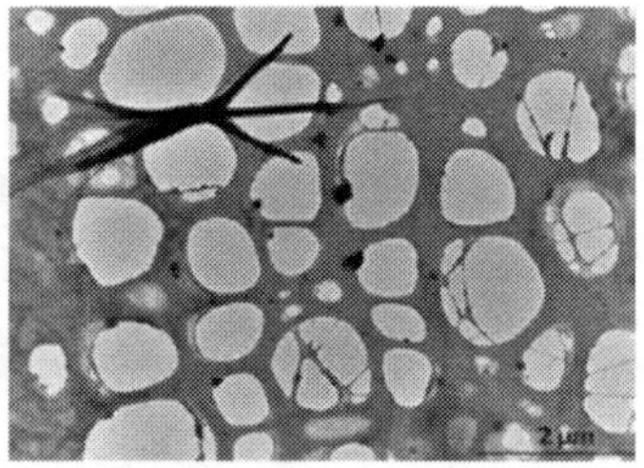 | Fleury *et al.* conducted extrusion of MWCNT-acrylonitrile-butadiene-styrene: complex polymer fumes, no free MWCNT[61] | At elevated temperatures, polymer vaporizes and re-condenses. Release of free MWCNTs was observed for thermal decomposition by TGA[66] and for dry-core drilling of MWCNT-epoxy composites,[47] but not for extrusion of MWCNT-polymer[61] |
| | **E** | | 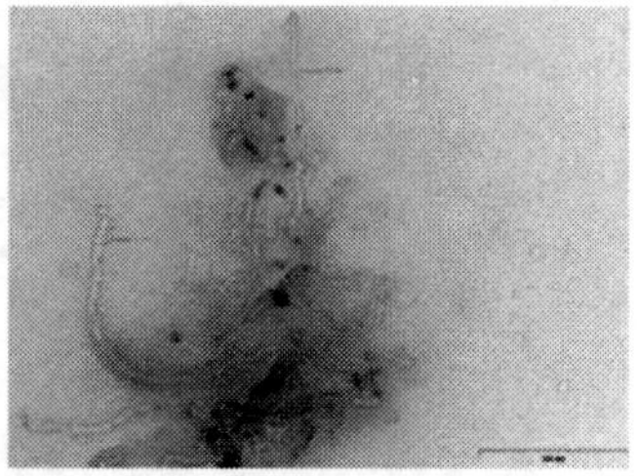 | Bouillard *et al.* found that low-temperature (400°C) combustion of the same composite can release MWCNT in flue gases[66] | Capacity to simulate real-world scenarios needs to be validated |
| Waste incineration (thermal degradation at controlled oxygen and temperatures above 1000°C) | **C** | Cone calorimeter ISO 5659-2 with off-gases monitored by CPC and sampled for TEM/SEM | 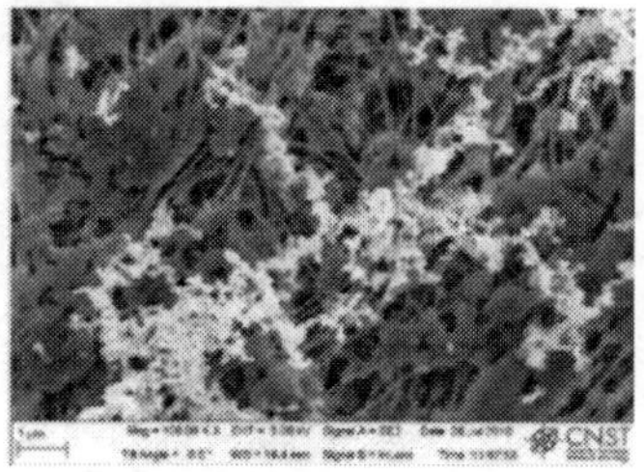 | Nyden *et al.* showed that fire smoke from carbon-nanofiber-polyurethanes is soot with no remaining fibers[6,41] | Existing reports carbon nanofiber-polymer find no free fibers in flue gas.[6,41] Release rate needs to be studied for end-of-life scenarios of MWCNT-polymer |

*Abbreviations:* APS, aerodynamic particle sizer; AUC, analytical ultracentrifugation; CPC, condensation particle counter; DLS, dynamic light scattering; EDX, energy dispersive X-ray analysis; ELPI, electrical low pressure impactor; ESP, electrostatic precipitator; FE-SEM, field emission scanning electron microscope; FMPS, fast mobility particle sizer; FTIR, Fourier transform infrared spectroscopy; ICP-MS, inductively coupled plasma mass spectrometry; LDPSA, laser diffraction particle size analyzer; LIBS, laser-induced breakdown spectroscopy; NAS, nano-aerosol sampler; OPC, optical particle counter; PAS, photoelectric aerosol sensor; PSSD, particle surface sensitive device (*e.g.*, NSAM, DiscMini, Nanocheck); SEM, scanning electron microscopy; SMPS, scanning mobility particle sizer; TEM, transmission electron microscopy ; TGA, thermogravimetric analyzer; TOF-SIMS, time-of-flight secondary ion mass spectrometry; TP, thermophoretic precipitator; UNPA, universal nano particle analyzer; WRASS, wide-range aerosol particle sampling system; XPS, X-ray photoemission spectroscopy.

[a]Images in this table are reprinted from the following sources in respective order: Nguyen *et al.*,[36] with permission from the authors; Hirth *et al.*,[7] with permission from Springer Science + Business Media; Bello *et al.*,[48] with permission from Springer Science + Business Media; Wohlleben *et al.*,[8] with permission from John Wiley and Sons; Bello *et al.*,[47] with permission from Maney Publishing; Fleury *et al.*,[61] with permission from Elsevier; Bouillard *et al.*,[66] with permission from Springer Science + Business Media; and Uddin and Nyden,[41] with permission from the Nano Science Technology Institute.

One issue to address in evaluation of the data regarding presence or absence of free MWCNTs in released material is that detection may be made difficult by the extreme polydispersity of particles, with typical particle sizes ranging from tens of micrometers for an elastic material such as polyurethane[9] to around 100 nm for epoxies[53,65] or acrylic coatings.[62] However, visual inspection of multiple scanning electron microscopy images typically reveals no free MWCNTs from machining of epoxy.[48,76] This is in contrast to detection for stiffer, longer fillers (carbon nanofiber)[76] and MWCNT agglomerates in epoxy composites sampled on filtered air grids.[54,55] The examination of released material from composites with two- and three-dimensional nanofillers (clays, pigments) began with similar expectations of a relationship between nanoparticle size, shear force, and bonding to composite. Again, no evidence of significant free particulate nanofillers has been reported.[5,49] Therefore, the available studies indicate that mechanical forces applied to MWCNT-composite materials tend to release matrix particles containing MWCNTs, with rare occurrence of unbound MWCNTs. The entity to which exposure occurs is thus expected to be a heterogeneous mixture of particles with embedded or protruding MWCNTs, for which very few toxicity studies have been conducted.

Conditions that may cause the release of free MWCNTs have been observed. Airborne clusters of unbound MWCNTs, for example, were observed during machining of MWCNT-epoxy with dry solid-core drilling accompanied (or caused) by thermal degradation of the matrix material,[47] which highlights the need for additional studies of synergistic stresses. Among the purely chemical degradation processes, the rate of matrix removal — and thus the potential rate of liberation of embedded MWCNTs — increases in the order of photolysis, hydrolysis (outdoor weathering), thermal degradation, combustion (accidental fires), and incineration (municipal waste treatment). Again, complete oxidation and destruction of MWCNTs is expected for incineration.[41,67]

Conditions that lead to apparent potential for release of free MWCNTs were confirmed for weathering, in which the matrix is hydrolyzed or oxidized from the top few hundred nanometers, leaving behind an entangled network of MWCNTs.[36,85] However, further research regarding life cycle release is also indicated by the finding that the same matt of fibers may act as an ultraviolet-protective layer that slows down the further polymer degradation and release.[36,39] It appears that, in contrast to particulate nanofillers that are releasable by run-off waters (*e.g.*, $TiO_2$[86] or $SiO_2$[36]), high aspect ratio MWCNTs collapse to an entangled network that minimizes their release unless ultrasonic shear forces overcome van der Waals forces.[7] However, the effects of solvents, acids, and bases such as those relevant for food contact or landfilling have not been studied.

## 5. Quantifying the Dose or Released Material Associated with Added Nanomaterial

The observations that a polydisperse mixture of embedded, free, and composite particles (Figure 2) is released and that the nature of the proportions among these types of release varies with release conditions raise the question of what should be measured and reported during the sustainable design and use of materials. What should be measured to assure that a "risk-relevant" release has not occurred? What should be measured to quantify dose in an exposure pathway relevant to the added nanomaterial at any point in the manufacture of the composite? Again, there is a limited amount of quantitative data available on release rates and mass flows relevant to evaluating materials of toxicological or epidemiological interest for modeling of exposure pathways.[30] Moreover, even the very first stages of problem formulation for risk assessment are made difficult due to the expected heterogeneity, other than for cases in which release can be physically precluded. It seems clear that a first step toward understanding what to measure will depend on a standardized description of release processes and released materials for scenarios of interest. Only then can standard methods be developed to measure the releases in those scenarios.

On an empirical level with the current state of research, the understanding of nanomaterial-specific release is also hindered by a paucity of attempts to quantitatively compare material released from nanomaterial composites to material released from nanomaterial-free composites. For example, for MWCNT-polymer composites, few studies[7] have attempted to develop quantitative data about

MWCNT release that go further than particle size distributions and examples of morphology seen through relatively few electron microscope grids. These data are not sufficient to estimate the dose of the MWCNT-specific addition. Furthermore, methods that seem applicable to nanomaterial-specific dose estimation would require tremendous effort to sample and evaluate morphology and composition of released particles. Without methodology to quantify differential dose for the addition of the nanomaterial, we cannot estimate the added risk of using the nanomaterial. We also cannot match toxicity data for the pristine nanomaterial to what is released in ways that inform needs for further toxicity data.

Risk management is still possible if the release can be characterized within worst-case boundaries relevant to exposure. For example, release scenarios in which there is a very low physical expectation of release could be verified through evaluation of the total mass of carbon released compared to reference standards for pure MWCNTs.[87,88] For this reason, the highest near-term practical priority for methods would be semi-quantitative release detection to provide upper bounds on the magnitude of release to assure sustainable development of many promising nanomaterial uses.

## 6. Required Research, Methods Development, and Standardization

While risk management based on surrogate measures and physical limits to exposure is possible, practical, quantitative measurement methods that would allow full risk assessment of uses of nanomaterial composites do not exist. Such methods must address the complexities of morphology, composition, and matrix-nanomaterial interactions, as well as the media into which release occurs in relation to those characteristics of the nanomaterials that are biologically deleterious.[89,90] In addition to variations in polymer formulations such as antioxidants and UV stabilizers, MWCNT formulations are often optimized for a specific polymer matrix through the modification or addition of functional groups on the nanotube surface. The multivariate sequences between nanomaterial addition and release create a formidable challenge for prioritizing efforts to address these methods requirements. Critical challenges include the following: i) filling vast knowledge gaps around nanocomposites and their potential degradation routes over life cycles that allow for nanomaterial release; ii) identifying a representative set of homogeneous and well-characterized materials that can serve as reasonable surrogates or reference materials of known degradation processes; iii) validating processes to develop composites that are also representative of systems in use; iv) establishing validated and eventually standardized methods and broadly available and applicable instrumentation for sampling, detection, and characterization of MWCNT release; and v) leveraging informatics and computational methodologies to allow for prediction of MWCNT release in the absence of empirical data.

For MWCNT measurement alone, there is a lack of validated, quantitative, and statistically relevant methods for measuring and reporting the characteristics of MWCNTs that are known to affect toxicity, such as MWCNT size and size distribution, morphology, surface functionalization, purity, tube alignment of the MWCNTs, and detection of the presence of different carbon allotropes.[91, 92] The most widely reported techniques for detecting CNTs and determining their morphology and size are scanning electron microscopy and transmission electron microscopy; however, these methods are semi-quantitative at best and cannot be used to provide representative results needed for science-based risk assessment. With regard to other key characteristics such as surface functionalization, methods such as X-ray photoelectronic spectroscopy, Raman spectroscopy,[93] and inductively coupled plasma mass spectrometry are sensitive analytic techniques that can elucidate specific elemental composition and physical structure; however, these techniques are not currently widely utilized apart from specific application evaluations.

For innovation to proceed with regard to safety for these valuable materials, there is an immediate need for the scientific community to adopt a systematic strategy for evaluating the critical determinants of release as well as a set of reference materials, sampling methods, and MWCNT-composites that can be used as a first step toward informing the release potential of nanotubes from MWCNT-nanocomposites. Testing materials and polymer systems selected as study systems should be as commercially relevant as possible and should preferably cover a representative range of characteristics for that material class. Laboratory testing strategies should strive to incorporate the multiple forms in which the MWCNT is found throughout the life cycle, including the pristine MWCNT (as synthesized), the formulated MWCNT (embedded in composite), and the transformed MWCNT (affected by environmental processes).[6]

As a result of the NanoRelease initiative, draft protocols to conduct sanding and weathering of nanofiller-polymer composites were presented to ISO TC/229 for consideration at the November 2013 meeting in Belo Horizonte, Brazil. Comparative testing by these harmonized protocols applied to a single batch of MWCNT polymer nanocomposites was initiated in November 2014 in laboratories across Europe, North America and Asia.

## 7. Conclusions

This state-of-the-science review for MWCNTs in polymers highlights the disconnect between as-produced MWCNTs with much studied hazards and the more realistic polydisperse fragments released from nano-enabled products made of MWCNT-composites. While releases do not necessarily correlate to the quantity of exposure because there may be other mitigating factors, the quantity and characteristics of the materials released from matrices containing nanomaterials do need to be addressed during risk assessment and risk management, including assessment of the mechanism of release for each phase of the product life cycle. Specifically, two scenarios were previously classified as having a greater possibility of release[3]: i) use of MWCNTs in tires and ii) recycling operations.

Despite a lack of best practices, and thus harmonization in measurements, measurement strategies, analysis, and reporting of measurement data, we were able to identify unifying principles across polymer classes. For most materials, degradation of the polymer matrix is associated with the greatest potential for release, with degradation and release rates dependent on the specific characteristics of the polymer, MWCNT, and environment. MWCNT-nanocomposites, as well as their susceptibility to degradation, may transform throughout the product life cycle.[4] Release during designed consumer use is projected to be quite low and composed of polydisperse fragments with only a small fraction of free MWCNTs. Direct handling of products containing nanomaterials could result in exposure; however, laboratory simulation of mechanical handling has demonstrated very low probability of release. Environmental releases of MWCNTs will undergo various reactions that could result in their destruction or alteration. In addition, misuse of products may expose MWCNT-nanocomposites to chemical environments that could alter rates of degradation. As such, we expected that thermolysis would represent a low potential for release given the increased thermal stability of many polymers with the addition of MWCNTs into the matrix, but available studies report divergent results.

We have identified an immediate need to use realistic release scenarios as the basis for coordinated research on nanotoxicology, occupational health and safety, and exposure assessments. A qualitative life cycle analysis or similar approach should be implemented to support sustainable nanotechnologies and inform risk management decisions. Tiered approaches based on releases measured by validated methods may provide practical ways to manage risk, with a relatively short timeline to reach predictive knowledge about nanomaterial release from composites. A benefit of these efforts may be the generation of data that inform improved design of products containing nanomaterials from performance, safety, and sustainability perspectives.

## Acknowledgments

The authors thank Debra Kaiser, Richard Zepp, and Aleksandre Stefaniak for review of the final drafts of this paper. Experts were convened and concepts were developed for this paper by the NanoRelease Consumer Products Steering Committee.[2] NanoRelease is funded by the US Environmental Protection Agency (EPA), the American Chemistry Council Nanotechnology Panel, Environment Canada, Health Canada, the American Cleaning Institute, the Society of Organic Chemical Manufacturers and Affiliates, the Adhesives and Sealant Council, and the ILSI Research Foundation. More than 60 experts listed on the NanoRelease Consumer Products website from government, academia, industry, and civil society organizations have also contributed time and expertise in support of the project. This state of the science evaluation considers publications and active MWCNT release measurement projects known to the authors as of November 2013. We especially thank Lyubov Tsytsikova, Center for Risk Science Innovation and Application of ILSI Research Foundation, for her contributions to this work under the NanoRelease project, and Christina West, for editorial assistance in completing this manuscript. This article has been reviewed in accordance with the US EPA's peer and administrative review policies and approved for publication. Mention of trade names or commercial products does not constitute an endorsement or recommendation for use by the US EPA. S. H. acknowledges support from the National Institute of Environmental Health Sciences (Grants ES017552-01A2, P30ES03850, and ES0166896-01) and the Air Force Research Laboratory (Grant FA8650-05-1-15041).

**References**

1. Froggett, S. J.; Clancy, S. F.; Boverhof, D. R.; Canady, R. A. A review and perspective of existing research on the release of nanomaterials from solid nanocomposites. *Part Fibre Toxico*l **2014**, 11, 17.
2. International Life Sciences Institute. ILSI Research Foundation NanoRelease Consumer Products. http://www.ilsi.org/ResearchFoundation/RSIA/pages/NanoRelease1.aspx (accessed August 10, 2013).
3. Nowack, B.; David, R. M.; Fissan, H.; Morris, H.; Shatkin, J. A.; Stintz, M.; Zepp, R.; Brouwer, D. Potential release scenarios for carbon nanotubes used in composites. *Environ. Int.* **2013**, *59*, 1.
4. Kingston, C.; Zepp, R.; Andrady, A.; Boverhof, D.; Fehir, R.; Hawkins, D.; Roberts, J.; Sayre, P.; Shelton, B.; Sultan, Y., et al. Release characteristics of selected carbon nanotube polymer composites. *Carbon* **2014**, *68*, 33-57.
5. Kuhlbusch, T. A.; Asbach, C.; Fissan, H.; Gohler, D.; Stintz, M. Nanoparticle exposure at nanotechnology workplaces: a review. *Part. Fibre. Toxicol.* **2011**, *8*, 22.
6. Petersen, E. J.; Zhang, L.; Mattison, N. T.; O'Carroll, D. M.; Whelton, A. J.; Uddin, N.; Nguyen, T.; Huang, Q.; Henry, T. B.; Holbrook, R. D., et al. Potential release pathways, environmental fate, and ecological risks of carbon nanotubes. *Environ. Sci. Technol.* **2011**, *45*, 9837-9856.
7. Hirth, S.; Cena, L.; Cox, G.; Tomovic, Z.; Peters, T.; Wohlleben, W. Scenarios and methods that induce protruding or released CNTs after degradation of nanocomposite materials. *J. Nanopart. Res.* **2013**, *15*, 1-15.
8. Wohlleben, W.; Brill, S.; Meier, M. W.; Mertler, M.; Cox, G.; Hirth, S.; von Vacano, B.; Strauss, V.; Treumann, S.; Wiench, K., et al. On the lifecycle of nanocomposites: comparing released fragments and their in-vivo hazards from three release mechanisms and four nanocomposites. *Small* **2011**, *7*, 2384-2395.
9. Wohlleben, W.; Meier, M. W.; Vogel, S.; Landsiedel, R.; Cox, G.; Hirth, S.; Tomovic, Z. Elastic CNT-polyurethane nanocomposite: synthesis, performance and assessment of fragments released during use. *Nanoscale* **2013**, *5*, 369-380.
10. David, R. M. Measuring engineered nanomaterials in the environment: a consortium view of how to address the problem. *Environ. Eng. Sci.* **2013**, *30*, 97-100.
11. McLeod, D. Nanotechnology's many unknowns raise risk management and liability concerns: environmental damage, bodily injury claims could materialize. www.businessinsurance.com/article/20130224/NEWS07/302249985# (accessed September 26, 2013).
12. Sargent, J. F. J. *Nanotechnology and Environmental, Health, and Safety: Issues for Consideration*. Congressional Research Service: Washington, DC, 2011.
13. US Government Accountability Office. *Nanotechnology: Nanomaterials Are Widely Used in Commerce but EPA Faces Challenges in Regulating Risk.* U.S. Government Accountability Office: Washington, DC, 2010; Vol. GAO-10-549.
14. Nowack, B.; Ranville, J. F.; Diamond, S.; Gallego-Urrea, J. A.; Metcalfe, C.; Rose, J.; Horne, N.; Koelmans, A. A.; Klaine, S. J. Potential scenarios for nanomaterial release and subsequent alteration in the environment. *Environ. Toxicol. Chem.* **2012**, *31*, 50-59.
15. Westerhoff, P.; Nowack, B. Searching for global descriptors of engineered nanomaterial fate and

transport in the environment. *Acc. Chem. Res.* **2013**, *46*, 844-853.
16. Burello, E.; Worth, A. P. QSAR modeling of nanomaterials. *Wiley. Interdiscip. Rev. Nanomed. Nanobiotechnol.* **2011**, *3*, 298-306.
17. Sun, B.; Li, R.; Wang, X.; Xia, T. Predictive toxicological paradigm and high throughput approach for toxicity screening of engineered nanomaterials. *Int. J. Biomed. Nanosci. Nanotechnol.* **2013**, *3*, 4-18.
18. International Council on Nanotechnology. Nano-EHS Database Analysis Tool. http://icon.rice.edu/report.cfm (accessed August 6, 2013).
19. Schnorr, J. M.; Swager, T. M. Emerging applications of carbon nanotubes. *Chem. Mater.* **2010**, *23*, 646-657.
20. Terrones, M. Science and technology of the twenty-first century: synthesis, properties, and applications of carbon nanotubes. *Annu. Rev. Mater. Res.* **2003**, *33*, 419-501.
21. Nanoposts.com. The global market for carbon nanotubes to 2015: A realistic assessment. http://www.Nanoposts.com (accessed July 2013).
22. Andrews, R.; Weisenberger, M. C. Carbon nanotube polymer composites. *Curr. Opin. Sol. State Mat. Sci.* **2004**, *8*, 31-37.
23. Baughman, R. H.; Zakhidov, A. A.; de Heer, W. A. Carbon nanotubes--the route toward applications. *Science* **2002**, *297*, 787-792.
24. Breuer, O.; Sundararaj, U. Big returns from small fibers: a review of polymer/carbon nanotube composites. *Polym. Compos.* **2004**, *25*, 630-645.
25. Coleman, J. N.; Khan, U.; Blau, W. J.; Gun'ko, Y. K. Small but strong: A review of the mechanical properties of carbon nanotube–polymer composites. *Carbon* **2006**, *44*, 1624-1652.
26. Coleman, J. N.; Khan, U.; Gun'ko, Y. K. Mechanical reinforcement of polymers using carbon nanotubes. *Adv. Mater.* **2006**, *18*, 689-706.
27. Endo, M.; Strano, M. S.; Ajayan, P. M. Potential applications of carbon nanotubes. *Top. Appl. Phys.* **2008**, *111*, 13-61.
28. Harris, P. J. F. Carbon nanotube composites. *Int. Mater. Rev.* **2004**, *49*, 31-43.
29. Wick, P.; Clift, M. J.; Rosslein, M.; Rothen-Rutishauser, B. A brief summary of carbon nanotubes science and technology: a health and safety perspective. *ChemSusChem* **2011**, *4*, 905-911.
30. Clark, K.; Tongeren, M.; Christensen, F.; Brouwer, D.; Nowack, B.; Gottschalk, F.; Micheletti, C.; Schmid, K.; Gerritsen, R.; Aitken, R., et al. Limitations and information needs for engineered nanomaterial-specific exposure estimation and scenarios: recommendations for improved reporting practices. *J. Nanopart. Res.* **2012**, *14*, 1-14.
31. Rillig, M. C. Microplastic in terrestrial ecosystems and the soil? *Environ. Sci. Technol.* **2012**, *46*, 6453-6454.
32. von Moos, N.; Burkhardt-Holm, P.; Kohler, A. Uptake and effects of microplastics on cells and tissue of the blue mussel Mytilus edulis L. after an experimental exposure. *Environ. Sci. Technol.* **2012**, *46*, 11327-11335.
33. Jones-Otazo, H. A.; Clarke, J. P.; Diamond, M. L.; Archbold, J. A.; Ferguson, G.; Harner, T.; Richardson, G. M.; Ryan, J. J.; Wilford, B. Is house dust the missing exposure pathway for PBDEs? An analysis of the urban fate and human exposure to PBDEs. *Environ. Sci. Technol.* **2005**, *39*, 5121-5130.
34. Webster, T. F.; Harrad, S.; Millette, J. R.; Holbrook, R. D.; Davis, J. M.; Stapleton, H. M.; Allen, J. G.; McClean, M. D.; Ibarra, C.; Abdallah, M. A.-E. Identifying transfer mechanisms and sources of decabromodiphenyl ether (BDE 209) in indoor environments using environmental forensic microscopy. *Environ. Sci. Technol.* **2009**, *43*, 3067-3072.
35. Institute of Nanotechnology. The world market for nanocomposites. http://www.nano.org.uk/nanotechnology-reports/132 (accessed August 10, 2013).
36. Nguyen, T.; Pellegrin, B.; Bernard, C.; Gu, X.; Gorham, J. M.; Stutzman, P.; Stanley, D.; Shapiro, A.; Byrd, E.; Hettenhouser, R. Fate of nanoparticles during life cycle of polymer

nanocomposites. *J. Phys. Conf. Ser.* **2011**, *304*, 012060.
37. Nguyen, T.; Pelligrin, B.; Mermet, L.; Gu, X.; Shapiro, A.; Chin, J.; Reichert, T. Degradation and nanofiller release of polymer nanocomposites exposed to ultraviolet radiation. *Nat. Artif. Ageing Polym.* **2009**, 149-162.
38. Nguyen, T.; Pelligrin, B.; Mermet, L.; Shapiro, A.; Gu, X.; Chin, J. Network aggregation of CNTs at the surface of epoxy/MWCNT composite exposed to UV radiation. In *Nanotechnology 2009: Fabrication, Particles, Characterization, MEMS, Elec tronics and Phototonics*, CRC Press, Nano Science and Technology Institute Houston, TX, 2009; Vol. 1, pp 90-93.
39. Asmatulu, R.; Mahmud, G. A.; Hille, C.; Misak, H. E. Effects of UV degradation on surface hydrophobicity, crack, and thickness of MWCNT-based nanocomposite coatings. *Prog. Org. Coat.* **2011**, *72*, 553-561.
40. Busquets-Fite, M.; Fernandez, E.; Janer, G.; Vilar, G.; Vazquez-Campos, S.; Zanasca, R.; Citterio, C.; Mercante, L.; Puntes, V. Exploring release and recovery of nanomaterials from commercial polymeric nanocomposites. *J. Phys. Conf. Ser.* **2013**, *429*, 012048.
41. Uddin, N. M.; Nyden, M. R. Characterization of nanoparticle release from polymer nanocomposites due to fire. In *Nanotechnology 2011: Advances Materials, CNTs, Particles, Films and Composites*, Nano Science Technology Institute: Danville, CA, 2011; pp 523-526.
42. Liu, X.; Hurt, R. H.; Kane, A. B. Biodurability of single-walled carbon nanotubes depends on surface functionalization. *Carbon N Y* **2010**, *48*, 1961-1969.
43. Liu, T.; Guo, S. Properties of polyurethane/carbon nanotube nanocomposites. In *Polymer Nanotube Nanocomposites: Synthesis, Properties and Applications*, Scrivener: Salem, MA, 2010; pp 141-176.
44. Kagan, V. E.; Konduru, N. V.; Feng, W.; Allen, B. L.; Conroy, J.; Volkov, Y.; Vlasova, II; Belikova, N. A.; Yanamala, N.; Kapralov, A., et al. Carbon nanotubes degraded by neutrophil myeloperoxidase induce less pulmonary inflammation. *Nat Nanotechnol* **2010**, *5*, 354-359.
45. Ashter, A.; Tsai, S. J.; Lee, J. S.; Ellenbecker, M. J.; Mead, J. L.; Barry, C. F. Effects of nanoparticle feed location during nanocomposite compounding. *Polymer Engineering & Science* **2010**, *50*, 154-164.
46. Tsai, S. J.; Ashter, A.; Ada, E.; Mead, J. L.; Barry, C. F.; Ellenbecker, M. J. Airborne nanoparticle release associated with the compounding of nanocomposites using nanoalumina as fillers. *Aerosol Air Qual Res* **2008**, *8*, 160-177.
47. Bello, D.; Wardle, B. L.; Zhang, J.; Yamamoto, N.; Santeufemio, C.; Hallock, M.; Virji, M. A. Characterization of exposures to nanoscale particles and fibers during solid core drilling of hybrid carbon nanotube advanced composites. *Int. J. Occup. Environ. Health.* **2010**, *16*, 434-450.
48. Bello, D.; Wardle, B.; Yamamoto, N.; Guzman deVilloria, R.; Garcia, E.; Hart, A.; Ahn, K.; Ellenbecker, M.; Hallock, M. Exposure to nanoscale particles and fibers during machining of hybrid advanced composites containing carbon nanotubes. *J. Nanopart. Res.* **2009**, *11*, 231-249.
49. Raynor, P. C.; Cebula, J. I.; Spangenberger, J. S.; Olson, B. A.; Dasch, J. M.; D'Arcy, J. B. Assessing potential nanoparticle release during nanocomposite shredding using direct-reading instruments. *J. Occup. Environ. Hyg.* **2012**, *9*, 1-13.
50. Gohler, D.; Stintz, M.; Hillemann, L.; Vorbau, M. Characterization of nanoparticle release from surface coatings by the simulation of a sanding process. *Ann. Occup. Hyg.* **2010**, *54*, 615-624.
51. Koponen, I. K.; Jensen, K. A.; Schneider, T. Comparison of dust released from sanding conventional and nanoparticle-doped wall and wood coatings. *J Expo Sci Environ Epidemiol* **2011**, *21*, 408-418.
52. Saber, A. T.; Jacobsen, N. R.; Mortensen, A.; Szarek, J.; Jackson, P.; Madsen, A. M.; Jensen, K. A.; Koponen, I. K.; Brunborg, G.; Gutzkow, K. B., et al. Nanotitanium dioxide toxicity in mouse lung is reduced in sanding dust from paint. *Part Fibre Toxicol* **2012**, *9*, 4.
53. Cena, L. G.; Peters, T. M. Characterization and control of airborne particles emitted during production of epoxy/carbon nanotube nanocomposites. *J. Occup. Environ. Hyg.* **2011**, *8*, 86-92.
54. Huang, G.; Park, J.; Cena, L.; Shelton, B.; Peters, T. Evaluation of airborne particle emissions

from commercial products containing carbon nanotubes. *J. Nanopart. Res.* **2012**, *14*, 1-13.
55. Golanski, L.; Guiot, A.; Pras, M.; Malarde, M.; Tardif, F. Release-ability of nano fillers from different nanomaterials (toward the acceptability of nanoproduct). *J. Nanopart. Res.* **2012**, *14*, 1-9.
56. Kohler, A. R.; Som, C.; Helland, A.; Gottschalk, F. Studying the potential release of carbon nanotubes throughout the application life cycle. *J Clean Prod* **2008**, *16*, 927-937.
57. Som, C.; Berges, M.; Chaudhry, Q.; Dusinska, M.; Fernandes, T. F.; Olsen, S. I.; Nowack, B. The importance of life cycle concepts for the development of safe nanoproducts. *Toxicology* **2010**, *269*, 160-169.
58. Upadhyayula, V. K. K.; Meyer, D. E.; Curran, M. A.; Gonzalez, M. A. Life cycle assessment as a tool to enhance the environmental performance of carbon nanotube products: a review. *J. Cleaner Prod.* **2012**, *26*, 37-47.
59. Tsai, S.-J.; Hofmann, M.; Hallock, M.; Ada, E.; Kong, J.; Ellenbecker, M. Characterization and evaluation of nanoparticle release during the synthesis of single-walled and multiwalled carbon nanotubes by chemical vapor deposition. *Environ. Sci. Technol.* **2009**, *43*, 6017-6023.
60. Yeganeh, B.; Kull, C. M.; Hull, M. S.; Marr, L. C. Characterization of airborne particles during production of carbonaceous nanomaterials. *Environ. Sci. Technol.* **2008**, *42*, 4600-4606.
61. Fleury, D.; Bomfim, J. A. S.; Vignes, A.; Girard, C.; Metz, S.; Muñoz, F.; R'Mili, B.; Ustache, A.; Guiot, A.; Bouillard, J. X. Identification of the main exposure scenarios in the production of CNT-polymer nanocomposites by melt-moulding process. *J. Cleaner Prod.* **2013**, *53*, 22-36.
62. Koponen, I. K.; Jensen, K. A.; Schneider, T. Comparison of dust released from sanding conventional and nanoparticle-doped wall and wood coatings. *J. Exposure Sci. Environ. Epidemiol.* **2010**, *21*, 408-418.
63. Le Bihan, O.; Shandilya, N.; Gheerardyn, L.; Guillon, O.; Dore, E.; Morgeneyer, M. Investigation of the release of particles from a nanocoated product. *Adv. Nanopart.* **2013**, *2*, 39-44.
64. Göhler D.; Nogowski A.; Fiala P.; Stintz M. Nanoparticle release from nanocomposites due to mechanical treatment at two stages of the life-cycle. *J. Phys. Conf. Ser.* **2013**, *429*, 012045.
65. Schlagenhauf, L.; Chu, B. T.; Buha, J.; Nuesch, F.; Wang, J. Release of carbon nanotubes from an epoxy-based nanocomposite during an abrasion process. *Environ. Sci. Technol.* **2012**, *46*, 7366-7372.
66. Bouillard, J.; R'Mili, B.; Moranviller, D.; Vignes, A.; Le Bihan, O.; Ustache, A.; Bomfim, J. S.; Frejafon, E.; Fleury, D. Nanosafety by design: risks from nanocomposite/nanowaste combustion. *J. Nanopart. Res.* **2013**, *15*, 1-11.
67. Mansfield, E.; Kar, A.; Hooker, S. A. Applications of TGA in quality control of SWCNTs. *Anal. Bioanal. Chem.* **2010**, *396*, 1071-1077.
68. Walser, T.; Limbach, L. K.; Brogioli, R.; Erismann, E.; Flamigni, L.; Hattendorf, B.; Juchli, M.; Krumeich, F.; Ludwig, C.; Prikopsky, K., et al. Persistence of engineered nanoparticles in a municipal solid-waste incineration plant. *Nat. Nanotechnol.* **2012**, *7*, 520-524.
69. Mueller, N. C.; Buha, J.; Wang, J.; Ulrich, A.; Nowack, B. Modeling the flows of engineered nanomaterials during waste handling. *Environ. Sci. Process Impact* **2013**, *15*, 251-259.
70. Bouillard, J. X.; Vignes, A.; Rmili, B. In *Nanowastes: risk assessment from the end-of-life combustion of nanomaterials*, Grenoble, France, International Conference on Safe Production and Use of Nanomaterials (NANOSAFE): Grenoble, France, **2010**.
71. Dahm, M. M.; Evans, D. E.; Schubauer-Berigan, M. K.; Birch, M. E.; Deddens, J. A. Occupational exposure assessment in carbon nanotube and nanofiber primary and secondary manufacturers: mobile direct-reading sampling. *Ann. Occup. Hyg.* **2013**, *57*, 328-344.
72. Golanski, L.; Guiot, A.; Tardif, F. New method for the characterization of abrasion-induced nanoparticle release into air from nanomaterials. In *Nanotechnology 2010: Advanced Materials, CNTs, Particles, Films and Composites*, NSTI: **2010**; pp 720-723.
73. Gupta, A.; Gaspar, D.; Yost, M.; Gross, G.; Rempes, P.; Clark, M.; Martin, J. In *Evaluating the potential for release of carbon nanotubes and subsequent occupational exposure during*

*processing of a nanocomposite*, Nanotechnology Occupational and Environmental Health and Safety, **2006**; pp 110-111.

74. Hellmann, A.; Schmidt, K.; Ripperger, S.; Berges, M.; . Freisetzung ultrafeiner Stäube bei der mechanischen Bearbeitung von Nanokompositen: Release of ultrafine dusts during the machining of nanocomposites. *Gefahrstoffe-Reinhaltung der Luft* **2012**, *72*, 473.
75. Meng, H.; Sui, G. X.; Xie, G. Y.; Yang, R. Friction and wear behavior of carbon nanotubes reinforced polyamide 6 composites under dry sliding and water lubricated condition. *Compos. Sci. Technol.* **2009**, *69*, 606-611.
76. Methner, M.; Crawford, C.; Geraci, C. Evaluation of the potential airborne release of carbon nanofibers during the preparation, grinding, and cutting of epoxy-based nanocomposite material. *J. Occup. Environ. Hyg.* **2012**, *9*, 308-318.
77. Nguyen, T. In *Quantitative studies of photo-induced surface accumulation and release of nanoparticles in polymer nanocomposites*, Grenoble, France, Nanosafe: Grenoble, France, 2012.
78. Ogura, I.; Kotake, M.; Shigeta, M.; Uejima, M.; Saito, K.; Hashimoto, N.; Kishimoto, A. Potential release of carbon nanotubes from their composites during grinding. *J. Phys. Conf. Ser.* **2013**, *429*, 012049.
79. Ono-Ogasawara, M.; Myojo, T. A proposal of method for evaluating airborne MWCNT concentration. *Ind. Health* **2011**, *49*, 726-734.
80. Orhan, T.; Isitman, N. A.; Hacaloglu, J.; Kaynak, C. Thermal degradation of organophosphorus flame-retardant poly(methyl methacrylate) nanocomposites containing nanoclay and carbon nanotubes. *Polym. Degrad. Stab.* **2012**, *97*, 273-280.
81. Schutz, J. A.; Morris, H. *Investigating the Emissions of Nanomaterials from Composites and Other Solid Articles during Machining Processes*. CSIRO: Clayton, South Victoria, Australia, 2013.
82. Scott, K.; Marvel, C.; Stranick, S.; Scott, J. H. In *Microscopy-based detection and analysis of carbon nanomaterials in commercially available baseball bats*, Nanotech Conference and Expo, Santa Clara, CA, June 19, 2012; Santa Clara, CA, **2012**.
83. Coleman, J. N.; Khan, U.; Blau, W. J.; Gun'ko, Y. K. Small but strong: a review of the mechanical properties of carbon nanotube-polymer composites. *Carbon* **2006**, *44*, 1624-1652.
84. Lachman, N.; Daniel Wagner, H. Correlation between interfacial molecular structure and mechanics in CNT/epoxy nano-composites. *Composites, Part A* **2010**, *41*, 1093-1098.
85. Bernard, C.; Nguyen, T.; Pellegrin, B.; Holbrook, R. D.; Zhao, M.; Chin, J. Fate of graphene in polymer nanocomposite exposed to UV radiation. *J. Phys. Conf. Ser.* **2011**, *304*, 012063.
86. Kaegi, R.; Ulrich, A.; Sinnet, B.; Vonbank, R.; Wichser, A.; Zuleeg, S.; Simmler, H.; Brunner, S.; Vonmont, H.; Burkhardt, M., et al. Synthetic TiO2 nanoparticle emission from exterior facades into the aquatic environment. *Environ. Pollut.* **2008**, *156*, 233-239.
87. NIOSH. *Occupational Exposure to Carbon Nanotubes and Nanofibers*. National Institute for Occupational Health and Safety: Cincinnati, OH, **2013**.
88. Pauluhn, J. Multi-walled carbon nanotubes (Baytubes®): Approach for derivation of occupational exposure limit. *Reg. Toxicol. Pharmacol.* **2010**, *57*, 78-89.
89. Plata, D. L.; Ferguson, P. L.; Westerhoff, P. Express it in numbers: efforts to quantify engineered nanoparticles in environmental matrices advance. *Environ. Sci. Technol.* **2012**, *46*, 12243-12245.
90. Plata, D. L.; Reddy, C. M.; Gschwend, P. M. Thermogravimetry-mass spectrometry for carbon nanotube detection in complex mixtures. *Environ. Sci. Technol.* **2012**, *46*, 12254-12261.
91. Lehman, J. H.; Terrones, M.; Mansfield, E.; Hurst, K. E.; Meunier, V. Evaluating the characteristics of multiwall carbon nanotubes. *Carbon* **2011**, *49*, 2581-2602.
92. Kaiser, D.; et al. **2104** www.ilsi.org/ResearchFoundation/RSIA/Documents/Methods for Measuring Release of MWCNTs from polymer composites-monographversion.pdf
93. Saito, R.; Hofmann, M.; Dresselhaus, G.; Jorio, A.; Dresselhaus, M. S. Raman spectroscopy of graphene and carbon nanotubes. *Adv. Phys.* **2011**, *60*, 413-550.

## Author biographies

Stacey Lynn Harper, PhD is an Assistant Professor of Nanotoxicology in a joint position between the Department of Environmental & Molecular Toxicology and the School of Chemical, Biological & Environmental Engineering at Oregon State University (OSU) and is a Signature Researcher of the Oregon Nanoscience and Microtechnologies Institute. She uses rapid assays and nanoinformatics to identify the specific nanomaterial features that govern its environmental behavior and biological interactions. Doctorate is from the University of Nevada Las Vegas with postdoctoral fellowships at the Environmental Protection Agency and through the National Institute of Environmental Health Sciences.

Wendel Wohlleben is Senior Scientist for Characterization of Nanomaterials at BASF. In the Dept. of Material Physics he leads research projects on advanced materials and on the safety of nanomaterials. Visiting scientist at the Harvard School of Public Health and at the Department of Materials and Interfaces at the Weizmann Institute (2012). Formerly responsible for the coordination between BASF's growth cluster nanotechnology and the European Technology Platform for Sustainable Chemistry. PhD 2003 (Munich University and Max-Planck-Institute) after studies of physics (Heidelberg University and Ecole Normale Supérieure, Paris).

Maria J. Doa, Ph.D. is the Director of the United States Environmental Protection Agency's Chemical Control Division in the Office of Chemical Safety and Pollution Prevention. She leads activities for the assessment and management of a wide range of industrial chemicals and nanoscale materials under the Toxic Substances Control Act. She is the United States lead for the Organization for Economic Cooperation and Development Working Party on Manufactured Nanomaterials. Previously, she led EPA programs to reduce risk to lead, PCBs, mercury and asbestos. She also led EPA's Toxics Release Inventory Program – a key community right-to-know program about toxic chemicals.

Prof. Dr. Bernd Nowack holds a MSc. (1992) and a PhD (1995) in environmental sciences from ETH Zürich. He is currently leading the "Environmental Risk Assessment and Management" group at Empa, the Swiss Federal Laboratories for Materials Science and Technology. His current research is about chances and risks of engineered nanomaterials for the environment: analysis of technological trends and opportunities; development and application of methods for material flow modeling, exposure modeling and environmental risk assessment; life cycle assessment; experimental studies about release of nanomaterials from products and about their transformation, fate and effects in the environment.

Shaun Clancy is Director of Product Regulatory Services for Evonik Corporation where his focus is on chemicals management; communication of their hazards, exposures and risks; and product stewardship. Shaun obtained his B.S. at University of Buffalo, SUNY and Ph.D. at Northwestern University. He is active in industry activities regarding the responsible development of chemicals and nanotechnology, International Organization for Standardization's Technical Committee 229 on Nanotechnologies and in the Nanotechnology Committee of the Business & Industry Advisory Committee to the OECD. Shaun participates on external advisory boards of the Nanomaterial Registry and the University of Michigan Risk Science Center.

Journal of Physics: Conference Series **617** (2015) 012026 doi:10.1088/1742-6596/617/1/012026

Richard Canady, PhD directs the Center for Risk Science Innovation and Application of the ILSI Research Foundation, which fosters collaborative research supporting risk management needs. Expert in regulatory risk assessment covering a wide range of cutting edge health risk management issues over a 25 year teaching and public policy career, including genomics, nanotechnology, biotechnology, obesity, and contaminants in the environment and in foods. Doctorate is from Rockefeller University. Positions include the US Centers for Disease Control, Food and Drug Administration, and the Office of Science and Technology Policy of the White House.

Andrew Maynard is the NSF International Chair of Environmental Health Sciences at the University of Michigan School of Public Health. He is Chair of the Environmental Health Sciences department at the University of Michigan, and Director of the University of Michigan Risk Science Center. Maynard is a leading expert on the responsible development and use of emerging technologies, and on innovative approaches to addressing emergent risks. His academic and professional work primarily addresses the potential health, environmental and social impacts and benefits of technology innovation, using integrated approaches to explore effective solutions to emerging challenges.

# Measurement of Nanoparticles Release during Drilling of Polymer Nanocomposites

**L Gendre, V Marchante Rodriguez, H Abhyankar, K Blackburn and J L Brighton**

Cranfield University, Centre for Automotive Technology, Bedfordshire, MK43 0AL, UK

E-mail: h.a.abhyankar@cranfield.ac.uk

**Abstract**. Nanomaterials are one of the promising technologies of this century. The Project on Emerging Nanotechnologies [1] reports more than 1600 consumer products based on nanotechnology that are currently on the market and advantages link to the reinforcement of polymeric materials using nano-fillers are not to demonstrate anymore. However, the concerns about safety and its consumer perception can slow down the acceptance of nanocomposites. Indeed, during its life-cycle, a nanotechnology-based product can release nano-sized particles exposing workers, consumers and environment and the risk involved in the use and disposal of such particles is not well known. The current legislation concerning chemicals and environment protection doesn't explicitly cover nanomaterials and changes undergone by nanoparticles during the products' life cycle. Also, the possible physio-chemical changes that the nanoparticles may undergo during its life cycle are unknown. Industries need a standard method to evaluate nanoparticles release during products' life cycle in order to improve the knowledge in nanomaterials risk assessment and the legislation, and to inform customers about the safety of nanomaterials and nanoproducts. This work aims to propose a replicable method in order to assess the release of nanoparticles during the machining of nanocomposites in a controlled environment. For this purpose, a new experimental set-up was implemented and issues observed in previous methods (background noise due to uncontrolled ambient environment and the process itself, unrepeatable machining parameters) were solved. A characterisation and validation of the chamber used is presented in this paper. Also, preliminary testing on drilling of polymer-based nanocomposites (Polyamide-6/Glass Fibre reinforced with nano-$SiO_2$) manufactured by extrusion and injection moulding were achieved.

## 1. Introduction

Nanomaterials are one of the most promising technologies of this century. They are defined as materials composed of several phases one of which has at least one dimension of less than 100 nanometers [2]. Usually, a nanocomposite is a matrix (like ceramic, metal or polymer) with an addition of nanofillers of varying shapes, like spheres, fibres, platelets, particles, or tubes, and of different chemical compositions.

Nowadays, industrial sectors, such as automotive or aerospace industry, include more and more nanocomposites materials in their products. In fact polymer-matrix nanocomposites seem to be a good alternative to replace metallic parts. They allow a considerable weight and cost reduction, and the use of nanofillers presents some advantages compared to traditional macro or microfillers: good mechanical properties, high energy absorption capability, recyclability, resistance to corrosion and chemical attack, high heat-distortion temperature, etc [3]–[5]. Compared to the traditional reinforcement, the addition of nanofillers in polymer implies a minor increase in the cost but reduces the weight. Actually, it is known that an addition of only 5wt.% of inorganic nano-particles in polymers is enough for a considerable improvement of the material's behaviour and properties compare to 20wt.% for a micro filler [6]–[10]. These improvements can be explained by the fact that fillers in nano-size allow a high volume-to-surface ratio of the nanoparticles, and so an increase of the contact surface between matrix and fibre [11]. It also allows a low inter-particles distance compare to microsize fillers and reduces stress concentrations around the fillers.

However, the risk involved in the use and disposal of such particles is not well known. The current legislation concerning chemicals and environment protection doesn't cover nanomaterials. Nevertheless the release of nano-sized particles from nanotechnology part can be a risk for human health and environment, and especially the physio-chemical properties of the nanoparticles embedded into the polymeric-matrix are unknown along the whole life cycle of the nanomaterials. Industries need a standard method to evaluate nanoparticles release during products' life cycle in order to improve the knowledge in nanomaterials risk assessment and the legislation, and to inform customers about the safety of nanomaterials and nanoproducts. It is safe to say that, given the explosive R&D and commercial uptake of nanomaterials (for example, the number of submissions per year to the Journal of Nanoparticle Research increased every year and reached 2149 in 2013 [12]), unsurprisingly, the regulations governing the use and disposal of nanomaterials during its life cycle is behind the curve. The wide acceptance of nanotechnology by the consumers depends on alleviating the perceived safety related concerns. In this context, many projects, aiming to understand the effects of nanomaterials usage on human health and environment, were and still are funded by the European Commission. Project Nanopolytox studied the "toxicological impact of nanomaterials derived from processing, weathering and recycling from polymer nanocomposites used in various industrial applications" [13], and SIRENA the SImulation of the RElease of NAnomaterials from consumer products for environmental exposure assessment, which is funding this study. We can also cite MARINA (Managing Risks of Nanomaterials) and NanoValid (Developing References Methods for Nanomaterials).

The SIRENA project aims to demonstrate and validate a methodology to simulate the unintended release of nanomaterials from consumer products by replicating different life cycle scenarios to be adopted by a wide number of industrial sectors in order to get the necessary information for exposure assessment [2]. In order to replicate different stages of products' life cycle, two types of experiments will be conducted: crashing (to simulate accidental or intended fractures), and drilling (which is a common procedure in different stages of product's usage phase). During these experiments, nanoparticles released have to be collected, sampled and characterised (chemical composition, shape, size, quantity, size distribution), in order to know if they present a risk to human health and environment. The main motivations of this work are:

- Provide different industrial sectors with a standard method to evaluate the release of nanoparticles from nanoproducts during their life-cycle, and so link to the potential risk on human health and environment.
- Increase and improve actual knowledge in nanomaterials risk assessment, in order to implement EU legislation in relation to chemicals and environmental protection.
- Inform consumers and general public about the safety of nanomaterials and nanoproducts, in order to allow a successful penetration into market and sustainability of Nanotechnology.

The work presented in this paper is part of the SIRENA project and aims to assess the release of nanoparticles during the machining of nanocomposites. The protocol developed in NEPHH project for the simulation of the release of nanoparticles was replicated and tested for this purpose. The results and deficiencies observed with this protocol led to the implementation of a new experimental set-up. A characterisation and validation of the chamber used for this work was done in order to assess the controllability of the environment and the replicability of the experiments. Also, preliminary testing on drilling of polymer-based nanocomposites (Polyamide-6/Glass Fibre reinforced by nano-SiO2) manufactured by extrusion and injection moulding were achieved.

## 2. Replication and assessment of the NEPHH protocol

The first step of this work was to assess the NEPHH protocol [14]. For this, we replicate the experiments with a similar method. The analysis of the results highlighted several deficiencies that are presented in the following part.

### *2.1. Replication of the NEPHH protocol: Materials & Methods*

*2.1.1. Materials.* The materials used for the replication of the NEPHH protocol are flat donut-shape rings of 4mm thick. The dimensions of the samples were 160mm for the external diameter and 100mm for the internal one. The materials studied were three-phase polymer matrix nanocomposites: Polyamide-6 (Durethan B30) reinforced by 30wt.% of Glass fiber (ThermoFlow 672) and different percentage of either nano-$SiO_2$ (Aerosil R 974) or organically modified Montmorillonite (OMMT, Dellite 43B). The nanocomposites were prepared at Fraunhofer – Institute of Chemical Technology (Germany), by direct melting extrusion in a twin-screw extruder at a maximum temperature of 280°C. The product was cooled in a water bath, pelletized and dried. The samples were injected moulded from the pellets produced. The composition of the different grades can be found *Table 1*.

**Table 1: Composition of the studied materials**

| | Type of Matrix | wt.% of PA6 | Type of Glass Fibre | wt.% of GF | Type of filler | wt.% of filler |
|---|---|---|---|---|---|---|
| PA/GF/OMMT 5wt.% | Durethan B30 | 65 | ThermoFlow 672 | 30 | Dellite 43B | 5 |
| PA/GF/OMMT 7.5wt.% | Durethan B30 | 62.5 | ThermoFlow 672 | 30 | Dellite 43B | 7.5 |
| PA/GF/OMMT 10wt.% | Durethan B30 | 60 | ThermoFlow 672 | 30 | Dellite 43B | 10 |
| PA/GF/SiO2 0.5wt.% | Durethan B30 | 69.5 | ThermoFlow 672 | 30 | Aerosil R 974 | 0.5 |
| PA/GF/SiO2 1wt.% | Durethan B30 | 69 | ThermoFlow 672 | 30 | Aerosil R 974 | 1 |
| PA/GF/SiO2 1.5wt.% | Durethan B30 | 68.5 | ThermoFlow 672 | 30 | Aerosil R 974 | 1.5 |
| PA/GF/SiO2 3wt.% | Durethan B30 | 67 | ThermoFlow 673 | 30 | Aerosil R 974 | 3 |

*2.1.2. Methods.* The method used replicated the protocol defined during the NEPHH project *[15]*. The generation of particles by drilling was carried out in a closed chamber, with the following dimensions: width of 690mm, depth of 330mm and height of 560mm. The samples were fixed into the chamber and the angle drill (Makita BDA351Z 18V LXT Angle Drill) was totally enclosed into the chamber during all the measurement cycle. The emissions of nano-particles released were measured with a portable aerosol sizer and counter SMPS+C (Grimm Aerosol) composed by a Condensation Particle Counter (CPC) model 5.403 with a classifier type Vienna, long U-DMA. It allows a particle size resolution of 44 channels over a size range of 11.1-1083.8nm. The SMPS+C was connected to the chamber by an antistatic hose. An overview of the installation can be found in Figure 1 and Figure 2.

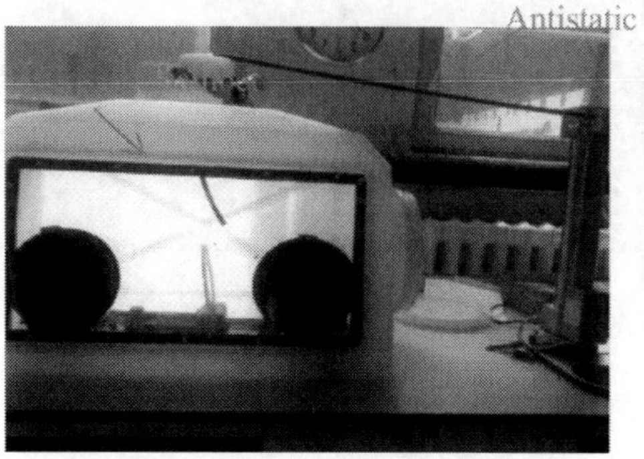

Figure 2: chamber connected to the

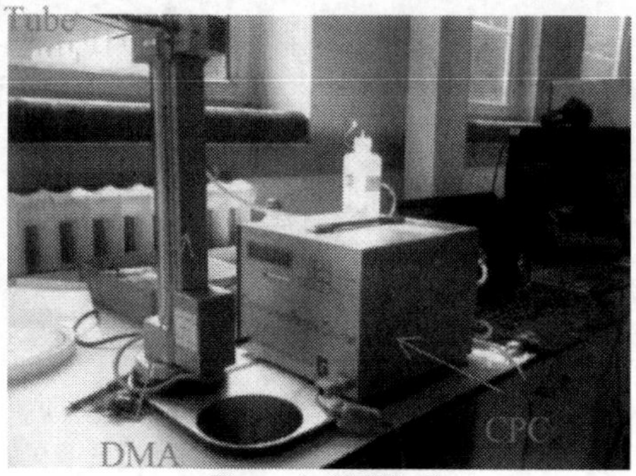


Figure 1: SMPS+C

The measurement cycle includes 20 minutes with the chamber open in order to purge it with lab air before the measurements. After, the chamber was closed and the measurements start with 30 minutes of record of the background noise, then a plate is drilled during 7 minutes, and the cycle finish with 60 minutes measurement of post-drilling.

The angle drill was used at its maximum speed: 1800 rpm; and two different sizes of drill bit were studied: 5 and 8 mm diameter. The experiment was repeated 3 times for each material composition and drill bit size. In addition, every morning one measurement cycle was conducted in order to record the noise of the drill itself.

### 2.2. *Replication of the NEPHH protocol: Results and Identification of the Deficiencies*

Results concerning the number concentration of particles along the time during the drilling of the different nanocomposites with a 5mm diameter drill bit are presented Figure 3. The background noise in the chamber recorded previous to drilling was around 10000 cm$^{-3}$. The concentration of particles was in every case at its maximum at the end of the 7 minutes of active drilling. Maximum airborne particles were comprised between 120000 cm$^{-3}$ to 520000 cm$^{-3}$.

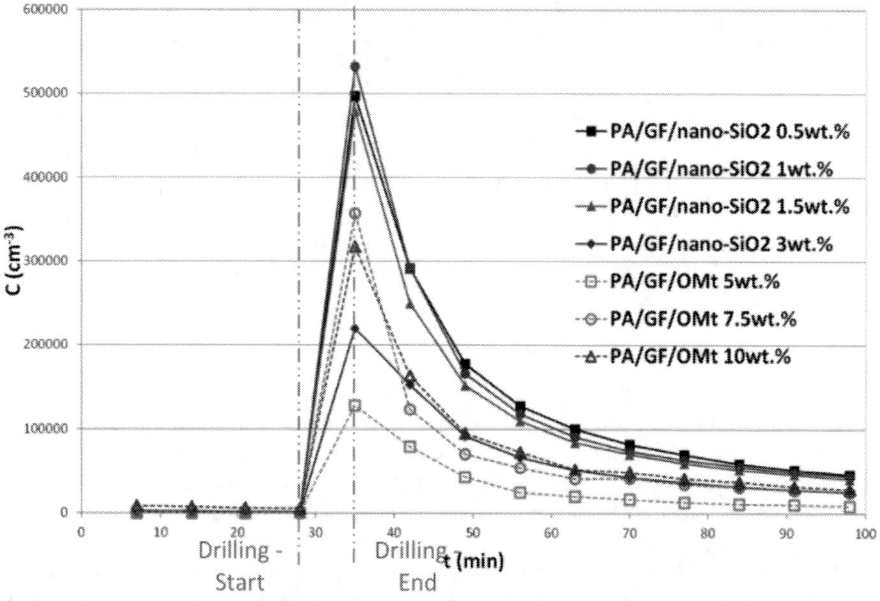


**Figure 3: Number concentration of particles (C) vs time during the replication of NEPHH protocol for the 5mmØ drill bit**

The replication of the experiments, these results and the analyses of several parameters showed some deficiencies of this protocol. The two main problems are the following:

- Variability of the process parameters: the spindle speed is controlled by an analogue switch with manual pressure. The speed is then only known when the pressure is maximal and cannot be controlled precisely. Also, the feed rate is determined by the pressure given by the operator on the manual drill. It is then dependant of the operator itself and totally variable during the experiment and not replicable. Figure 4 shows the difference in the particle size distribution between a manual pressure pushed around 4mm/min (corresponding to 7holes drilled in 7 minutes), and a manual pressure pushed around 1.14 mm/min (corresponding to 2 holes drilled in 7 minutes). The difference is significant and so the feed rate is a non-negligible parameters influencing the release of nanoparticles and needs to be controlled to insure replicable data.

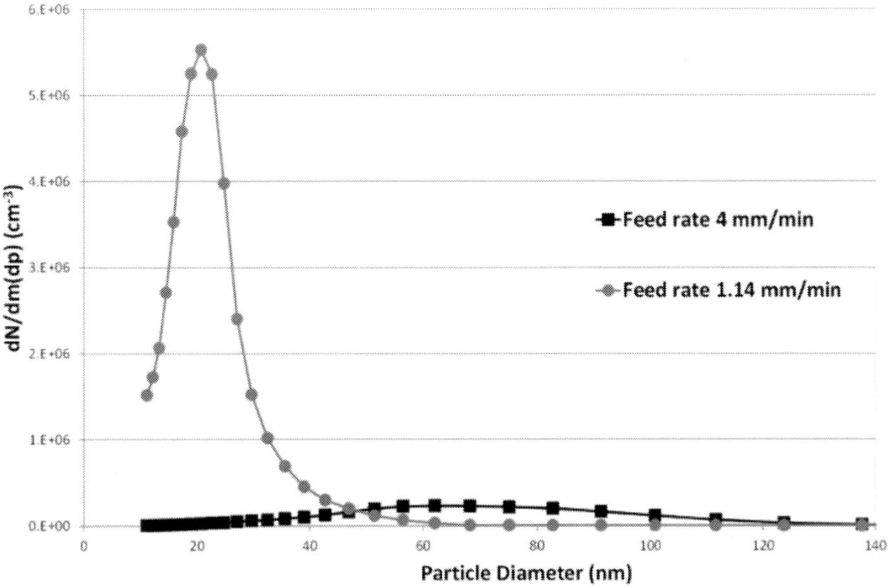


Figure 4: Normalised particle size distribution (dN/dm(dp)) inside the chamber at time 35 minutes - Study of the feed rate influence

- Influence of the background noises: first of all, in NEPHH protocol the chamber is purged with lab air during 20 minutes prior to drilling. The environment is dependant of the quality of the air in the lab and of the activities carried out in the same room. Also, every day, a blank test was done recording the level of particles when the manual angle drill was on, but no materials were drilled. Figure 5 presents the results of these tests for different day. It is clear that the manual angle drill itself is producing a significant level of particles (up to 700000 $cm^{-3}$). Also, the level of particles generated by the drill is variable according to the day. Again, these tests show that the protocol used is not replicable, and the results of these experiments are not exploitable.

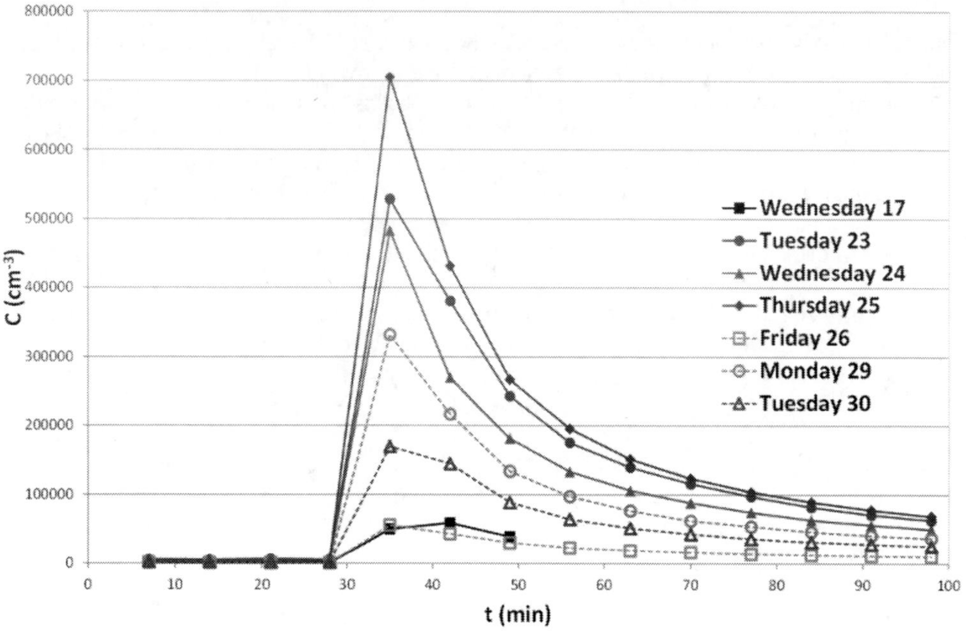


**Figure 5: Number concentration of particles (C) versus time (t) for the particles released by the manual drill without sampling in different days (Blank tests)**

## 3. Presentation and validation of the new protocol

### *3.1. Presentation of the prototype*

Following the previous results, it was decided to implement a new system, presented Figure 6, in order to address these deficiencies and propose a robust and replicable method to assess the nanoparticle release during the machining of nanocomposites. The main features and elements of this system are:

- *Environmental control*: The environmental control system comprising of a sealed chamber with a fan, BenchVent I100-4, has been implemented. In addition, a pre-filter and HEPA filter (category H14) are used to clean the air inside the chamber. An air recirculation system was also implemented in order to reduce the amount of 'dirty' air from the room to enter the chamber. Also, the chamber was transformed into a glove box in order to reduce the time between the opening and closing of the chamber. This configuration ensures a good control of the environment inside the chamber, as well as protection for the operator
- *Automatic machining system:* A CNC machine was designed and built at Cranfield University (not of-the-shelf), which allows the precise control of drilling parameters (feed rate, spindle speed, etc.). This system makes it possible to have reproducible and repeatable experiments in a controlled environment. Additionally, a water cooled spindle drill is used in order to avoid background noise or particles produced by the motor, as the motor is totally sealed
- *Dust collection system:* A fixture system composed of a base plate made of antistatic polymer (Tecafine HDPE), with a pattern to drill the holes in the samples was implemented. In addition, a Petri dish with lid, adapted for the drilling process, will be located on the surface of the sample. Therefore, the deposited fraction of particles could be easily collected into the Petri dish and will not be blown away because of the fan. Furthermore, the Petri dish will be sealed and used as a container. This way, the collection and storage of generated dust will be reduced to a single step
- *Instrumentation:* The scanning mobility particle sizer plus particle counter ('SMPS+C') from Grimm Aerosol was used to monitor the nanoparticles released. The' SMPS+C'

comprises of a Condensation Particle Counter (CPC) model 5.403 with a classifier type Vienna, long U-DMA, for the measurement of the airborne particles. This equipment is connected to the chamber using antistatic hoses. The particle size range measured is from 11.1 to 1083.8nm distributed in 44 channels

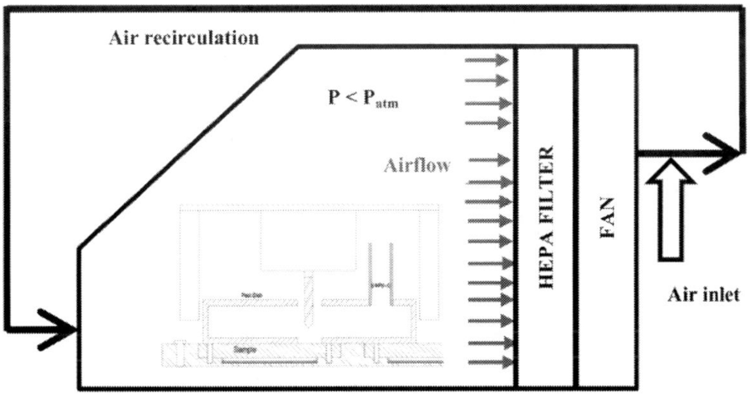


Figure 6: Scheme of the new configuration for the drilling prototype

### *3.2. Validation of the new protocol: Characterisation of the environmental background in the chamber*

A baseline experiment was conducted in order to characterise the air inside the chamber. Results are presented Figure 7. We can see that the air in the room is usually around 6000 $cm^{-3}$. Then, when the fan is on, and the air recirculated, it takes around 2 hours to reach an acceptable level of particles inside the chamber under 1000 $cm^{-3}$. The environment is then stabilized and the average number of particles inside the chamber is 312 $cm^{-3}$, which is a first improvement compared to the chamber used previously.

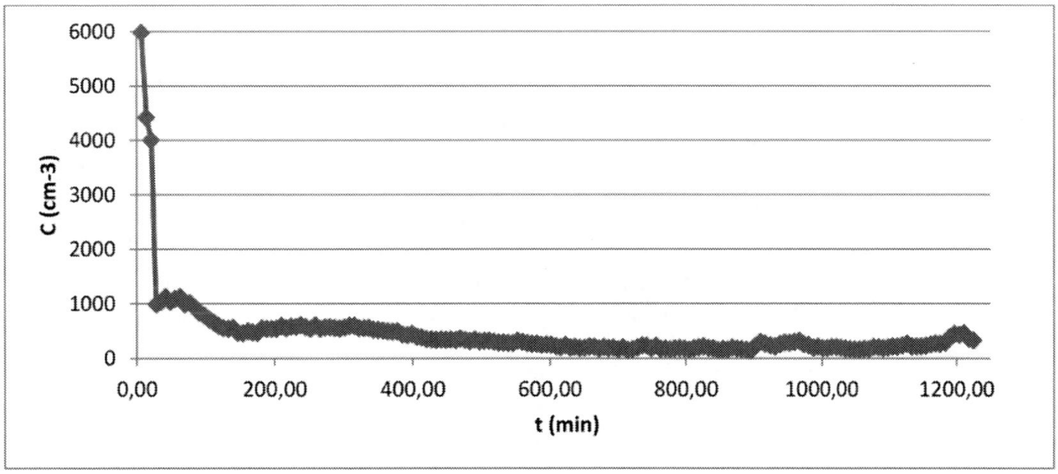


Figure 7: Baseline test of the air inside the chamber prior to cutting or drilling activities

Journal of Physics: Conference Series **617** (2015) 012027 doi:10.1088/1742-6596/617/1/012027

### *3.3. Validation of the new protocol: Comparison between a manual angle drill and the CNC drill*

In order to compare the new protocol with the protocol developed in NEPHH, the manual drill, used in the NEPHH experiments, was monitored. The manual drill was placed completely inside the chamber with the air recirculation system working ("fan on"). The manual drill was working for 7 minutes with a drill bit of 8 mmØ at maximum speed (1400 rpm). But no sample was machined. The air inside the chamber was monitored before using the manual drill and the airborne particles released by the drill (Figure 8). The hose for air inlet to the 'SMPS+C' was placed near the drill bit.
Inside the chamber, the average of C was about 590 ± 75 $cm^{-3}$. In case of the manual drill, it was switch on for 7 minutes in 3 occasions, but drilling no sample. These occasions correspond to scans 9, 13 and 21, in which, C increased to 8688, 8066 and 4609 $cm^{-3}$ respectively. Under similar operating conditions (7 minutes working but drilling no sample), for the CNC machine (scan 32) C increased only to 905 $cm^{-3}$. This experiment proved that, unlike the manual drill from NEPHH protocol, the CNC machine was not a contamination source. These particles are probably metallic ones produced by the engine of the manual drill which are metal brushes.

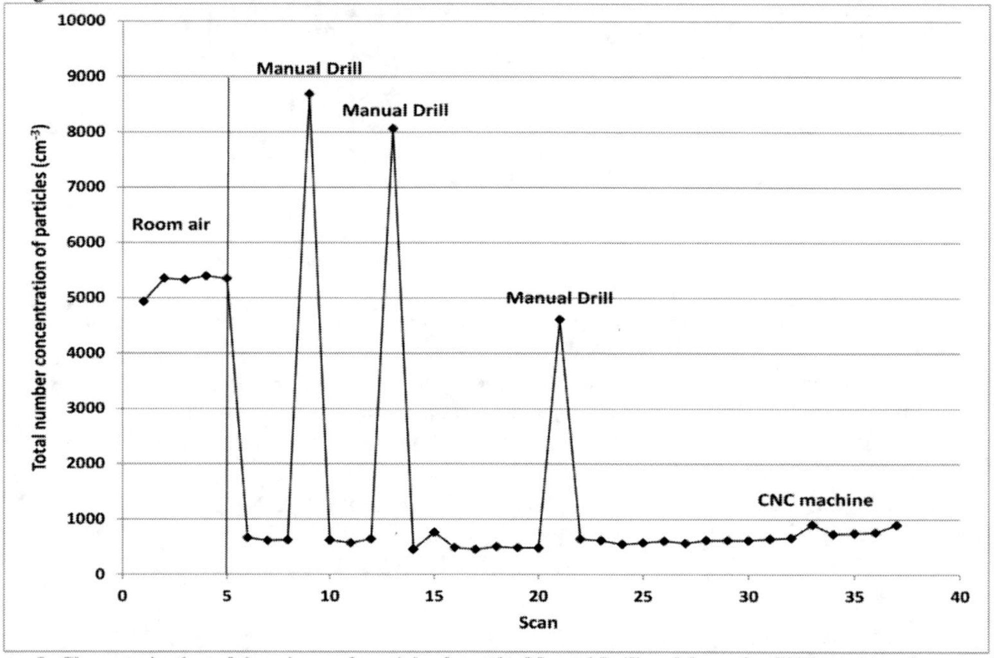


**Figure 8: Characterisation of the release of particles from the Manual Drill and from the CNC machine. In scans 1 to 5 the room air is measured. Scans 5 to 38 measure the air inside the chamber. Manual drill on in scans 9, 13 and 21; and CNC machine on in scan 3**

## 4. Preliminary testing on drilling of polymer-nanocomposites

After assessing the level of particles inside the chamber as background noise, first drilling experiments were conducted in order to validate the method for dust collection.

### *4.1. Preliminary Testing: Materials & Methods*

Three holes were drilled in a Polyamide-6/Glass Fiber/3wt.% nano-$SiO_2$ at three different feed rate: 2mm/min; 20mm/min and 200mm/min. Every hole was drilled using a different Petri dish, which was sealed following the drilling. The spindle speed was set to 10000 rpm according to industrial guidelines.

### *4.2. Preliminary Testing: Results & Discussions*

Pictures of the deposited particles collected in the three cases are available in Figure 9. Differences in shape and size could be noticed. From the hole drilled at 2mm/min, the particles were long and thin as filament, around 200μm diameter and 5mm long and several (around 30) particles could be found in the Petri dish. At a higher speed, 20mm/min, particles were fewer (less than 10) but bigger in size, 1mm large and few mm long. Size was more difficult to assess as the particles were shaped as remaining from a pencil sharpener. Finally, at high speed drilling, 200 mm/min, only one big piece of material was remaining after drilling.

**Figure 9: SEM pictures of the deposited particles at a micro scale**

SEM images of the deposited particles generated by drilling at a nano scale can be found Figure 10. Nano-$SiO_2$ particles are provided as spherical particles with a diameter from 7nm to 50μm for the agglomerates. On the three images, spherical nanoparticles with a diameter under 100nm can be found. It is noticeable that more particles can be found at the surface of the deposited dust generated by drilling at higher feed rate. This can probably be explain by the fact than drilling at high speed produced big bites but accompanied with a large quantity of nanoparticles, as at lower speed, large amount of small micro-sized particles are produced but only a few quantity of nanoparticles is released from this mechanism. These assumption need to be correlated with a record of the airborne nanoparticles released during the experiments.

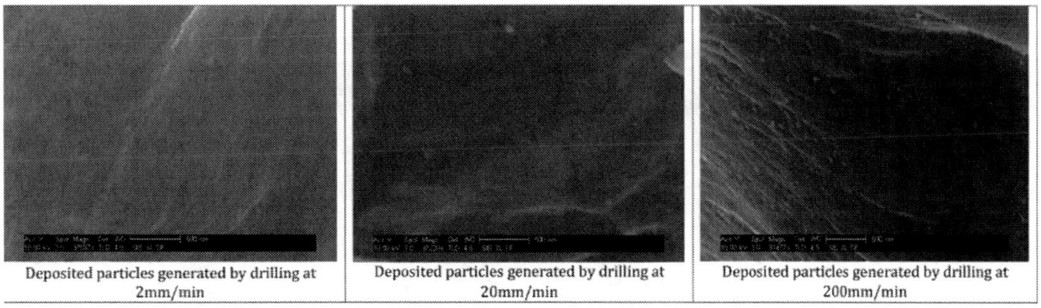


**Figure 10: SEM pictures of the deposited particles at a nano scale**

## 5. Conclusions

A new prototype and protocol have been developed to assess the release of nanoparticles from nanocomposites. An automated system controls the process parameters and there is precise control over the environment in the chamber were the experiments takes place. In addition, safety measurements are considered to protect the operator. In general, it can be concluded that the new prototype provides reproducibility and reliability, overcoming issues in the previous protocol like contamination from the manual drill, precise control of process parameters and reduction of the contamination from background.

**Aknwoledgment**

This work is funded by SIRENA ('Simulation of the release of nanomaterials from consumer products for environmental exposure assessment') Project (LIFE11 ENV/ES/596). SIRENA project is the continuation of NEPHH project. The author would like to thank the LIFE program and SIRENA project (LIFE 11 ENV/ES/596) for supporting this study.

**References**

[1] PEN, "CPI Home." [Online]. Available: http://www.nanotechproject.org/cpi/. [Accessed: 17-Jun-2014].
[2] ISO Standards, "ISO/TS 27687:2009 - Nanotechnologies - Terminology and definitions for nano-objects - Nanoparticle, nanofibre and nanoplate." .
[3] G. Dorairaju, *Additive Synergy in Flexible PVC Nanocomposites for Wire and Cable Applications*. ProQuest, 2008.
[4] Y. Kojima, A. Usuki, M. Kawasumi, A. Okada, Y. Fukushima, T. Kurauchi, and O. Kamigaito, "Mechanical properties of nylon 6-clay hybrid," *J. Mater. Res.*, vol. 8, no. 05, pp. 1185–1189, 1993.
[5] J. Xiong, Z. Zheng, H. Jiang, S. Ye, and X. Wang, "Reinforcement of polyurethane composites with an organically modified montmorillonite," *Compos. Part Appl. Sci. Manuf.*, vol. 38, no. 1, pp. 132–137, Jan. 2007.
[6] Claude Duval, "Plastiques et automobile - D'aujourd'hui à demain." Techniques de l'ingenieur, 10-Jul-2007.
[7] J. M. Garcés, D. J. Moll, J. Bicerano, R. Fibiger, and D. G. McLeod, "Polymeric Nanocomposites for Automotive Applications," *Adv. Mater.*, vol. 12, no. 23, pp. 1835–1839, 2000.
[8] Damien M. Marquis, Eric Guillaume, and Carine Chivas-Joly, "Properties of Nano fillers in Polymer," in *Nanocomposites and Polymers with Analytical Methods*, InTech, 2011, pp. 261–284.
[9] M. Rafiee, F. Yavari, J. Rafiee, and N. Koratkar, "Fullereneepoxy nanocomposites-enhanced mechanical properties at low nanofiller loading," *J. Nanoparticle Res.*, vol. 13, no. 2, pp. 733–737, 2011.
[10] Anne-Lise Goffin, "Polymer bionanocomposites reinforced by functionalized nanoparticles: impact of nanofiller size, nature and composition," PhD Thesis, Universite de Mons, Belgium, 2010.
[11] J.-J. Luo and I. M. Daniel, "Characterization and modeling of mechanical behavior of polymer/clay nanocomposites," *Compos. Sci. Technol.*, vol. 63, no. 11, pp. 1607–1616, Aug. 2003.
[12] "Journal of Nanoparticle Research ( JNR )," *springer.com*. [Online]. Available: http://www.springer.com/materials/nanotechnology/journal/11051. [Accessed: 06-May-2014].
[13] "nanopolytox « Toxicological impact of nanomaterials." [Online]. Available: http://www.nanopolytox.eu/. [Accessed: 29-Jan-2013].
[14] "Nephh: Home." [Online]. Available: http://www.nephh-fp7.eu/. [Accessed: 29-Jan-2013].
[15] S. Sachse, F. Silva, A. Irfan, H. Zhu, K. Pielichowski, A. Leszczynska, M. Blazquez, O. Kazmina, O. Kuzmenko, and J. Njuguna, "Physical characteristics of nanoparticles emitted during drilling of silica based polyamide 6 nanocomposites," *IOP Conf. Ser. Mater. Sci. Eng.*, vol. 40, no. 1, p. 012012, Sep. 2012.

Journal of Physics: Conference Series **617** (2015) 012028 doi:10.1088/1742-6596/617/1/012028

# Particle release from single-wall and multiwall carbon nanotubes in polystyrene-based composites during grinding

**I Ogura[1,2], M Shigeta[2], M Kotake[2], M Uejima[2] and K Honda[1,2]**

[1] National Institute of Advanced Industrial Science and Technology (AIST), 16-1 Onogawa, Tsukuba, Ibaraki 305-8569, Japan
[2] Technology Research Association for Single Wall Carbon Nanotubes (TASC), 1-1-1 Higashi, Tsukuba, Ibaraki, 305-8565, Japan

E-mail: i-ogura@aist.go.jp

**Abstract**. The aerosol particles released during the grinding of polystyrene (PS)-based composites with well- and poorly dispersed single-wall carbon nanotubes (CNTs) and multiwall CNTs were measured using real-time aerosol measuring instruments. Increases in the concentration of aerosol particles were recorded during the grinding of the samples. However, similar increases were observed even when CNT-free polystyrene was ground. Electron microscopic analysis of the released particles revealed that particles with protruding CNTs were observed for the well-dispersed CNT–PS composites, but free-standing CNTs were not found. On the other hand, particles like agglomerated CNTs were found for the poorly dispersed CNT–PS composites.

## 1. Introduction

Carbon nanotubes (CNTs) exhibit unique properties, and are promising candidates as filler materials in composites. Although composites based on CNTs are expected to be used in a wide range of industrial applications and consumer products, the potential impact of CNTs on human health remains a concern. Therefore, in order to control exposure to CNTs, an awareness of the potential release of CNTs throughout their life cycle is important.

In our previous study [1], we investigated the particle release during the grinding of polystyrene (PS)-based composites with well-dispersed single-wall CNTs (SWCNTs). In this study, we have broadened the range of sample types to include PS-based composites with both well- and poorly dispersed SWCNTs and multiwall CNTs (MWCNTs).

## 2. Methods

### *2.1. Test samples*

We used SWCNTs (AIST/TASC Super-growth SWCNTs) with a tube diameter of approximately 3 nm and a carbon purity of 99.9% [2], and MWCNTs (Nanocyl NC7000) with a tube diameter of approximately 10 nm and a carbon purity of 90% [3]. As one of the most widely used plastics, PS was used as a composite matrix. Although PS has poor heating resistance and impact durability, it is expected that the addition of CNTs can modify its properties.

Five types of test plate samples were prepared: PS (CNT-free), 5wt% well- and poorly dispersed SWCNT–PS composites, and 7.5wt% well- and poorly dispersed MWCNT–PS composites. The well-dispersed CNT–PS composites were prepared as follows: (1) CNTs were added into an aqueous solution with 1wt% sodium dodecyl sulfate (SDS) and dispersed using a jet mill (JN20, Jokoh, Tokyo, Japan); (2) PS dissolved in toluene was mixed with 1wt% SDS aqueous solution using a mixer (Filmix Model 56-50, Primix, Osaka, Japan) to obtain a latex solution with 4wt% toluene; (3) the CNT and latex solutions were mixed using a magnetic stirrer, and the mixture was dripped into 2-propanol; (4) precipitating composites composed of PS, and well-dispersed CNTs were obtained by filtration and dried under vacuum to remove 2-propanol and water; and (5) the composite material was formed into a plate by pressing. On the other hand, the poorly dispersed CNT–PS composites were prepared as follows: (1) as-produced CNTs were directly mixed with PS using a compounding machine (DSM Xplore, Geleen, The Netherlands); (2) the compounded composites were pelletized; and (3) the pellets were pressed into a plate.

Figure 1 shows photographs of the test plate samples. For the poorly dispersed SWCNT-PS composite, agglomerated CNTs were visually observed, whereas the other samples were entirely black.

**Figure 1.** Photographs of test plate samples.

*2.2. Experimental setup*

The composites were ground by hand using a micro-grinder in a conductive-antistatic plastic box that had an air supply opening with a high efficiency particulate air (HEPA) filter (see figure 2). The rotating speed was set at 20,000 rpm; grinding time was 30 s; and the air change rate was approximately 1 $min^{-1}$.

Prior to the test, purified air was passed through the box to eliminate background particles, and the aerosol particles released into the box by the grinding process were then measured using real-time aerosol measuring instruments: a scanning mobility particle sizer (SMPS, model 3936L75, TSI Inc., USA), an optical particle sizer (OPS, model 3330, TSI Inc., USA), and a condensation particle counter (CPC, model 3776, TSI Inc., USA). Each sample was tested at least three times.

In addition, the released aerosol particles were collected on Nuclepore membrane filters (25-mm diameter, 0.080-μm pore size, Nomura Micro Science Co., Ltd., Japan) and holey carbon film-coated transmission electron microscope (TEM) grids (Quantifoil R1.2/1.3 on 200 mesh Cu, Agar Scientific, Stansted, UK) with a TEM grid holder (Mini Particle Sampler, MPS, Ecomesure, Janvry, France) [4, 5] for analysis using a scanning electron microscope (SEM; Quanta 250 FEG, FEI Company, USA) and a TEM (JEM-1010, JEOL, Tokyo, Japan), respectively.

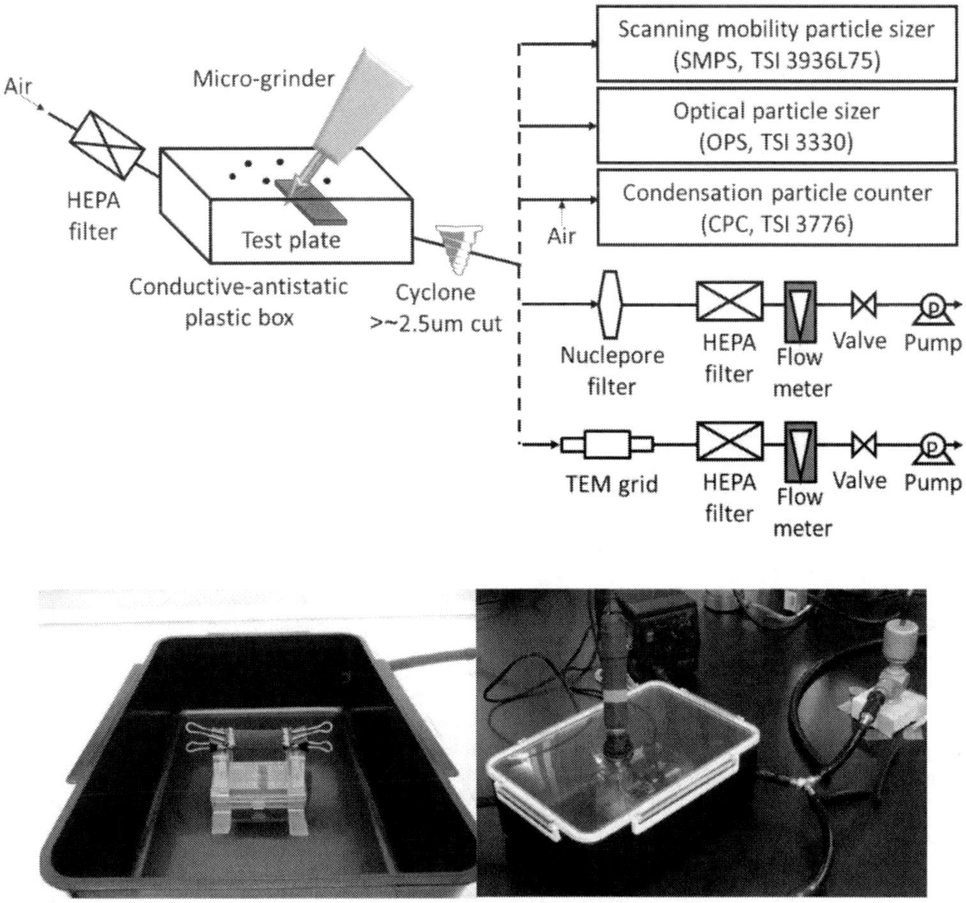


**Figure 2.** Schematic of experimental setup.

## 3. Results and discussion

### *3.1. Real-time aerosol measurements*

Figure 3 shows the particle size distribution of the released aerosol particles measured using real-time aerosol measurements. The results for idle running of the micro-grinder (without grinding) indicate that particles were generated from the micro-grinder itself. Increases in the concentration of nanometer- and micrometer-sized aerosol particles were recorded during the grinding of the samples, regardless of the sample type used, and similar increases were observed even when the CNT-free PS was ground. It is likely that the nanometer-sized particles were volatile particles released by frictional heat produced during the grinding of the composites [1].

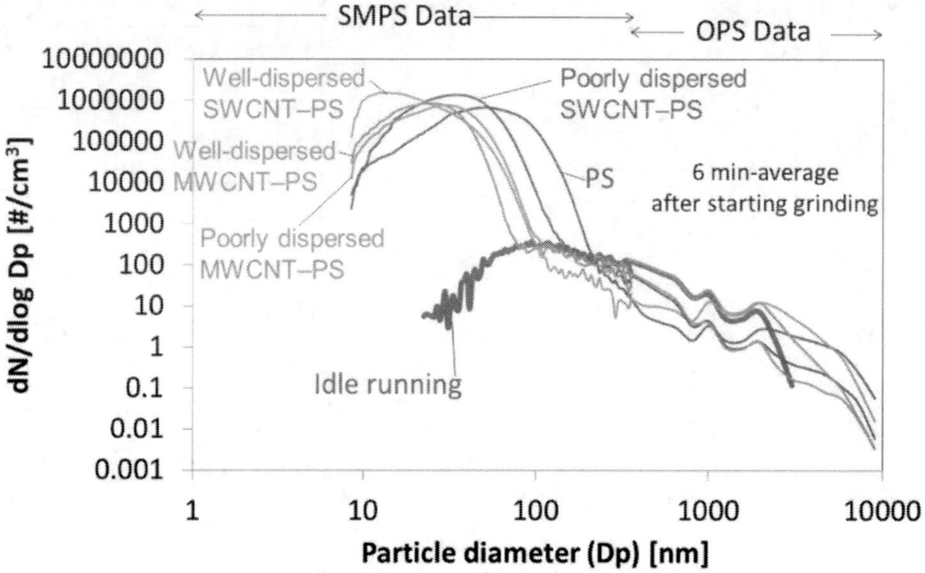


**Figure 3.** Number-based particle size distributions of released particles measured using SMPS and OPS.

### *3.2. Electron microscopic observation*

Figures 4–8 show SEM and TEM images of particles released into the air during the grinding of the test samples. Although particles with protruding CNTs were observed for the well-dispersed SWCNT–PS and MWCNT–PS composites (see figures 5 and 7), free-standing CNTs were not found. However, for the aerosol particles released from the poorly dispersed SWCNT–PS and MWCNT–PS composites, particles like agglomerated CNTs were found (see figures 6 and 8).

Journal of Physics: Conference Series **617** (2015) 012028 doi:10.1088/1742-6596/617/1/012028

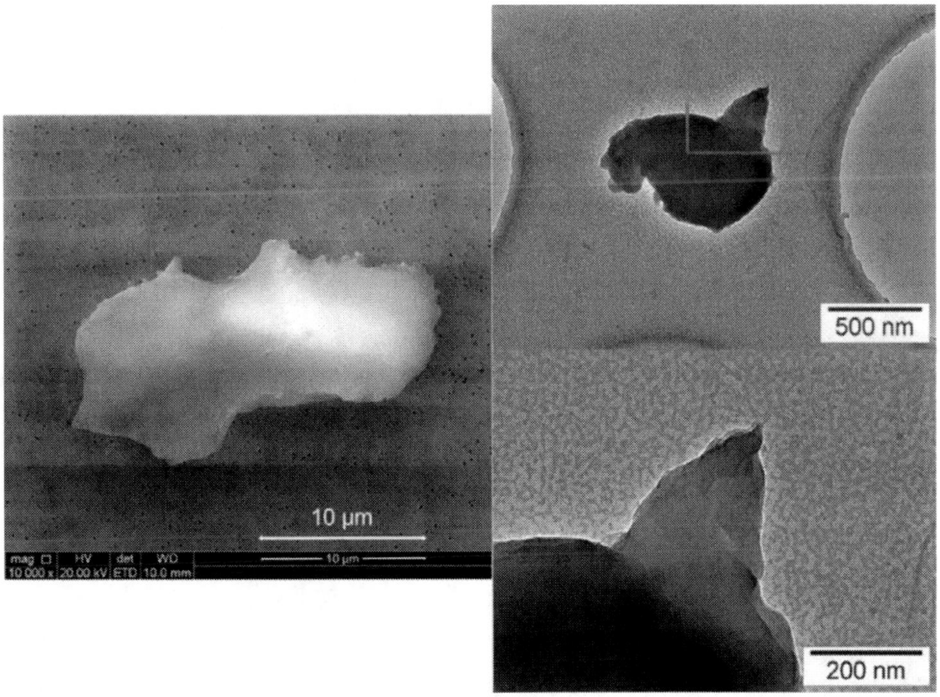


**Figure 4.** SEM and TEM images of particles released into the air during the grinding of CNT-free PS.

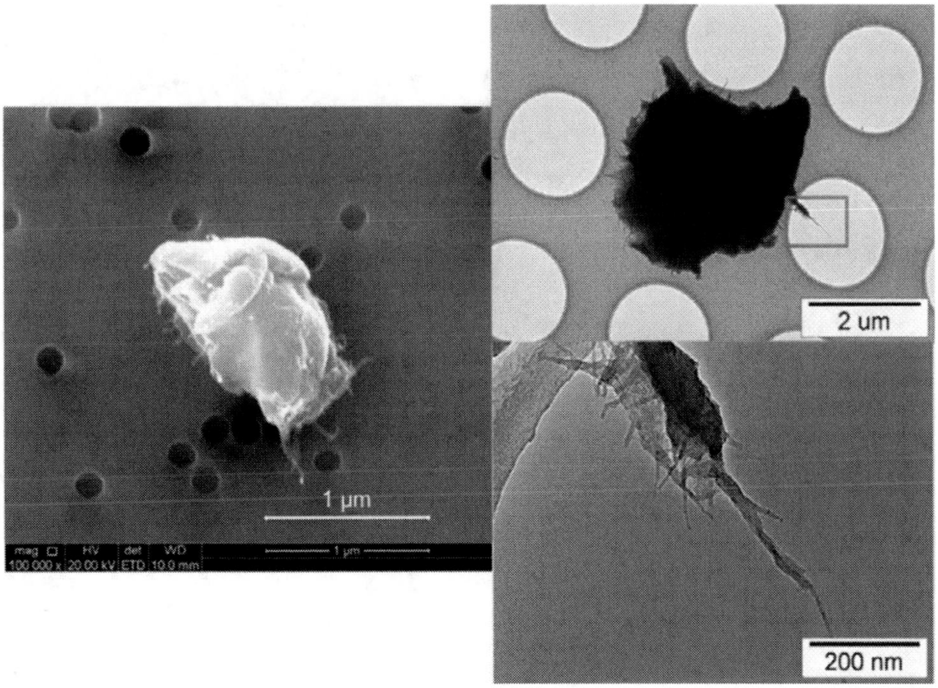


**Figure 5.** SEM and TEM images of particles released into the air during the grinding of well-dispersed SWCNT–PS composite.

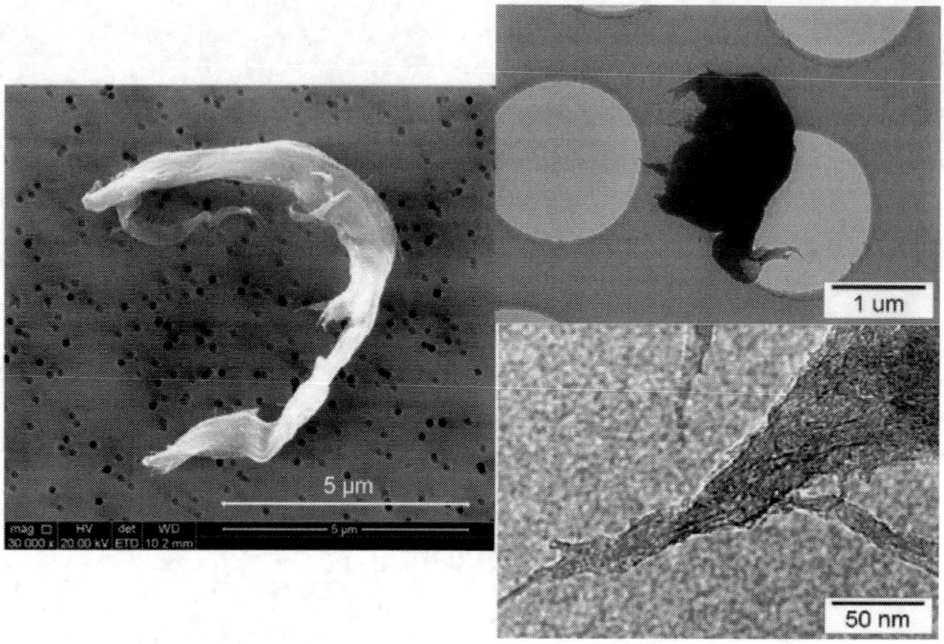


**Figure 6.** SEM and TEM images of particles released into the air during the grinding of poorly dispersed SWCNT–PS composite.

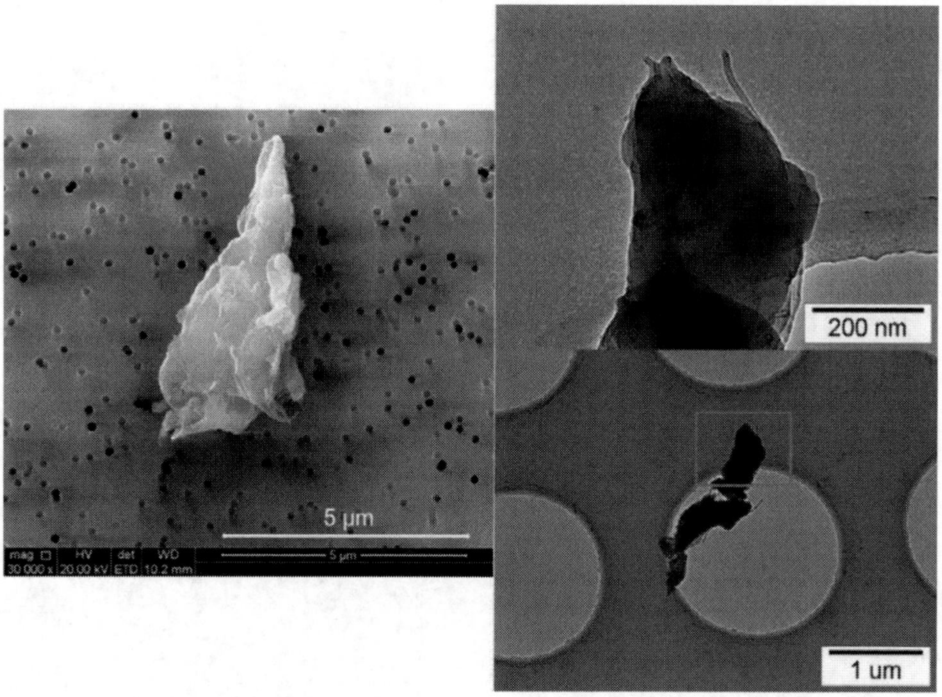


**Figure 7.** SEM and TEM images of particles released into the air during the grinding of well-dispersed MWCNT–PS composite.

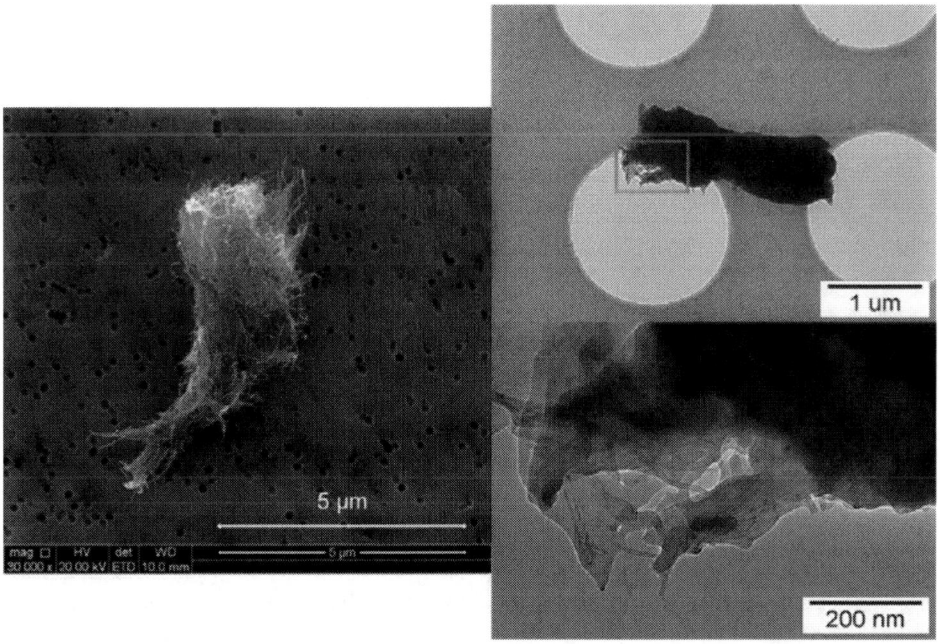


**Figure 8.** SEM and TEM images of particles released into the air during the grinding of poorly dispersed MWCNT–PS composite.

**Acknowledgment**
This work is based on results obtained from a project commissioned by the New Energy and Industrial Technology Development Organization (NEDO).

**References**

[1] Ogura I, Kotake M, Shigeta M, Uejima M, Saito K, Hashimoto N and Kishimoto A 2013 Potential release of carbon nanotubes from their composites during grinding *J. Phys.: Conf. Ser.* **429** 012049

[2] Hata K, Futaba D N, Mizuno K, Namai T, Yumura M and Iijima S 2004 Water-assisted highly efficient synthesis of impurity-free single-walled carbon nanotubes *Science* **306** 1362–1364

[3] Nanocyl Nanocyl™ NC 7000 Thin Multiwall Carbon Nanotubes http://www.nanocyl.com/de/Products-Solutions/Products/Nanocyl-NC-7000-Thin-Multiwall-Carbon-Nanotubes

[4] R'mili B, Le-Bihan OLC, Dutouquet C, Aguerre-Charriol O and Frejafon E 2013 Particle sampling by TEM grid filtration *Aerosol Sci. Technol.* **47** 767–775

[5] Ogura I, Hashimoto N, Kotake M, Sakurai H, Kishimoto A and Honda K 2014 Aerosol particle collection efficiency of holey carbon film-coated TEM grids *Aerosol Sci. Technol.* **48** 758–767

Journal of Physics: Conference Series **617** (2015) 012029 doi:10.1088/1742-6596/617/1/012029

# Nanoparticle release quantification during weak and intense dry dispersion of nanostructured powders

**Daniel Göhler and Michael Stintz**

Research Group Mechanical Process Engineering, Institute of Process Engineering and Environmental Technology, Technische Universität Dresden, D-01062 Dresden, Germany.

E-mail: daniel.goehler@tu-dresden.de

**Abstract**. Currently, the characterisation of particle release from nanostructured powders is realised based on different release scenarios, which were typically designed to describe real powder handling processes. Most of these scenarios base on short term stress situations (e.g. drop down, mixing) and are thus accompanied with dynamic changes of the particle number concentration, which complicates both the process characterisation and the release characterisation. In this study, two continuous release processes for weak and intense dispersion of powder agglomerates were analysed based on considerations regarding energetic and instrumental impact.

## 1. Introduction

Nanostructured powders are the basic components of numerous nano-products and were thus the subject of several particle release investigations for risk estimation [1]. Performed studies examined different dispersion procedures, which can by classified to the acting stress intensity level. Three groups may be distinguished: drop down procedures [2-5], fluidized bed approaches [6-8] and high energy dispersion treatments [9].

The amount of the dispersion energy affects both the particle size distribution (PSD) and the particle number concentration (PNC) and can thus lead for one powder to a large data diversity regarding the particle release. Typical release scenarios for powder handling are short term stress situation (e.g. drop down, mixing), which lead in most cases to dynamic changes on the particle release that complicates tasks regarding the basic process characterisation or the selection of suitable measurement devices. This in turn limits the comparability among the results of different release studies.

This paper aims at a uniform methodological approach to the quantification of (nano)-particle release from dry powder due to weak and intense dispersion processes. These processes are intended to provide a time-constant release signal, as is already feasible with commercial aerosol generators for intense dispersion like the well-known rotating brush generator (RBG). A new release process was developed that ensures time-constant size and concentration values due to weak dispersion. The so-called weak dispersion unit (WDU) was designed by considering previous experiences and the possible impact of analytic instrumentation on the state of dispersion, i.e. the measurement results.

## 2. Characteristics of dispersion processes

### 2.1. Technical power/energy consumption

Today it is common practise to quantify the progress of dispersion via any kind of technical power or energy input into the dispersed system, i.e. an indirect but measureable or accessible power/energy input of the operated disperser.

For example, wet dispersion processes by ultrasonication are frequently described by the volume-specific calorimetric energy input $e_{V,CAL}$ based on the calorimetric power $P_{CAL}$ related to the suspension volume $V_{SUS}$ and dispersion duration $t_{DIS}$ [10]:

$$e_{V,CAL} = \frac{P_{CAL} \cdot t_{DIS}}{V_{SUS}} = \left( \frac{m_F \cdot c_p \cdot \Delta T}{t} \right) \cdot \frac{t_{DIS}}{V_{SUS}} \tag{1}$$

In contrast to $e_{V,CAL}$ employed for wet dispersion, the mass-specific pneumatic energy consumption $e_{M,PNE}$ based on the pneumatic power $P_{PNE}$ and the supplied powder mass flow could be used for describing dry dispersion processes [11,12]:

$$e_{M,PNE} = \frac{P_{PNE}}{\dot{m}_P} = (Q \cdot \Delta p) \frac{1}{\dot{m}_P} \tag{2}$$

The pressure drop $\Delta p$ is affected by the acceleration of the particles within the volumetric air flow $Q$, and also by the geometry of the piping system. The pressure drop due to particle acceleration is often too low for measurement, whereas the geometric-based one differs considerably among existing devices. Thus, there is a need for additional characteristic parameters for dry dispersion processes.

### 2.2. Dispersion stress

Powder dispersion usually relies on several interacting forces (resulting from drag, inertia, gravity, impacts, etc.), which act on the particles/agglomerates and cause tension, compression, bending and shear. This results in deagglomeration, i.e. agglomerate erosion and/or fragmentation [13], when the corresponding stresses exceed the agglomerate strength:

$$\sigma_{STRENGTH} < \sigma_{STRESS} \tag{3}$$

A comparison of existing dispersion models [11, 14-17] shows that tensile stress caused by particle acceleration dominates the dispersion of submicron agglomerates in air flows at similar and low particle load (< 0.1 vol.-%). Hence, the decisive parameter for pneumatic dispersion is then the maximum relative velocity $v_{rel,max}$ between the undisturbed air flow and the particle within the dispersion unit.

In order to demonstrate this statement, three different commercial aerosol generators (see Table 1) were operated with two test powders (see Table 3) at comparable low powder feeding (powder flow rate of 30 $mm^3 \cdot min^{-1}$).

**Table 1.** Commercial aerosol generators operated for statement demonstration.

| | | RBG 1000 | RODOS | SAG 410/U |
|---|---|---|---|---|
| manufacturer | [-] | Palas GmbH, Karlsruhe, Germany | Sympatec GmbH, Clausthal-Zellerfeld, Germany | Topas GmbH, Dresden, Germany |
| dispersion unit | [-] | dispersion cover type A | ejector | ISO-injector (ISO 5011) |
| $A_{DIS}$ [a] | [$mm^2$] | 18.5 | 32.9 | 12.6 |
| Re [b] | [-] | $8.8 \cdot 10^3$ - $2.8 \cdot 10^4$ | $1.0 \cdot 10^4$ - $2.1 \cdot 10^4$ | $6.2 \cdot 10^3$ - $1.4 \cdot 10^4$ |
| reference | | [18] | [19] | [20] |

[a]... dispersion cross section used for the calculation of maximal air velocity
[b]... Reynolds number for the analysed system pressure range (0.5 bar - 3.5 bar)

The corresponding number-weighted PSDs ($PSD_0$) were measured by means of an electrical mobility spectrometer (SMPS, Model 3934, TSI Inc., USA) after sufficient dilution and bipolar neutralisation within a flow channel.

Figure 1 a shows determined agglomerate number median diameters $x_{50,0}$ of the analysed two test powders over the system pressure, whereas Figure 1 b gives $x_{50,0}$ over $v_{rel,max}$.

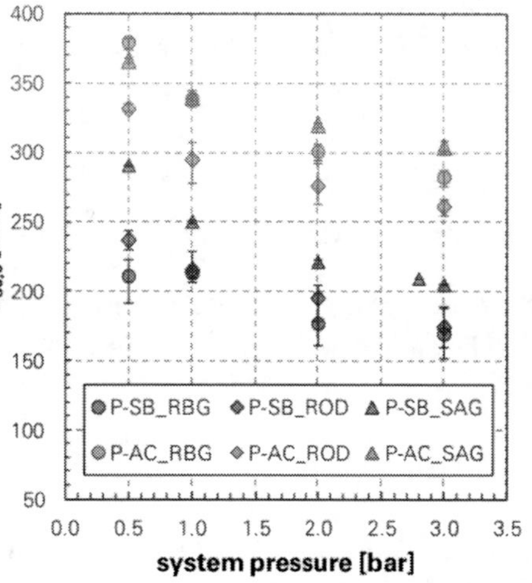


(a) $x_{50,0}$ over system pressure

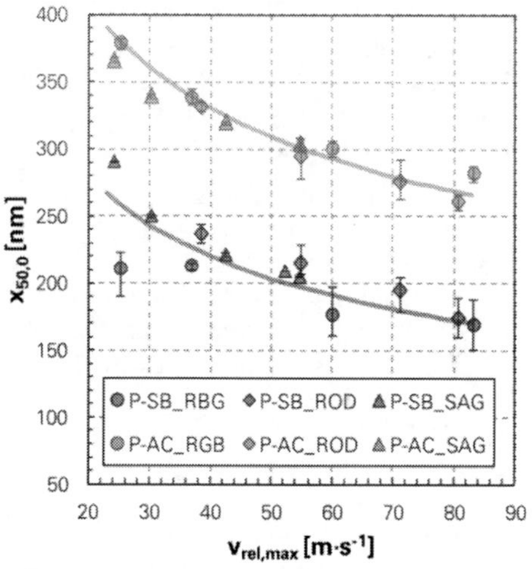


(b) $x_{50,0}$ over $v_{rel,max}$

**Figure 1**. Progress of dry powder dispersion for 3 different aerosol generators (RBG, ROD, SAG) and two filter test powders (P-SB, P-AC) at similar powder feeding as function of system pressure (a) and $v_{rel,max}$ (b).

It is evident, that $v_{rel,max}$ is more appropriate to compare the progress of dry powder dispersion than the typical used system pressure or the volumetric aerosol flow (not shown here). According to this, the agglomerate number median diameter $x_{50,0}$ of powder dispersion is a function of $v_{rel,max}$:

$$x_{50,0} = f\left(v_{rel.\max}\right) \tag{4}$$

2.3. Conclusion for the design of weak dispersion test-scenarios

The relationship for $x_{50,0}$ was the basis for our considerations regarding the design of a suitable test scenario. Additionally, attention has to be paid also to the employed aerosol measurement instruments, which have often critical zones where a post-dispersion of airborne agglomerates is possible. Such zones are particle separators at the inlet (e.g. impactors, aerocyclones), bends and constrictions (e.g. nozzles) in the piping system or the measurement zone. Table 2 provides approximated values of $v_{rel.max}$ for typical operation conditions of selected aerosol measurement instruments.

**Table 2.** Estimation of $v_{rel,max}$ within common used aerosol measurement instruments.

| instrument [-] | critical zones [-] | volumetric flow [L·min$^{-1}$] | inner diameter [mm] | $v_{rel.max}$ [m·s$^{-1}$] |
|---|---|---|---|---|
| aerodynamic particle sizer (APS) | staged acceleration nozzle | 1.0 to 5.0 | 0.81 to 1.04 | ≈ 60 |
| scanning mobility particle sizer (SMPS) | inlet impactor | 0.3 | 4.5 to 0.71 | ≈ 12.3 |
| | | | 4.5 to 0.46 | ≈ 30.2 |
| condensation particle counter (CPC) | bends | 0.3 | 4.5 | ≈ 0.3 |

It is obvious that $v_{rel,max}$ in the instruments must not exceed that of the intended dispersion process. Otherwise particle characterisation would solely reflect the post-dispersion within the measurement instrument. Moreover, since losses of coarse agglomerates in the experimental setup are unavoidable and measures of dilution are mandatory, the particle mass flow through the critical zones of the instruments is considerable lower than the stressed particle mass flow by the simulated process. Consequently, the process should warrant considerably higher $v_{rel,max}$ than those prevailing in the analytical instruments.

## 3. Experimental details for the characterisation of particle release

### 3.1. Materials

Different types of powders were used within this study for dry dispersion process characterisation and particle release characterisation (see Table 3). Two typical filter test powders (P-SB, P-AC) were employed for the characterisation of the dispersion process (see subchapter 2.2).

**Table 3.** Materials of interest: identification key of analysed powders; data on bulk density and compacted apparent density (for operation of RBG).

| Powder | | Pural SB | AC Fine 2[a)] | Iron oxide pigment, transparent | | | Iron oxide pigment, opaque | | |
|---|---|---|---|---|---|---|---|---|---|
| batch | [-] | A | A | A | B | C | A | B | C |
| sample ID | [-] | P-SB-A | P-AC-A | P-TE-A | P-TE-B | P-TE-C | P-DE-A | P-DE-B | P-DE-C |
| $\rho_{bulk}$ | [kg·m$^{-3}$] | 794 | 764 | 779 | 756 | 754 | 695 | 690 | 670 |
| $\rho_{comp}$ | [kg·m$^{-3}$] | 872 | 1515 | 889 | 902 | 892 | 1615 | 1609 | 1439 |
| $k_{POW}$[b)] | [-] | 0.09 | 0.50 | 0.12 | 0.16 | 0.16 | 0.57 | 0.57 | 0.53 |
| water content | [%] | 2.98 | 0.03 | 1.70 | 1.31 | 1.46 | 0.06 | 0.07 | 0.07 |
| BET | [m$^2$·g$^{-1}$] | 250 | n.a. | 85 | 85 | 85 | n.a. | n.a. | n.a |

[a)] … Arizona test dust type ISO 12103-A2 with 68-76 wt.-% $SiO_2$ and 10-15wt.-% of $Al_2O_3$
[b)] … powder compressibility based on $\rho_{bulk}$ and $\rho_{comp}$

Beside the test powders, two iron oxide pigments, a nanostructured one (P-TE) and a non-nano-structured one (P-DE), were focused for release characterisation. The former pigment is the same as used in [21]. In order to estimate the influence of sample age, three different batches (A, B, C) were analysed for each pigment.

To approximate the $PSD_0$ with those of the smallest dispersible units for very intense dispersion, wet dispersion series were performed by ultrasonication (UDS 751/S7, Topas GmbH, Germany) in combination with dynamic light scattering (HPPS 5002, Malvern Instruments GmbH, Germany) for different $e_{V,CAL}$ according to eq. 1 with $P_{CAL,UDS}(P_{EL} = 100\ \%, V_{AQU} = 50\ mL) = 31$ W.

The aqueous suspensions at $e_{V,CAL} = 367$ MJ·m$^{-3}$ were than aerosolized (ATM 220, Topas GmbH, Germany) and dried to determine the corresponding $PSD_0$ by an electrical mobility aerosol spectrometer (SMPS, Model 3938, TSI Inc., Shoreview, USA). As typical for this procedure, salt crystals of the pure dispersion medium (ultrapure water at 18.3 MΩ, UPW) were already measurable. To distinguish the fraction of salt crystals from the pigments, a data fitting was performed assuming a lognormal density distribution (see Figure 2). According to this, the approximated $PSD_0$ of smallest dispersible units of the nanostructured pigment P-TE can be described by $x_{50,0} = 63$ nm, $\sigma_{ln} = 0.43$ with $Q_0(100\ nm) = 86\ \%$, the one for the non-nanostructured P-DE by $x_{50,0} = 168$ nm, $\sigma_{ln} = 0.49$ with $Q_0(100\ nm) = 15\ \%$.

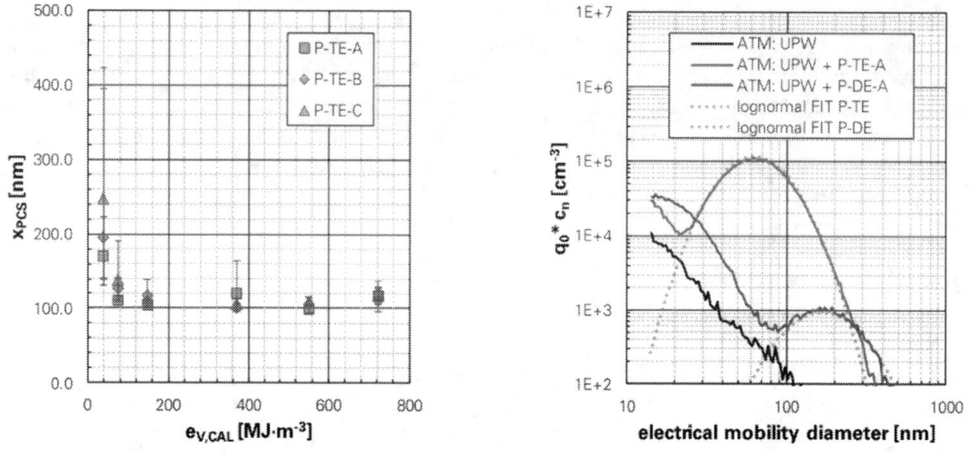


(a) wet dispersion series on P-TE (b) aerosolized and dried suspensions

**Figure 2**. Mean intensity-weighted particle size $x_{PCS}$ of P-TE over $e_{V,CAL}$ (a) and $PSD_0$ of the aerosolized suspensions at $e_{V,CAL} = 367$ MJ·m$^{-3}$ with fitted log-normal density distribution (b).

3.2. Instrumentation

*3.2.1. Dry dispersion systems.* Two different test-scenarios were examined within this study for a quantitative release characterisation. The intense dispersion system is based on a commercial rotation brush generator (RBG 1000, Palas GmbH, Germany), whereas for the low energetic consideration a weak dispersion unit (WDU) was developed that offers continuous powder feeding and low but adjustable dispersion energy input.

The WDU (Figure 3 a) consists of the linear-driven translation unit as described in [22]. A base plate with notches of different geometries and dimensions serves for the purpose of continuous volumetric powder feeding. The powder removal from the notch and the dispersion process is achieved by a stain-less steel capillary suction nozzle arranged 0.5 mm above the notch. Due to the separation of the dosing system from the dispersion system, the supplied volumetric powder flow and the dispersion intensity can be varied independently. Within this study, the WDU was operated at a volumetric powder flow rate of 30 mm$^3$·min$^{-1}$ and $v_{rel.max} \approx 25$ m·s$^{-1}$ ($P_{PNE,WDU}$(4.6 L·min$^{-1}$) = 0.2 W). The process parameters were chosen with respect to the operating conditions of the measurement devices as discussed in subchapter 2.3.

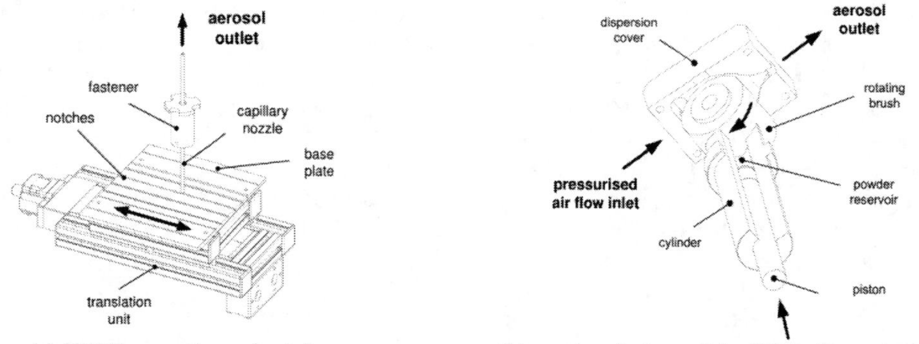


(a) WDU operation principle (b) sectional view of the RBG dispersion unit

**Figure 3**. Operated dispersion units for the characterisation of particle release from powders due to weak and intense dry dispersion of powder agglomerates.

The RBG consists of a cylinder-piston-system for a constant powder feeding into a rotating brush (Figure 3 b). The brush transports the powder into the dispersion cover, were the actual dispersion process is caused by a pressurised air flow [18]. The RBG was operated with dispersion cover type A at a dispersion pressure of 3.5 bar ($v_{rel,max} \approx 100\ m \cdot s^{-1}$), a powder flow of $25.7\ mm^3 \cdot min^{-1}$ and a rotational steel brush frequency of $1210\ min^{-1}$ ($P_{PNE,RBG}(3.5bar) \approx 134$ W). Note, the lower the system pressure is during RBG operation, the higher is the relative impact of mechanical stress by the rotating brush during dry powder dispersion.

*3.2.2. Particle measuring instruments.* The generated aerosols were analysed from the nanoscale up to the microscale using a condensation particle counter (CPC, Model 3022A, TSI Inc., USA), an electrical mobility aerosol spectrometer (SMPS, Model 3938, TSI Inc., USA) and a time-of-flight spectrometer (APS, Model 3321, TSI Inc. USA).

To avoid post-dispersion and re-entrainment of agglomerates during the analyses with the WDU, the SMPS was operated without inlet impactor. Furthermore, SMPS-data were evaluated without multiple charge correction (MCC) for both release scenarios. The occasional high fractions of large particles outside the SMPS detection limit (see Figure 7) would otherwise lead to considerable artefacts in the curve shapes. Note, that the omission of the MCC leads to finer $PSD_0$s and higher PNCs as present [25]. Based on the absence of detailed information regarding the effective agglomerate density, the given APS-data were evaluated without Stokes correction. In the case of WDU dispersion, a post-dispersion within the APS acceleration nozzle could not be ruled out. The most robust results based on the determined PNC measured by the CPC.

### 3.3. Experimental setups and procedures

*3.3.1. WDU setup.* The experimental setup that was operated for the characterization of the WDU-induced particle release is shown in Figure 4. The experimental setup was operated within a laminar flow bench (Model LF-VM-K0615, STEAG Laminarflow Prozesstechnik GmbH, Germany) to avoid particle contamination from the background.

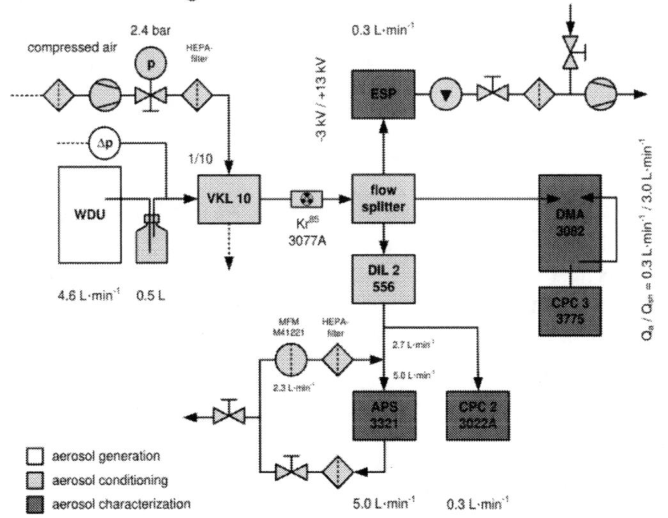


**Figure 4.** Experimental setup for weak dry powder dispersion by operating the WDU.

The aerosols generated by the capillary tube were firstly passed through a 0.5 L compensation tank to decrease fluctuations in PNC and to deposit surviving coarse agglomerates. The suction flow through the nozzle and the compensation tank was realized by a compressed air driven dilution unit

Journal of Physics: Conference Series **617** (2015) 012029 doi:10.1088/1742-6596/617/1/012029

(Model VKL10, Palas GmbH, Germany), which served furthermore for a defined pre-dilution and a pressure-compensated aerosol sampling. All measuring devices received their aerosol sample through a flow splitter. An additional dilution system (Model DIL 556, Topas GmbH, Germany) was placed before APS and CPC. The sample flow of the APS was generally post-diluted by partial back-mixing with HEPA-filtered device exhaust. During the analyses, a mean relative humidity of 19.5 % ± 3.3 % and a mean temperature of 25.0°C ± 1.2°C were logged.

*3.3.2. RBG setup.* The experimental setup that was operated for the characterisation of the RBG-induced particle release is shown in Figure 5. To attain a defined pre-dilution, the generated aerosol particles were passed parallel to an adjustable and particle-free air flow into an atmospheric decoupled mixing channel. For this purpose, both flows were turbulently mixed at the channel entrance. The aerosol flow to be analysed was firstly passed through a 5.0 L compensation tank to reduce fluctuations in the PNC, which can arise e.g. by the incidence of the stick-slip-effect. Afterward, the aerosol passed a commercial dilution unit (Model VKL10, Palas GmbH, Germany) and a bipolar neutraliser (Model 3077A, TSI Inc., USA) for a further defined reduction of PNC and to improve the charge condition of the aerosols. The measuring devices received their aerosol sample through a flow splitter. The sample flow of the APS was additionally diluted by partial back-mixing with HEPA-filtered device exhaust.

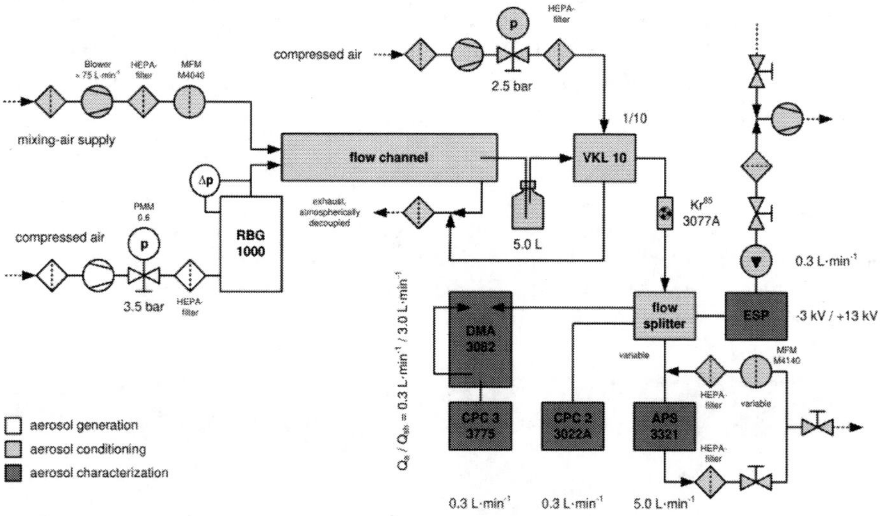


**Figure 5**. Experimental setup for intense dry powder dispersion by operating the RBG.

For the operation of the experimental setup, non-conductive tubing was installed before the neutralisers, whereas conductive tubing was used behind the neutralisers. In order to attain charge saturation of aerosol-facing components and thus to reduce particle losses and ensure maximum PNCs, the measurements were started not earlier than 60 min after feeding the RBG 1000 with a new powder sample. During the analyses, a mean relative humidity of 19.4 % ± 2.1 % and a mean temperature of 27.9°C ± 1.7°C were logged.

## 4. Results for particle release for weak and intense dispersion

The electron microscopic images (SEM, Model Gemini 982, Karl Zeiss AG, Germany) of equal magnification given in Figure 6 show the dispersion energy impact on the agglomerate size of the nanostructured Pigment P-TE. The first image based on simple powder dusting with a brush onto a carbon-conductive tab and shows thus the "unstressed" agglomerates.

Figure 6 b and Figure 6 c show electrostatic precipitated agglomerates after the dry dispersion procedures, whereas Figure 6 d refers to the agglomerates after atomization of the ultra-sonicated suspension ($e_{V,CAL} = 367\ MJ \cdot m^{-3}$). Despite the comparatively high energy input during RBG operation, released agglomerates are significantly coarser then the ones from the wet dispersion process by ultrasonication.

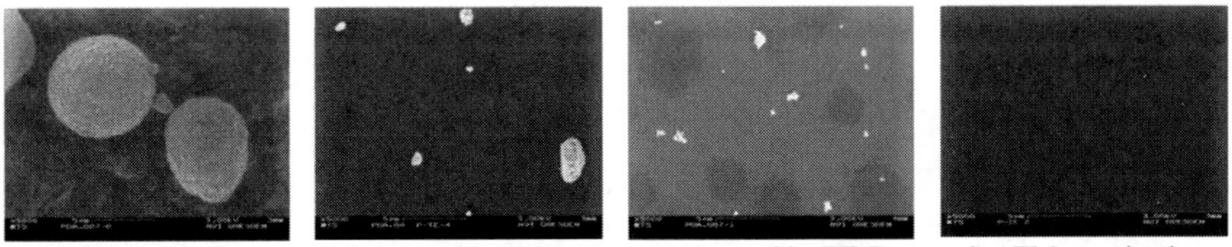

a) unstressed b) stressed by WDU c) stressed by RBG d) ATM atomization

**Figure 6.** SEM-images of P-TE agglomerates for the different dispersion procedures; bar ≙ 5 µm.

The corresponding $PSD_0$ of P-TE and P-DE given in Figure 7 were derived solely from the SMPS-data. Results show that the dry dispersion processes (WDU, RBG) were not able to reach the $PSD_0$ of the smallest dispersible units (based on the wet dispersion process, i.e. ATM) despite intense dispersion, immediate aerosol dilution and bipolar neutralisation.

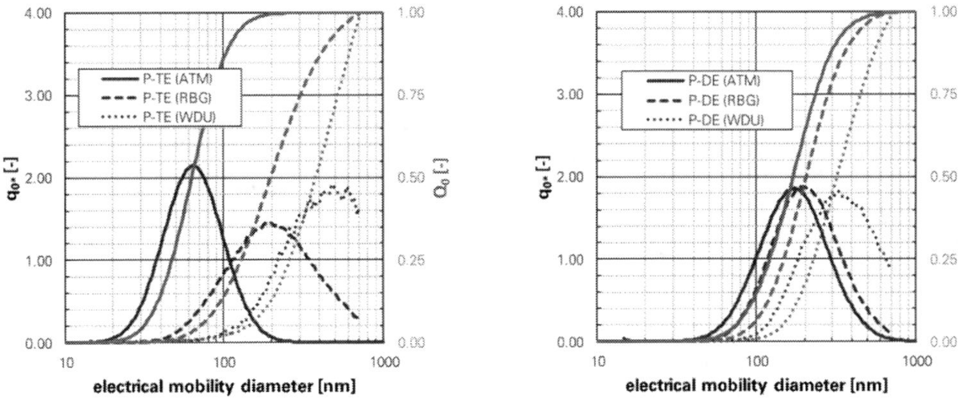


(a) P-TE (b) P-DE

**Figure 7.** Comparison of the $PSD_0$ determined for the simulated dry powder dispersion procedures in comparison to the $PSD_0$ smallest dispersible units determined by atomization (ATM) of the ultra-sonicated suspensions ($e_{V,CAL} = 367\ MJ \cdot m^{-3}$).

An overview on the nanoparticle fractions and the fractions of agglomerates x ≤ 200 nm determined by SMPS during the weak and intense dry dispersion procedures for the analysed powder samples and batches are given in Figure 8.

Results show that the nanoparticle fraction does not exceed a value of 20 %. Taking into account that solely the SMPS measurement range was used for the determination of $Q_0(100\ nm)$ and $Q_0(200\ nm)$, the quantities would be significantly lower than the determined ones by considering the whole $PSD_0$.

According to our previous studies regarding the particle release characterization from nanostructured materials [21, 24], fractional particle release numbers were determined from the measurement data and the analytical conditions and related to the dispersed powder sample mass.

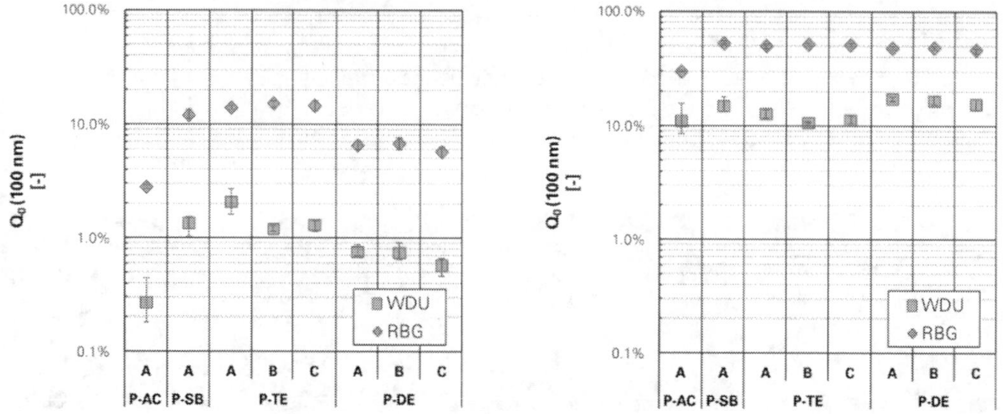


(a) $Q_0$(100 nm) (b) $Q_0$(200 nm)

**Figure 8**. Selected number quantity-fractions based on SMPS-Data of the analysed powders and their batches, error bars = data span of 3-5 repeated dispersion procedures with at least 3 measurements.

The fractional and mass-specific numbers of released particles are given in Figure 9 for total particle release (i.e. x < 10 μm based on CPC-Data), the nanoparticle release (i.e. x < 0.1 μm based on SMPS-Data) and the micrometer particle release (i.e. 1.0 μm ≤ x ≤ 20.0 μm based on APS-Data).

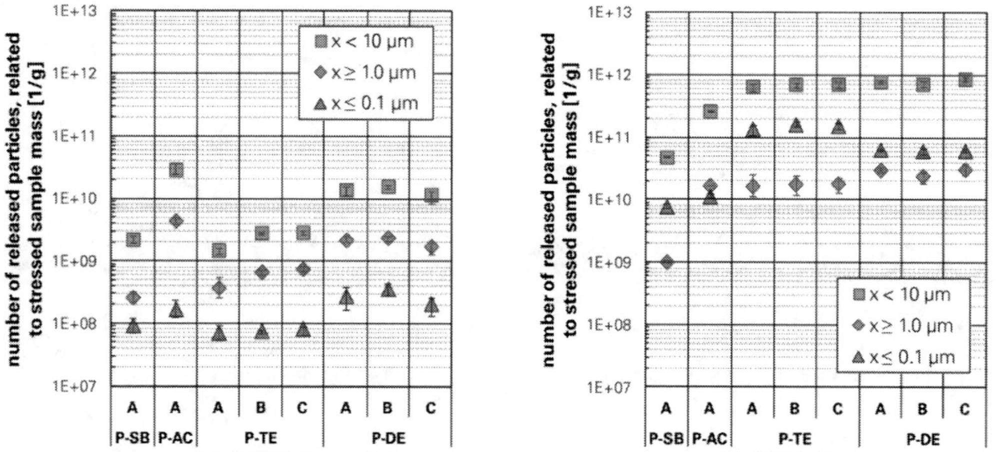


(a) WDU-dispersion (b) RBG-dispersion

**Figure 9**. Fractional particle release numbers related to the dispersed sample mass; error bars = data span of 3-5 repeated dispersion procedures with at least 3 measurements.

Depending on the applied energy input, the particle release varied over magnitudes. For example, the simulated intense release scenario led to particle release numbers (x < 10 μm) up to 400 times higher in the case of P-TE and up-to 80 times higher for P-DE in comparison to the simulated weak one. The powder sample age (i.e. comparison between the different batches A, B and C) shows a comparable minor impact on the particle release.

## 5. Summary and conclusion

The paper addressed test-scenarios for particle release from nano-structured powders, which simulates dispersion processes with low and high energy input and provides time-constant release signals. In this context it could be shown that the relative particle velocity $v_{rel}$ is the crucial agglomerate dispersion parameter in air flows. This set a serious limitation to the measurement technique, which can be used to study the particle release: aerosol sample feeding systems frequently expose the airborne particles to a maximum $v_{rel}$, which are larger than in the test-scenario, thus induce a second dispersion step, which distort the result of the particle release test. The designed "weak dispersion unit" is operated under such conditions ($v_{rel} = 25\ m{\cdot}s^{-1}$) that post-dispersion can be excluded for most aerosol characterisation instruments. Note that for powders with sufficient flowability, the operation of some commercial aerosol generators at similar conditions is also possible. The WDU is supplemented by an "intense dispersion unit" (rotating brush generator) operated at similar powder mass flow and at $v_{rel} = 100\ m{\cdot}s^{-1}$.

The combined test-scenario for low and high energy dispersion was employed to study the particles release from two iron oxide powders (a nano-structured one and a non-nanostructured one) and two filter test powders. The results allow to conclude that the combined test-scenario yields reproducible results on particle release, clearly demonstrates the impact of energy input and facilitates the grouping of powders according to their release characteristics.

Finally, the authors would like to initiate a discussion on how to deal with the fact, that current state-of-the-art sizing instrumentation does not allow a quantitative release study for very weak dispersion energies or - more generally - implicate remarkable dispersion stress for particles to be measured.

## Acknowledgement
This work is supported by the German Federal Ministry of Economic Affairs and Energy (BMWi) under grant IGF 17785/BR/1.

## References
[1] Kuhlbusch T A J, Asbach C, Fissan H, Göhler D and Stintz M 2011 *Part. Fibre Toxicol.* **8** 22
[2] Schneider T and Jensen KA 2008 *Ann. Occup. Hyg.* **52** 23-34
[3] Jensen K, Koponen I, Clausen P and Schneider T 2009 *J. Nanopart. Res.* **11** 133-46
[4] Tsai CJ, Wu CH, Leu ML, Chen SC, Huang CY, Tsai PJ and Ko FH 2009 *J. Nanopart. Res.* **11** 121-31
[5] Dahmann D and Monz C 2011 *Gefahrstoffe - Reinhalt. Luft* **71** 481-87
[6] Maynard A 2002 Ann. Occup. Hyg. 46 197-202
[7] Ogura I, Sakurai H and Gamo M 2009 *J. Phys. Conf. Ser.* **170** 012003
[8] Plitzko S, Gierke E, Dziurowitz N and Broßell D 2010 Gefahrstoffe - Reinhalt. Luft **70** 31-35
[9] Stahlmecke B, Wagener S, Asbach C, Kaminski H, Fissan H and Kuhlbusch TAJ 2009 *J. Nanopart. Res.* **11** 1625-35.
[10] Pohl M, Schubert H and Schuchmann HP 2005 *Chem. Ing. Tech.* **77** 258-62
[11] Brenner H 1958 *The Physics of Fluids* **1** 338-46
[12] Naue G, Liepe F, Mascheck HJ, Reher EO and Schenk R 1983 *Technische Strömungslehre I.* (Leipzig: Deutscher Verlag für Grundstoffindustrie).
[13] Zeidan M, Xu B, Jia X and Williams R 2007 *Chem. Eng. Res. Des.* **85** 1645-54
[14] Fuchs NA 1964 The mechanics of aerosols (Oxford: Pergamon Press)
[15] Kousaka Y, Okuyama K, Shimizu A and Yoshida T 1979. *J. Chem. Eng. Jpn.* **12** 152-59.
[16] Bagster D and Tomi D 1974 *Chem. Eng. Sci.* **29** 1773-83
[17] Zahradnicek A 1976 *Untersuchungen zur dispergierung von Quarz- und Kalksteinfraktionen im Korngrößenbereich 0.5 - 10 µm in strömenden Gasen.* (PhD thesis, Universität Karlsruhe)
[18] Zahradnicek A and Löffler F 1976 *Staub - Reinhalt. Luft* **36** 425-27
[19] Leschonski K, Röthele S and Menzel U 1984 Part. Part. Syst. Charact. **1** 161–66
[20] Bürgelt T, Adam R, Rudolph A and Große S 2013 *Filtrieren und Separieren* **27** 72-77
[21] Göhler D, Nogowski A, Fiala P and Stintz M 2013 *J. Phys. Conf. Ser.* **429** 01204
[22] Göhler D, Stintz M, Hillemann L and Vorbau M 2010 *Ann. Occup. Hyg.* **54** 615-24
[23] He M and Dhaniyala S 2013 *J. Aerosol Sci.* **61**, 13-26
[24] Göhler D and Stintz M 2014 *J. Nanopart. Res.* **16** 2520

Journal of Physics: Conference Series **617** (2015) 012030 doi:10.1088/1742-6596/617/1/012030

# Towards understanding the mechanisms and the kinetics of nanoparticle penetration through protective gloves

**L Vinches[1], C Peyrot[2], L Lemarchand[2], N Boutrigue[1], M Zemzem[1], K J Wilkinson[2], S Hallé[1] and N Tufenkji[3]**

[1] École de technologie supérieure – 1100 Notre-Dame Ouest, Montréal, QC, H3C 1K3 Canada

[2] Dept. de chimie, Université de Montréal – C.P. 6128, succ. Centre-ville, Montréal, QC, H3C3J7 Canada

[3] Dept. of chem. Engineering, McGill University – 3610, rue University, Montreal, QC, H3A0C5 Canada

E-mail: ludwig.vinches.1@ens.etsmtl.ca

**Abstract.** Parallel to the increased use of engineered nanoparticles (ENP) in the formulation of commercial products or in medicine, numerous health & safety agencies have recommended the application of the precautionary principle to handle ENP; namely, the recommendation to use protective gloves against chemicals. However, recent studies reveal the penetration of titanium dioxide nanoparticles through nitrile rubber protective gloves in conditions simulating occupational use. This project is designed to understand the links between the penetration of gold nanoparticles (nAu) through nitrile rubber protective gloves and the mechanical and physical behaviour of the elastomer material subjected to conditions simulating occupational use (i.e., mechanical deformations (MD) and sweat). Preliminary analyses show that nAu suspensions penetrate selected glove materials after exposure to prolonged (3 hours) dynamic deformations. Significant morphological changes are observed on the outer surface of the glove sample; namely, the number and the surface of the micropores on the surface increase. Moreover, nitrile rubber protective gloves are also shown to be sensitive to the action of nAu suspension and to the action of the saline solution used to simulate sweat (swelling).

## 1. Introduction

Engineered nanoparticles, particularly colloidal gold nanoparticles, have attracted enormous scientific and technological interest due to their ease of synthesis, chemical stability, and unique optical properties [1-3]. These characteristics allow the use of nAu for many applications as photocatalysts [4], as Raman sensors [5], in electronics components and in medicine [6, 7]. In the field of bioscience, nAu are used in chemical sensing, biological imaging, drug delivery, and cancer treatment [8-11]. For example, Ducan et al. (2010) have shown how gold nanoparticles can be employed as platforms for drug and biomacromolecule delivery [8]. In 2004, James et al measured a significant enhancement in radiotherapy treatment in mice using gold nanorods [11]. However, a number of studies have also highlighted the toxicity of functionalized or bare nAu [12-15]. In 2009 and 2012, two studies demonstrated the effect of gold nanoparticles on human and mouse spermatogenesis, respectively [16, 17]. Wiwanitkit et al showed that when human semen was placed in contact with a suspension of gold nanoparticles, 25% of the sperm cells became non-motile. Penetration of nAu into the sperm heads and tails was observed. Moreover, a study conducted on rat skin after dermal exposure to 15 nm, 102 nm and 198 nm bare nAu shows that the nAu can be observed in the deep layer of the skin [18]. Transmission electron microscopy (TEM) analysis reveals an accumulation of larger nAu mainly in dermis and epidermis layer, whereas the smaller nAu are observed in a deeper region. Other studies indicated the penetration of ENP through intact or damaged human skin [19-22].

To handle ENP, and particularly nAu, the use of protective gloves against chemicals has been recommended by several health and safety government agencies [23, 24]. However, recent studies have shown the poor efficiency of disposable protective nitrile rubber gloves against titanium dioxide nanoparticles in suspension and in powder form [25, 26].

This paper reports some physicochemical mechanisms which facilitate the penetration of nAu through disposable nitrile rubber gloves. Moreover, it also reports correlations between the penetration of nAu suspensions and the mechanical and physical behaviour of the elastomer material used in conditions simulating occupational use.

## 2. Materials

### *2.1. Protective gloves*

For this work, two models of disposable 100% nitrile rubber gloves (Nitri-care®3005PF and N-Dex®6005PF) from Showa Best Glove were selected. Both are non-latex, powder-free and silicon-free. In this paper, the gloves are identified as NBR-1 for Nitri-care®3005PF and NBR-2 for N-Dex®6005PF, and have a thickness of 73.2±3.0 µm and 117.0±6.5 µm, respectively. All samples were taken from the palm section or back of the gloves.

### *2.2. Nanoparticles*

Nanoparticle suspension in water was prepared containing 50 nm gold particles (nAu-50) coated with polyvinylpyrrolidone (Nanocomposix, San Diego, CA). The gold concentration is 0.05 mg/mL (NanoXact grade).

## 3. Methods

### *3.1. ENP penetration experimental setup*

This project used the same experimental test setup developed by Vinches et al. (Figure 1) [27]. It includes an exposure chamber to expose the outer surface of the glove samples to colloidal suspensions and a sampling chamber to collect the nAu with a sampling solution (saline solution) after their passage through the glove samples. Both chambers are separated by the sample. The test setup is also equipped with a probe (diameter 33.5mm × length 80mm) used to apply mechanical deformations via an electronically controlled interface. The probe head used corresponds to a conical-spherical geometry which simulates MD produced by a flexing hand [26]. The time profile of the sample MD consists in 50% out-of-plane deformations applied every 10 seconds [28].

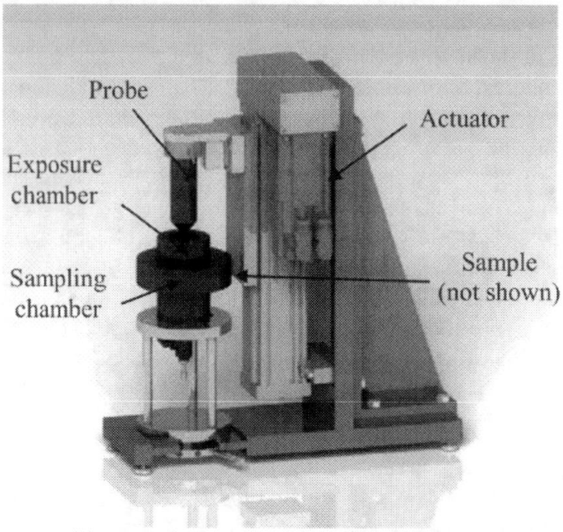


Figure 1. Isometric view of the test setup

### *3.2. Characterization of the nAu suspension*
The size distribution of the nAu was verified by statistical analysis of more than 100 particles imaged by TEM (Hitachi JEM-2100F). The hydrodynamic diameters were also measured by dynamic light scattering (DLS) (Mobius-Wyatt). To obtain statistically significant data, triplicate measurements were performed for all analyses. Thermogravimetric analysis (TGA) (Diamond TGA/DTA Perkin Elmer) was used to evaluate the presence of additives in the liquid carrier. Gradual evaporation of the liquid carrier occurred between 25 and 150 °C at a heating rate of 5°C/min.

### *3.3. Morphological and chemical characterization of the protective gloves surface*
The outer surface morphology of three samples for each nitrile rubber glove was analysed by scanning electron microscopy (SEM) (Hitachi S3600N – $V_{acc}$ = 15 kV – magnification × 1000). ImageJ image processing software was used to perform statistical analyses (more than 600 pores for NBR-1 and 700 pores for NBR-2) on the measured diameter and the surface area of the pores [28]. Energy dispersive spectroscopy (EDS) (Oxford INCA X-sight 200) analyses were conducted to determine the presence of gold in the outer surface of both glove materials.

### *3.4. nAu sampling protocol and detection method*
For all the tests, physiological solution was used as the sampling solution. This solution is composed of 0.5 % (w/w) sodium chloride, 0.1 % (w/w) lactic acid and 0.1 % (w/w) urea [29]. The pH of the solution is adjusted to 6 with 1% ammonia solution. The sampling chamber is filled with saline solution up to the level where it makes contact with the inner surface of the glove sample. Before each test, the absence of contamination of the sampling chamber by nAu was verified with a control sample produced by rinsing the sampling chamber with the sampling solution. Gold concentrations were analysed by inductively coupled plasma mass spectrometry (ICP-MS, PerkinElmer NexION 300X). To decrease the sodium chloride concentration, all samples were diluted 10 times in MilliQ water (resistivity 18.2 MΩ·cm and TOC < 2 $\mu gC \cdot L^{-1}$) before the ICP-MS analysis.

### *3.5. Swelling measurements*
In occupational use, the outer surface of the protective glove is in contact with the ENP and the inner surface with the skin. During the wearing time of the glove, a microclimate is produced inside the glove (sweat, temperature). To evaluate the effect of the nAu suspension and the physiological solution on the material properties, length change measurements were performed on glove samples.

Rectangular samples (5 × 60 mm) were taken from the palm section or the back of the gloves. The samples were immersed in a suspension of nAu in water or in saline solution and at regular intervals, samples were removed from the liquid and their length measured using a calliper (±0.01 mm). The length change data were analyzed using Eq. (1):

$$\Delta L(t) = \frac{L_t - L_0}{L_0} \quad (1)$$

with $L_t$ being the length at $t$ time and $L_0$ the corresponding value before immersion. To obtain statistically significant data, triplicate measurements were carried out for each type of glove.

## 4. Results and discussion

### *4.1. Characterization of the nAu suspension*

The size distribution of the nAu-50 was measured by TEM (Figure 2). The particle sizes measured were 47.9±4.7 nm. These results confirmed the manufacturer's data. DLS analysis yielded hydrodynamic diameters for the nAu-50 of 77.7±0.7 nm (Figure 3).

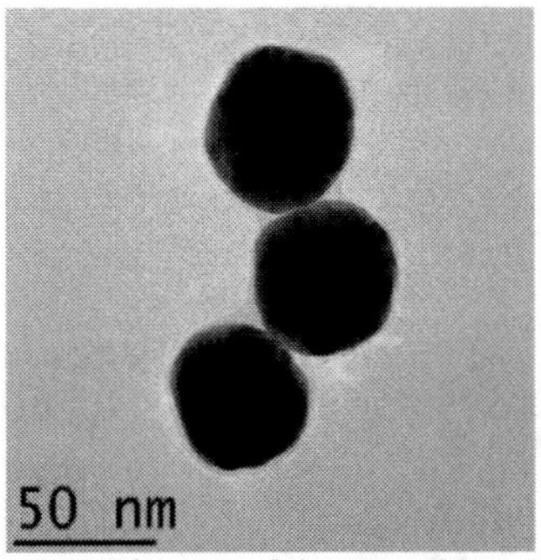


Figure 2. TEM image of the nAu-50 suspension

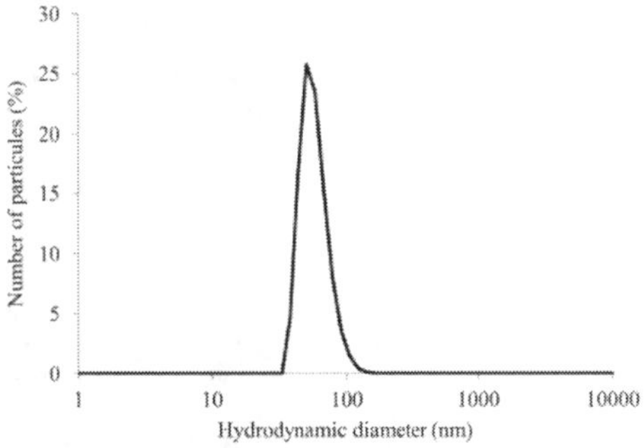


Figure 3. Distribution in number of the hydrodynamic diameter of nAu-50

Finally, thermogravimetric analyses were carried out to verify the presence of additives such as stabilizing agents in the stock colloidal suspension. Figure 4 compares the spectrum obtained with nAu-50 and that obtained with MilliQ water. A change in slope is observed near 80°C (black arrow) on the nAu-50 spectrum corresponding to the presence of one or more additives in the carrier liquid. These additives may play a significant role in the penetration of nAu through the protective gloves. Moreover, the shift towards higher temperatures is due to the presence of the nAu to which the solvent molecules adsorb, thus increasing the vapour pressure [30].

Journal of Physics: Conference Series **617** (2015) 012030 doi:10.1088/1742-6596/617/1/012030

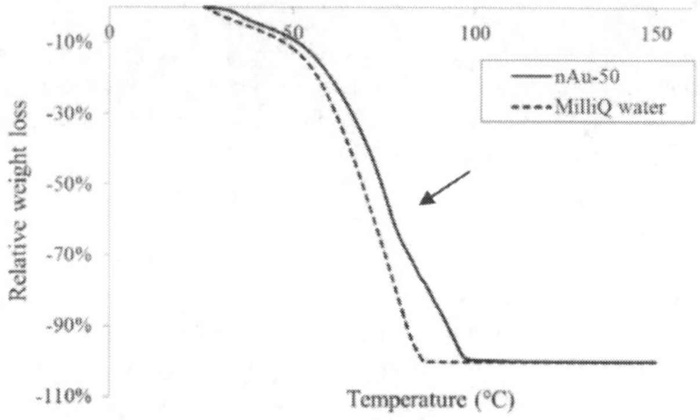


Figure 4. TGA spectra of nAu-50 in water and milliQ water

*4.2. Characterization of the protective gloves surface*

On the outer glove surface which will be placed in contact with the ENP and the probe, micrometer-size pores can be observed for both native nitrile rubber gloves (Figure 5).

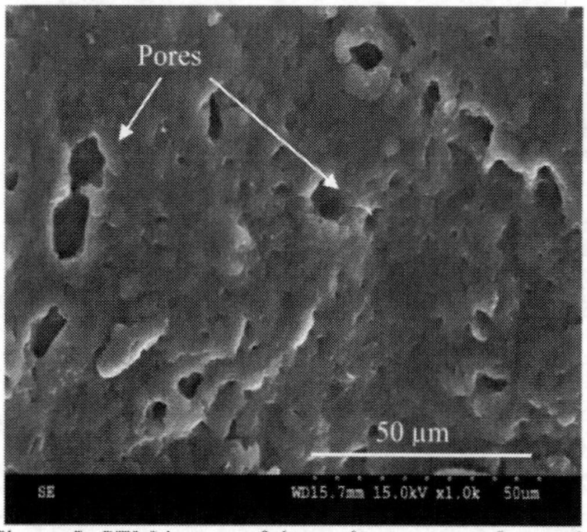


Figure 5. SEM image of the native outer surfaces of NBR-1

Statistical analysis provides three important data: (i) the number of pores as a function of pore diameter (Figure 6), (ii) the mean number of pores per unit area and, (iii) the area occupied by the pores per unit area of the glove (Table 1). Figure 6 displays that the diameter distribution of native NBR-2 pores is centered on 2 µm. In the case of native NBR-1, the distribution is also centered on 2 µm (data not shown). The number of pores per unit area is 17% greater for NBR-2 than for NBR-1. Similarly, the area occupied by the pores is 21% greater for NBR-2 (Table 1).

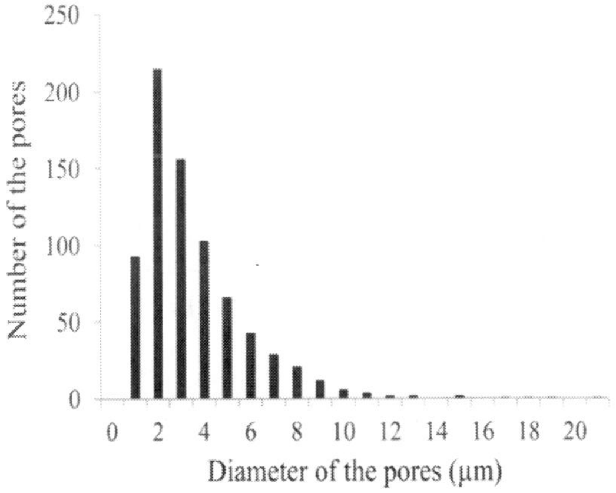


Figure 6. Statistical distribution of pore diameters for native NBR-2

Table 1. Number of pores per unit area and area occupied by the pores of both native gloves

| | NBR-1 | NBR-2 |
|---|---|---|
| **Number of pores/mm²** | 5654 | 6839 |
| **Area occupied by the pores (%)** | 9.6 ± 2.1 | 12.1 ± 3.4 |

The chemical elements present on the outer were determined by EDS. Few differences can be noted between both glove types (data not shown), but neither contains gold.

*4.3. nAu penetration: ICP-MS analyses*

Glove samples were exposed to the nAu-50 in water while being simultaneously subjected to a 50%-MD every 10 seconds for up to 3h (for a total of 620 MD). The same tests were performed without nAu and in this case, the gold concentration is lower than the limit of detection (LOD) of the ICP-MS (0.048 $\mu g \cdot L^{-1}$). For NBR-1, a clear increase in gold concentration in the saline solution can be observed after 3h of 50%-MD as a result of the presence of nAu in the sampling chamber (Figure 7).

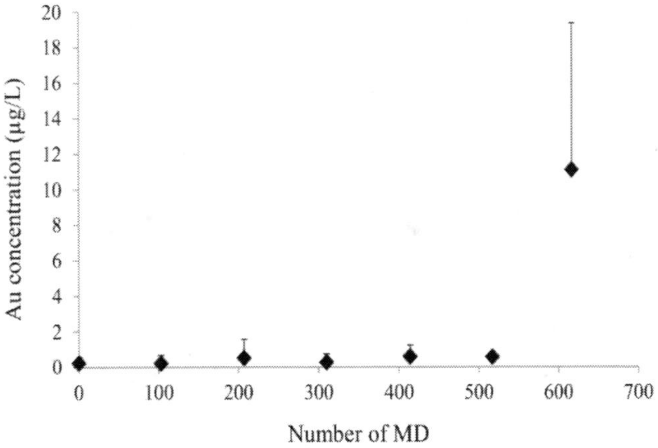


Figure 7. Variation in gold concentration as a function of test duration for NBR-1 samples exposed simultaneously to 50%-MD and nAu-50

Journal of Physics: Conference Series **617** (2015) 012030 doi:10.1088/1742-6596/617/1/012030

When NBR-2 was exposed to nAu-50 in the presence of 50%-MD, no significant gold penetration was measured. This result is probably due to the greater thickness of this protective glove material.

*4.4. Behaviour of the protective glove outer surface submitted to MD*

Statistical analysis was performed to determine the effects of the MD on the outer surface of both protective nitrile rubber gloves. Firstly, the pore size distribution for NBR-1 native gloves and the same model after 620 50%-MD (3 hours of testing) is compared (Figure 8). After 620 50%-MD, the pore size distribution is centered on 3 µm. This represents a shift of 1 µm compared to the NBR-1 native glove. Thus, mechanical deformations increase the pore diameters but their number by mm² and the occupied area are smaller after 3h 50%-MD (Table 2). Interestingly, for NBR-2, all pores are completely missing are exposure to 3h of 50%-MD (Figure 9). This phenomenon is due to abrasion induced by the contact between the probe and the glove sample [28]. For NBR-2, abrasion likely removes the first layer of the glove material.

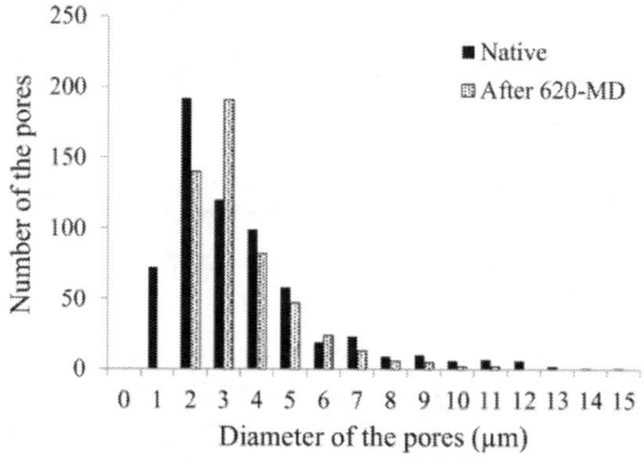


Figure 8. Statistical distribution of pore diameters for native NBR-1 after 620 50%-MD

Figure 9. SEM images of the outer surfaces of NBR-2 a) native and b) after 620 50%-MD

Table 2. Number of pores per unit area and area occupied by the pores after 620 50%-MD

| | NBR-1 | NBR-2 |
|---|---|---|
| **Number of pores/mm²** | 4632 | N/D |
| **Area occupied by the pores (%)** | 6.7 ± 1.6 | N/D |

N/D: Not detectable

### *4.5. Swelling measurements*

Swelling measurements were carried out by recording the length of glove samples after immersion in nAu-50 suspension, in physiological solution or in MilliQ water. Figure 10 compares the length change ratio (LCR) recorded for NBR-1 samples immersed in the three different solutions. For each case, a steady increase in sample length is observed until a plateau is reached. The LCR after 3h of immersion reaches 3.8 % in the three liquids. Thus, the swelling behaviour is the same for an immersion time lower than 3h (maximum duration of penetration tests). However, following longer immersion times, differences are observed in the LCR for the different soak solutions. After 175 hours of immersion, the maximum LCR is similar for nAu suspension and MilliQ water (between 13.6% and 14.6% for MilliQ water and nAu-50, respectively), but it is considerably lower for the saline solution (8.5%). This difference in swelling behavior can be attributed to differences in the solvents' ability to penetrate (diffuse) into the elastomer. According to these results, gold nanoparticles (in nAu suspension) do not have a significant effect on the swelling of the glove material when compared to MilliQ water, the liquid carrier for the nanoparticles. On the other hand, the chemical composition of the physiological solution seems to decrease the swelling phenomenon compared to MilliQ water. Ongoing studies are underway to confirm this hypothesis.

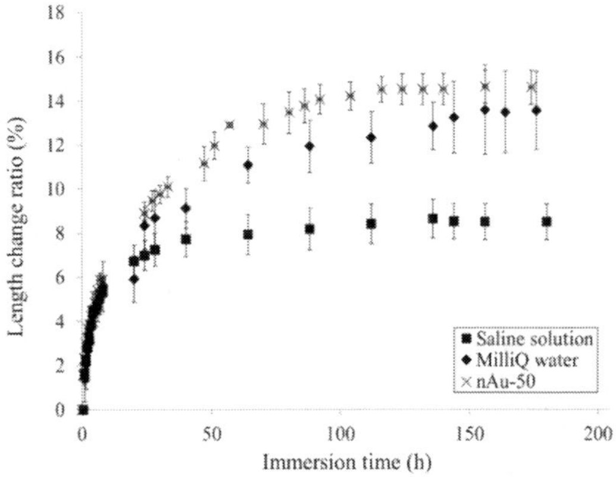


Figure 10. Length change ratio for NBR-1 as a function of immersion time in nAu suspension in water, saline solution or MilliQ water

Table 3 summarizes the LCR after 3h of immersion and the maximum LCR (value reached at the plateau) for both protective nitrile rubbers. Unlike NBR-1, a difference can be observed for LCR after 3 hours of immersion between nAu-50 suspension (2.2%) and the other solutions (3.7% and 3.6% in milliQ water and in saline solution respectively). The maxima of LCR are increasingly important for NBR-2 than NBR-1. These results are consistent with those obtained by Vinches et al. [28].

Table 3. LCR after 3h of immersion and the maximum LCR for NBR-1 and NBR-2 as a function of immersion time in nAu-50 suspension in water, in saline solution and in MilliQ water

| | *NBR-1* | | *NBR-2* | |
|---|---|---|---|---|
| | LCR after 3 hours (%) | Maximum of LCR (%) | LCR after 3 hours (%) | Maximum of LCR (%) |
| MilliQ water | 3.7 ± 1.0 | 13.6 ± 1.7 | 3.7 ± 0.5 | 19.0 ± 0.8 |
| Saline solution | 3.9 ± 0.5 | 8.5 ± 0.8 | 3.6 ± 0.2 | 13.0 ± 0.6 |

| | | | | |
|---|---|---|---|---|
| nAu-50 | 3.8 ± 0.2 | 14.6 ± 0.7 | 2.2 ± 0.4 | 13.0 ± 1.1 |

## 5. Conclusion

This study has confirmed the penetration of ENP – in this case nAu – through nitrile protective gloves after repeated deformations simulating occupational use. Nanoparticle penetration has been recorded after 620 50%-MD corresponding to 3h of wearing gloves. Secondly, mechanical and physicochemical phenomena facilitating ENP penetration have been quantified. Significant changes have been observed on the outer surface of the glove material. Indeed, the number and the area occupied by the pores decrease after 3h of 50%-MD and the first layer of glove material were torn from the surface. Moreover, the swelling of the elastomeric membrane has been measured in the presence of nAu-50 as well as in physiological solution simulating the sweat produced inside the glove during their use. To conclude, the disappearance of the first layer of glove materials after repeated deformations, the swelling of elastomeric membrane in contact with the colloidal suspension, exposure to sweat and the measured decrease of the material resistance due to glove use may lead to the weakening of the efficiency of protective gloves used to handle ENP. Ongoing studies are being carried out to further identify the mechanisms facilitating nanoparticle penetration through glove materials.

## Acknowledgements

The authors would like to acknowledge the Institut de Recherche Robert-Sauvé en Santé et Sécurité du Travail (IRSST) and NanoQuébec for financial support. Moreover, we would like to thank the contribution of M. Ben Salah (École de technologie supérieure) to the project as well as the contribution of Madjid Hadioui (Université de Montréal).

## References

1. Zhou, J., et al., *Functionalized gold nanoparticles: Synthesis, structure and colloid stability.* Journal of Colloid and Interface Science, 2009. **331**(2): p. 251-262.
2. Balasubramanian, S.K., et al., *Characterization, purification, and stability of gold nanoparticles.* Biomaterials, 2010. **31**(34): p. 9023-9030.
3. Zhao, P., N. Li, and D. Astruc, *State of the art in gold nanoparticle synthesis.* Coordination Chemistry Reviews, 2013. **257**(3–4): p. 638-665.
4. Hillyer, J.F. and R.M. Albrecht, *Gastrointestinal persorption and tissue distribution of differently sized colloidal gold nanoparticles.* Journal of Pharmaceutical Sciences, 2001. **90**(12): p. 1927-1936.
5. Gratton, S.E.A., et al., *Nanofabricated particles for engineered drug therapies: A preliminary Biodistribution study of PRINT (TM) nanoparticles.* Journal of Controlled Release, 2007. **121**(1-2): p. 10-18.
6. Tiwari, P., et al., *Functionalized Gold Nanoparticles and Their Biomedical Applications.* Nanomaterials, 2011. **1**(1): p. 31-63.
7. Rojanathanes, R., et al., *Gold Nanoparticle as an Alternative Tool for a Urine Pregnancy Test.* Taiwanese Journal of Obstetrics and Gynecology, 2008. **47**(3): p. 296-299.
8. Duncan, B., C. Kim, and V.M. Rotello, *Gold nanoparticle platforms as drug and biomacromolecule delivery systems.* Journal of Controlled Release, 2010. **148**(1): p. 122-127.
9. Dykman, L. and N. Khlebtsov, *Gold nanoparticles in biomedical applications: recent advances and perspectives.* Chemical Society Reviews, 2012. **41**(6): p. 2256-2282.

10. Huang, X., et al., *Cancer Cell Imaging and Photothermal Therapy in the Near-Infrared Region by Using Gold Nanorods.* Journal of the American Chemical Society, 2006. **128**(6): p. 2115-2120.
11. James, F.H., N.S. Daniel, and M.S. Henry, *The use of gold nanoparticles to enhance radiotherapy in mice.* Physics in Medicine and Biology, 2004. **49**(18): p. N309.
12. Goodman, C.M., et al., *Toxicity of Gold Nanoparticles Functionalized with Cationic and Anionic Side Chains.* Bioconjugate Chemistry, 2004. **15**(4): p. 897-900.
13. Tedesco, S., et al., *Oxidative stress and toxicity of gold nanoparticles in Mytilus edulis.* Aquatic Toxicology, 2010. **100**(2): p. 178-186.
14. Alkilany, A. and C. Murphy, *Toxicity and cellular uptake of gold nanoparticles: what we have learned so far?* Journal of Nanoparticle Research, 2010. **12**(7): p. 2313-2333.
15. Das, S., et al., *Comparative analysis of stability and toxicity profile of three differently capped gold nanoparticles for biomedical usage.* BioMetals, 2012. **25**(5): p. 1009-1022.
16. Zakhidov, S.T., et al., *Effect of gold nanoparticles on mouse spermatogenesis.* Biology Bulletin, 2012. **39**(3): p. 229-236.
17. Wiwanitkit, V., A. Sereemaspun, and R. Rojanathanes, *Effect of gold nanoparticles on spermatozoa: the first world report.* Fertility and Sterility, 2009. **91**(1): p. e7-e8.
18. Sonavane, G., et al., *In vitro permeation of gold nanoparticles through rat skin and rat intestine: Effect of particle size.* Colloids and Surfaces B: Biointerfaces, 2008. **65**(1): p. 1-10.
19. Labouta, H.I., et al., *Mechanism and determinants of nanoparticle penetration through human skin.* Nanoscale, 2011. **3**(12): p. 4989-4999.
20. Larese, F.F., et al., *Human skin penetration of silver nanoparticles through intact and damaged skin.* Toxicology, 2009. **255**(1–2): p. 33-37.
21. Rouse, J.G., et al., *Effects of mechanical flexion on the penetration of fullerene amino acid-derivatized peptide nanoparticles through skin.* Nano Letters, 2007. **7**(1): p. 155-160.
22. Sadrieh, N., et al., *Lack of Significant Dermal Penetration of Titanium Dioxide from Sunscreen Formulations Containing Nano- and Submicron-Size TiO2 Particles.* Toxicological Sciences, 2010. **115**(1): p. 156-166.
23. Ostiguy, C., et al., *Les nanoparticules de synthèse - Connaissances actuelles sur les risques et les mesures de prévention en SST - 2e édition.* 2009, Institut de recherche Robert-Sauvé en santé et en sécurité au travail p. 159.
24. OECD, *Current developments/activities on the safety of manufactured nanomaterials - Tour de table at the 7th meeting of the working party on manufactured nanomaterials.*, in *Series on the safety of manufactured nanomaterials.* 2010: Paris.
25. Vinches, L., et al., *Experimental evaluation of the penetration of TiO2 nanoparticles through protective clothing and gloves under conditions simulating occupational use.* Nanoscience Methods, 2013: p. 1-15.
26. Vinches, L., et al., *Which Gloves Are Efficient To Protect Against Titanium Dioxide Nanoparticles In Work Conditions?* International Journal of Theoretical and Applied Nanotechnology, 2014. **2**: p. 24-27.
27. Vinches, L., et al., *Experimental evaluation of the resistance of nitrile rubber protective gloves against TiO2 nanoparticles in water under conditions simulating occupational use.* Journal of Physics: Conference Series, 2013. **429**(1): p. 012056.
28. Vinches, L., et al., *Physical phenomena facilitating the penetration of solutions of TiO2 nanoparticles through protective gloves.* International Journal of Theoretical and Applied Nanotechnology, 2013. **1**(2): p. 1-16.
29. E.C.F., *Reference test method for release of nickel from product intended to come in direct and prolonged contact with the skin: BS EN 1811:1998+A1:2008.* 1998: Brussels CEN.
30. Vinches, L., et al., *Swelling of Protective Gloves in Commercial TiO2 Nanoparticles Colloidal Solutions.* International Journal of Theoretical and Applied Nanotechnology, 2012. **1**(1): p. 45-51.

Journal of Physics: Conference Series **617** (2015) 012031 doi:10.1088/1742-6596/617/1/012031

# Nanomaterials in construction and demolition – how can we assess the risk if we don't know where they are?

**Wendy Jones[1], Alistair Gibb[1], Chris Goodier[1], Phil Bust[1], Jie Jin[2] and Mo Song[2]**

[1] School of Civil and Building Engineering, Loughborough University, LE11 3TU, UK

[2] Department of Materials, Loughborough University, LE11 3TU, UK

E-mail: w.jones2@lboro.ac.uk

**Abstract**. This research, funded by the Institution of Occupational Safety and Health in the United Kingdom, has used a combination of literature review, web searching and unstructured interviews with a range of industry professionals to compile a list of products used in construction and the built environment which might contain nanomaterials. Samples of these products have been analysed using Scanning Electron Microscopy and Energy Dispersive X-Ray Spectroscopy to investigate whether nanomaterials are actually present and to what extent. Preliminary results of this testing are presented here. It is concluded that there is a discrepancy between the academic literature and the reality regarding the current application of nanomaterials in the construction industry and the built environment. There are also inaccuracies and deficiencies in the information provided by manufacturers which makes it difficult to accurately assess the location and application of nanomaterials within the industry. Further testing is planned to evaluate the risk of nanoparticle release from nano-enabled building products at their end of life by reproducing common demolition and recycling processes such as crushing, grinding, burning and melting. Results of this will form the basis of practical guidance for the construction, demolition and recycling industries to help them identify where particular protection or control measures may be appropriate as well as providing reassurance where no additional action is required.

## 1. Introduction

Construction is an important arena for the application of nanomaterials as the high volume of concrete, steel and glass used offers a substantial market for novel products. It is, unfortunately, also a sector where there is a high potential for causing harm, as the industry employs almost 15 million people across Europe who typically have high exposure to biological, chemical and ergonomic hazards [1]. Risks from the physical demands of the work are compounded by inadequate training in some parts of the sector and there are additional challenges associated with the highly varied and rapidly changing nature of the work environment. There is particular anxiety within the industry regarding the possible health risks of developments in nanotechnology. The construction and demolition industries deal on a daily basis with the legacy of asbestos, which was once lauded as a 'magical' material and used widely before its association with mesothelioma, lung cancer and asbestosis was recognised. There is concern that nanomaterials may represent a similar scenario, particularly in the light of findings that some carbon nanotubes may affect the body in similar ways to asbestos fibres due to their high aspect ratio and low solubility [2].

This paper describes a research project, funded by the Institution of Occupational Safety and Health (IOSH) in the UK, which addresses the application of nanomaterials in the built environment, in particular their impact in the demolition sector. Preliminary results are presented from the first year of this three year study, and the planned later stages of work are described.

The aims of the research include:

- to identify where nanomaterials are used in the built environment

A number of useful summary papers have explored this issue [3-8]. However, these are based largely on published academic literature and web searches, and are generally more effective at describing where nanomaterials have the potential to be used rather than where they are actually used in practice. There is no current requirement to label building materials which are nano-enabled, although recent legislation in France, Belgium and Denmark requires central reporting or registration of nano-enabled products.

- to assess the risks of nanoparticle release during refurbishment, demolition and recycling

Demolition and refurbishment tasks introduce particular challenges as dust generation is common and the exact provenance of building materials may not be known. High levels of recycling (95% or higher in the UK) [9] increase the exposure potential for workers stripping building materials out in an enclosed environment and also create additional exposure potential downstream.

- to generate guidance on how to manage this

The demolition industry in particular has had to adapt its practices to accommodate the presence of asbestos and other toxic materials in the built environment. This is only possible where there is a good understanding of where such materials might be found. User-friendly guidance on the level of risk posed by nano-enabled products is needed to support informed decisions regarding the demolition of buildings which may contain these materials. At the same time, there is a benefit for the construction industry as a whole in understanding the level of risk involved, to ensure that disproportionate anxieties do not prevent the industry from taking full advantage of the benefits of these new technologies.

## 2. Methods

The literature review has encompassed wider academic literature in addition to the summary papers mentioned above. It has also used web-based data, in particular the nano product inventory run by the Project on Emerging Nanotechnologies [10], The Nanodatabase in Denmark [11], and the specific inventory of nano-enabled construction products hosted by The Center for Construction Research and Training (CPWR) in the USA [12]. A further source of data is manufacturers' product sites, identifying construction materials either labelled 'nano' or having novel properties which might indicate nano-enablement.

Based on the literature, a summary table has been produced of construction products which might be nano-enabled. Samples of many of these have been collected either by requesting products directly from manufacturers or suppliers, buying them, or collecting offcuts from users. Preliminary testing of samples with Scanning Electron Microscopy (SEM) and Energy Dispersive X-Ray Spectroscopy (EDX) has been used to identify the presence of nanoparticles and the chemical composition of materials. Both the identification of possible nano-enabled products and the collection and analysis of samples are ongoing.

Interviews with builders and project managers, visits to construction sites and discussions with manufacturers are providing data regarding the current use of nano-enabled construction products. Interviews with demolition contractors and recycling specialists are being used to identify the common processes involved in recycling and demolition.

## 3. Preliminary findings

*3.1 .Overview of nanomaterials in construction*

Interviews with those working in the construction industry have confirmed the relatively conservative nature of the sector. There is a very strong focus on cost effectiveness, with the suggestion that 'market forces' are the key consideration when selecting materials. Projects generally include the cheapest materials which will fulfil a particular requirement and, as one manager observed, *"they just want us to put bricks on top of bricks."* The implication is that novel materials are not widely used, nor are they likely to be unless there are financial or regulatory reasons to do so.

Set against this is a quote from Bill Looney of AECOM who has suggested that *"By 2025, over 50 percent of building materials are expected to contain nanomaterials as more take advantage of these lighter, stronger, and more energy efficient materials"* [13]. If this prediction is accurate, we can expect to see a massive expansion of nano-use in the sector over the next 10 years. Although this seems unlikely given the generally slow evolution of materials and techniques, we may be at a tipping point where the use of new products will become more practical and/or cost effective, perhaps for the regulatory reasons mentioned above. For example, discussion with those working in the glazing sector has found that the use of soft-coated (i.e. nano-enabled) glass has increased rapidly in the UK in recent years as it is the most effective way of fulfilling the insulation requirements of the building regulations.

This research to date has highlighted the difficulties of accurately identifying nano-enabled products. Materials which contain nanoparticles are not required to be labelled as such. Material safety data sheets have been gathered for many products and rarely specify that the product contains nanoparticles. Email exchanges with manufacturers have met with a variety of responses – some are open regarding the presence of nanomaterials, some are reluctant to provide detailed information and some simply fail to respond.

In addition, many materials and products which are labelled as 'nano' do not actually contain nanoparticles, as this extract from a manufacturers' product literature illustrates: *"[the] name derives from its nanotechnology,... [the company] does not use Nano particles in any of its products."*

In the current study, a supplier was contacted after microscopy of a purchased sample failed to identify the presence of nanomaterials; the response given was that *"Nanocoating is purely a name given to the treatment we offer for the [product], rather than a description of the material itself."*

*3.2. Common categories of nanomaterials in construction*

The products identified by this research as being potentially nano-enabled are summarised in table 1. Samples of 35 of these products have been acquired. This section will proceed by providing further details regarding the four most common product types identified and illustrating these with initial results of microscopy. This table will be developed further during the course of the research, and sample collection will also continue.

**Table 1**. Number of construction products identified as being potentially nano-enabled.

| Product type | Number of potential construction products |
|---|---|
| Surface coatings | 58 |
| Glass | 23 |
| Concrete | 21 |
| Insulation | 10 |
| Metal | 8 |
| Composites | 1 |
| Other (roofs, flooring, lighting) | 16 |

*3.2.1. Concrete.* Microsilica (also known as silica fume) has been used as a concrete additive for over 20 years. Although the average particle size is around 150 nm, it is classed as a nanomaterial under the EU definition as more than 50% of the particles are less than 100 nm in size (although these account for only a small proportion of the material by mass). The small spherical particles contribute to concrete being self-compacting, dense (and therefore stronger at lower thicknesses), gaining rapid strength during curing, or having good flow for pumping to high levels. Most of the major concrete producers include microsilica-enabled concretes in their portfolio, although these are not always labelled as such. It was identified from interviews with industry professionals that microsilica-enabled concretes account for only a small proportion of the concrete market in the UK, largely because there are many other additives which provide similar properties and are cheaper or easier to use. Nanosilica (which is finer than microsilica) can also be added to concrete to produce a very high performance material but appears to be even less widely used.

The SEM results in figure 1 are from a sample of microsilica concrete and show nanoparticles to be present, but only a relatively small number. Discussions with professionals in the concrete industry have highlighted that much of the microsilica added at the mixing stage participates in the concrete reaction and is therefore no longer present as original microsilica particles once the material is cured. This would explain the small number of nanoparticles visible in figure 1.

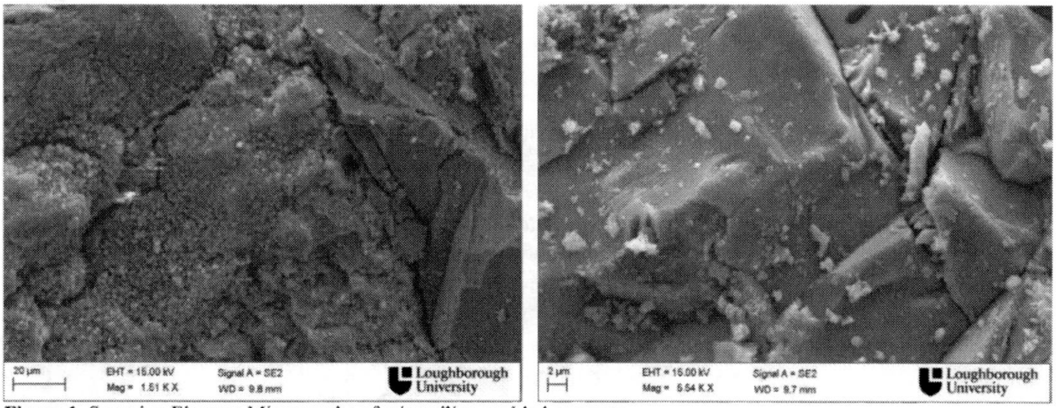


**Figure 1.** Scanning Electron Micrographs of microsilica-enabled concrete.

A second additive which can be used in concrete in its nano form is titanium dioxide, which is typically used to keep concrete clean and white or to absorb pollutants from the environment. It is used even less commonly than microsilica due to its high cost; where it is used, it is as a surface layer rather than being present throughout a concrete structure.

As figure 2 shows, only a very small number of nanoparticles are visible in the sample of nanotitanium concrete tested. Further work will be required to assess whether this unexpected result reflects the true state of nanotitanium concrete, or whether it demonstrates a poorly prepared sample. It is widely acknowledged that dispersion of nanoparticles is a major challenge in construction materials.

**Figure 2.** Scanning electron micrographs of nanotitanium-enabled concrete.

In conclusion, concrete is currently used to a limited extent in nano-enabled forms. Preliminary results suggest that the presence of nanoparticles in the cured concrete is also relatively low. Further testing will be required to confirm this and to assess how the material behaves under demolition-related processes. However, given that destruction of ordinary (non-nano) concrete generates large numbers of nano-sized particles [14], it is possible that the inclusion of nanomaterials such as microsilica (and maybe titanium) do not significantly increase any risk: and will remain less important than the existing risks from quartz silica, and can be managed through similar control measures.

*3.3.2. Glass.* Window glass appears to be the most widely used nano-enabled construction material. Almost all new buildings include windows which have nano coatings to improve thermal insulation, with these having displaced the more traditional 'hard' coatings in recent years. In addition, many conservatories and other glass roofed buildings have self-cleaning coatings, and the use of nanosilica intumescent layers has contributed to high grade fire safety glass for around 30 years. Other nano-enabled properties are available such as solar protection, anti-fogging, and electro chromic capability (which allows the user to vary light penetration through a window) but are less frequently used.

Glazing nanolayers are extremely thin (e.g. the Pilkington website gives a figure of 15 nm) [15]. As a consequence, the quantities of nanomaterials involved are low. This is demonstrated by the willingness of some glass manufacturers to reuse offcuts of coated glass within their float glass production, which is a highly sensitive process. Testing of nano-coated glass within this project is due to begin in the near future.

*3.3.3. Insulation materials.* The most commonly reported nano-enabled insulation materials are those based on silica aerogels. These materials are 96% air and have been described as *"the most effective thermal insulation on earth"* [16]. They are generally constructed as blankets which are wrapped around pipes or other structures or applied to plasterboard. They may also be used as insulation in translucent walls such as Kalwall. They are commonly described in manufacturers' and academic literature as being 'nanoporous' [3] or nanostructured and as not generally containing nanoparticles. Two different brands of aerogel blanket have been tested to date, the results are shown in figures 3 and 4.

Insulation blanket A, as shown in figure 3, is constructed of fibres which then have micro and nano sized particles on their surface. The material is very dusty and releases particles when touched. Insulation blanket B is much less dusty in use. It is still constructed of particles and fibres, but the particles are not at the nanoscale (figure 4).

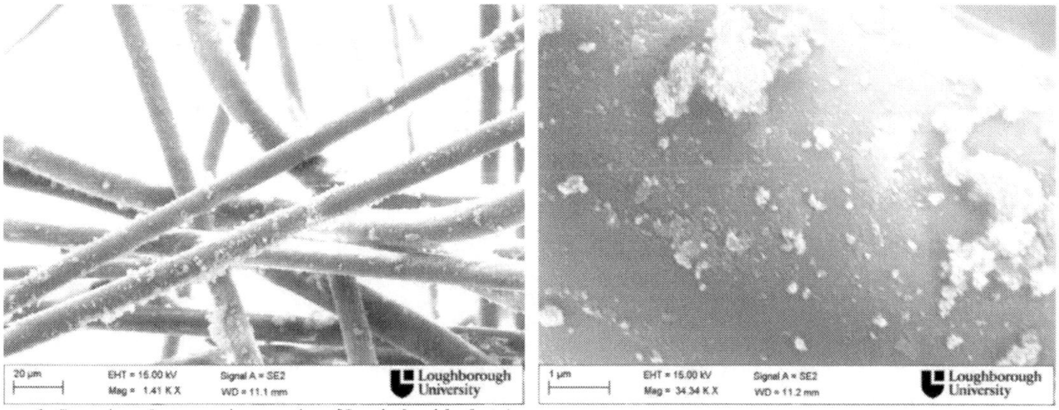


**Figure 3.** Scanning electron micrographs of Insulation blanket A.

**Figure 4.** Scanning electron micrographs of Insulation blanket B.

In conclusion, not all insulation blankets are the same in their construction or their potential to release nanoparticles, despite being generally considered as a homogenous group within the literature. This topic hence warrants further consideration. Although amorphous nanosilica is commonly considered to be a relatively low risk material, it is important to assess the quantity and nature of particles released through use/removal of these products and compare these to the levels associated with possible health effects in the literature, for example those discussed by Napierska [17].

*3.3.4. Surface coatings.* Surface coatings are the most diverse nano-enabled products in the building industry. They are available for domestic use, are included in professionally applied coatings and paints, and can be incorporated on the surfaces of products such as ceiling tiles or toilets during manufacture. Many are silica based, providing water resistant or anti-graffiti properties. Others include titanium to enable easy cleaning and pollution absorption. A third category are those which include silver as an anti-microbial, although it can be difficult to distinguish between products which contain silver at the nanoscale and those which are based on different forms.

Figure 5 shows the structure of a surface render which is marketed as being easy to clean and thermally insulating, and as containing 'nano-quartz'. As the micrographs show, the material contains both nano and micro sized particles. Further investigation is required to assess whether the particles are actually quartz, or whether they are amorphous silica, which seems more likely.

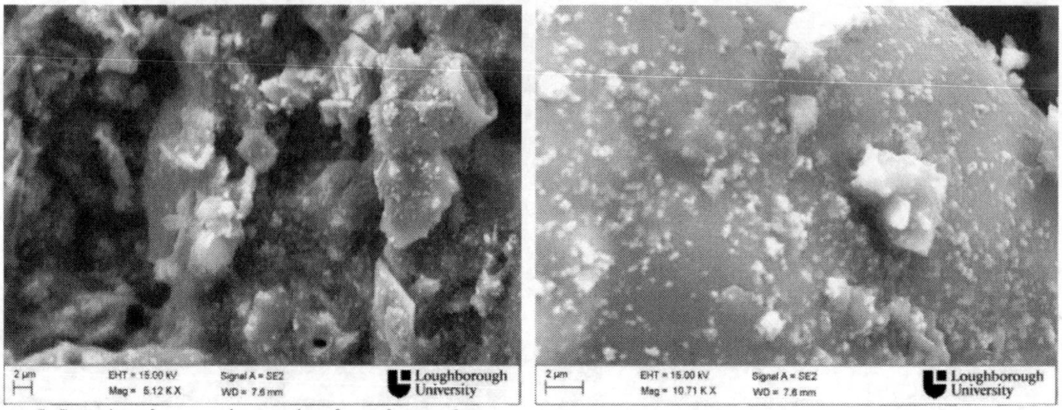


**Figure 5.** Scanning electron micrographs of a surface render.

The material shown in figure 6 is marketed for worksurfaces and other internal building uses which require scratch resistance and easy clean properties. It is described as a nanotechnology material but the micrographs suggest that it does not contain nanoparticles.

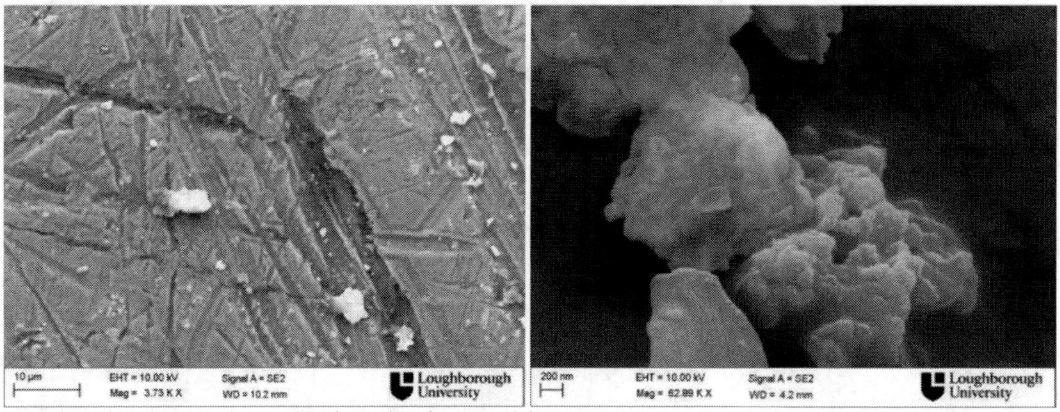


**Figure 6.** Scanning electron micrographs of a worksurface material.

In conclusion, there are a wide variety of nanomaterials used in surface coatings. It is possible that many products which are sold as being nano-enabled do not contain nanomaterials or do so in extremely small quantities. It is difficult at this stage to determine which, if any, might present potential health risks and drawing generic conclusions is likely to remain difficult due to the wide variations between products.

### *3.4. Carbon Nanotubes in construction*

Carbon nanotubes (CNT) are widely discussed within the academic literature with reference to construction, identifying that they have the potential to enable concretes which are particularly strong or self-healing; or which are self-sensing and can transmit information regarding the state of deterioration of the material [3-8]. The papers which summarise the use of nano in construction generally refer to the benefits of adding 1% CNT to concrete, although it is not clear what the source of this figure is, as the wider literature explores a variety of concentrations as well as many different shapes and sizes of tube.

The literature also acknowledges that there are difficulties with the incorporation of CNT into commercial products due to high cost, low availability and difficulties dispersing tubes effectively. Interviews for the current research with those working in the industry have similarly indicated that the technology is some way from commercial use. However in recent months there have been developments in this field, with a recent report by Eden Energy stating that they are close to production and that *"[a] preliminary trial in USA during the next 3 months of Eden's CNT enriched concrete on a suitable roadway or similar area is scheduled"* [18].

A second market for the use of carbon nanotubes in construction is in coatings or thin films. These take advantage of the material's conductive capability and have the capacity to be used for heatable paints and to shield rooms from radio and microwaves. It is unclear how extensive this use is and in fact whether these materials are currently being sold to users or to interim manufacturers who will carry out further development and testing.

## 4. Ongoing research

Continued development of the list of potential nano-enabled products is ongoing, together with the accumulation and testing of further product samples to identify and characterise the nanomaterials. It is hoped to extend the range of products considered by gathering samples of CNT enabled products as well as composites and nanoclays. These materials have scope to be nano-enabled but do not yet appear widely in the marketplace.

The second stage of the research will involve a subset of the samples already tested. These will be subjected to a range of simulated demolition processes such as crushing and grinding (which are the most common end points for concrete and other aggregates and also currently for glass), as well as melting and burning and other processes involved in recycling.

## 5. Conclusions

This research has sought to identify where nanomaterials are located in practice in the construction industry. It has identified that there is a mismatch between the developments in the technology (as described in the literature) and availability in reality in the marketplace. It has also demonstrated that there is sometimes a disparity between the information provided by manufacturers regarding the nano-content of products and the reality; even where correct information is provided it is generally incomplete. There is no legal requirement to provide details to the user regarding the nanocontent of construction products.

It is difficult currently to draw firm conclusions regarding the risk potential from nanomaterials to those working in the demolition industry. It is anticipated that further testing will provide clarification in at least some areas and hence enable pragmatic guidance for the industry to be developed to indicate where particular protection or control measures may be appropriate and to provide reassurance where no additional action is required.

## Acknowledgements

This research is funded by the Institution of Occupational Safety and Health (IOSH) and their support is gratefully acknowledged.

Additional support from the Institute of Demolition Engineers and the National Federation of Demolition Contactors is also acknowledged.

**References**

[1] OSHA *Construction* Available at https://osha.europa.eu/en/sector/construction/index_html Accessed November 26, 2014
[2] Donaldson K, Poland CA, Murphy FA, MacFarlane M, Chernova T and Schinwald A. Pulmonary toxicity of carbon nanotubes and asbestos - similarities and differences 2013 *Adv Drug Deliv Rev* Dec 65[15] 2078-86
[3] van Broekhuizen F and Van Broekhuizen P 2009 *Nano-products in the European Construction Industry State of the Art* Available at www.efbww.org/pdfs/Nano%20-%20final%20report%20ok.pdf
[4] Teizer J, Teizer W, Venugopal M and Sandha B 2009 *Nanotechnology and its impact on construction* CII report 251-11
[5] Mann S *Nanotechnology and Construction* 2006 Available at http://nanotech.law.asu.edu/Documents/2009/10/Nanotech%20and%20Construction%20Nanoforum%20report_259_9089.pdf Accessed April 17, 2014
[6] Lee J, Mahendra S and Alvarez PJ 2010 Nanomaterials in the construction industry: a review of their applications and environmental health and safety considerations *Acs Nano* 4[7] 3580-90.
[7] Greßler S and Gazsó A 2012 Nano Trust dossier No. 032en *Nano in the Construction Industry*. 2012 Available at epub.oeaw.ac.at/ita/nanotrust-dossiers/dossier032en.pdf
[8] Hanus MJ and Harris AT 2013 Nanotechnology innovations for the construction industry *Progress in Materials Science* 8 58[7] 1056-1102
[9] NFDC *NFDC yearbook* 2014 Available at http://epublishing.theconstructionindex.co.uk/magazine/nfdcyearbook2014/ Accessed November 26, 2014
[10] Project on Emerging Nanotechnologies *Consumer Product Inventory* Available at http://www.nanotechproject.org/cpi/ Accessed November 26, 2104
[11] Nanodatabase 2014 Available at http://nanodb.dk/ Accessed November 26, 2014
[12] elCOSH Nano. *Construction Nanomaterial Inventory* Available at http://nano.elcosh.org/. Accessed November 26, 2014
[13] AECOM *The Blue Book: Property and Construction Handbook International Edition* 2014 Available at http://www.aecom.com/deployedfiles/Internet/Geographies/Australia-New%20Zealand/Documents/Blue%20Book%202014.pdf Accessed November 26, 2014
[14] Kumar P, Mulheron M, Fisher B and Harrison RM 2012 New Directions: Airborne ultrafine particle dust from building activities – A source in need of quantification *Atmos Environ* 9 56(0) 262-4
[15] Pilkington website Available at https://www.pilkington.com/en-gb/uk/householders/types-of-glass/self-cleaning-glass/faqs/technical-faqs Accessed December 8, 2014
[16] Aspen Aerogels Available at http://www.aerogel.com/ Accessed November 26, 2014
[17] Napierska D, Thomassen L, Lison D, Martens JA and Hoet PH 2010 The nanosilica hazard: another variable entity *Part Fibre Toxicol* 7[1] 39
[18] Eden Energy Ltd *Annual Report for the Year Ended 30 June 2014* Available at http://www.edenenergy.com.au/pdfs/9655b8bc99f0f15e9fa9c3ac46d9ec92.pdf Accessed 8 December, 2014

Journal of Physics: Conference Series **617** (2015) 012032 doi:10.1088/1742-6596/617/1/012032

# Nanomaterials in Food – Current and Future Applications and Regulatory Aspects*

**K Aschberger[1†], S Gottardo[1], V Amenta[1], M Arena[1], F Botelho Moniz[1], H Bouwmeester[2], P Brandhoff[2], A Mech[1], L Quiros Pesudo[1], H Rauscher[1], R Schoonjans[3], M Vittoria Vettori[3], R Peters[2],**

[1]IHCP-Nanobiosciences, JRC Ispra, Via Fermi 2749, 21027 Ispra, Italy
[2]RIKILT Wageningen UR, Akkermaalsbos 2, 6700 AE Wageningen, The Netherlands
[3]EFSA, Via Carlo Magno 1a, 43126 Parma, Italy

E-mail: Karin.aschberger@ec.europa.eu

**Abstract.** Nanotechnology can contribute to the development of innovative applications in the agriculture, food and feed sector by e.g. enabling improved delivery of nutrients or increased efficacy of agrichemicals. It is expected that applications will increase in the near future and may therefore become a relevant source of human exposure to nanomaterials (NM). To gain more up-to date information, RIKILT and the Joint Research Centre (JRC) were commissioned by the European Food Safety Authority (EFSA) to prepare an inventory of currently used and reasonably foreseen applications of NM in agriculture and food/feed production and carried out a review of regulatory aspects concerning NM in both EU and non-EU countries. An analysis of the information records in the inventory shows that nano-encapsulates, silver and titanium dioxide are the most frequent type of NM listed and that food additives and food contact materials are the most frequent types of application. A comparison between marketed applications and those in development indicates a trend from inorganic materials (e.g. silver) towards organic materials (nano-encapsulates, nanocomposites). Applications in novel food, feed additives, biocides and pesticides are currently mostly at a developmental stage. The review of EU and non-EU legislation shows that currently a few EU legal acts incorporate a definition of a nanomaterial and specific provisions for NM, whereas in many non-EU countries a broader approach is applied, which mainly builds on guidance for industry.

## 1. Introduction

Nanotechnology can contribute to the development of innovative applications in the agriculture, food and feed sector (hereinafter referred to as agri/feed/food) with new and enhanced properties [1, 2]. Applications include nano-encapsulated agrochemicals or nutrients, antimicrobial nanoparticles and active and intelligent food packaging. It is expected that applications will increase in the future and thereby represent a relevant source of direct exposure of humans to nanomaterials (NM). To address this potential concern and to gain more up-to date information on the exposure potential to NM applied in agri/feed/food products, RIKILT and JRC have prepared for the European Food Safety

* Any opinions expressed in this publication are those of the authors only and do not represent EFSA's, JRC or RIKILT- Wageningen UR's official position.

† To whom any correspondence should be addressed

Journal of Physics: Conference Series **617** (2015) 012032 doi:10.1088/1742-6596/617/1/012032

Authority (EFSA) an inventory of currently used in commercial products and in development applications of NM in agri/feed/food [3].

## 2. Nano Inventory

Relevant information for the Nano Inventory was retrieved through an extensive literature search performed in bibliographic databases such as Scopus, Web of Science, PubMed Central, NANOnetBase and SciFinder Scholar, and complemented by ad-hoc searches in the internet, at company websites and through questionnaires sent to different stakeholders. The retrieved data and information were evaluated based on selection and exclusion criteria and, if considered relevant, inserted into the Nano Inventory. The Nano Inventory was developed in a Microsoft® Access environment and organised in 12 different application fields related to agri/feed/food (Figure 1). From the database, three predefined queries were developed to extract the most relevant data and information.

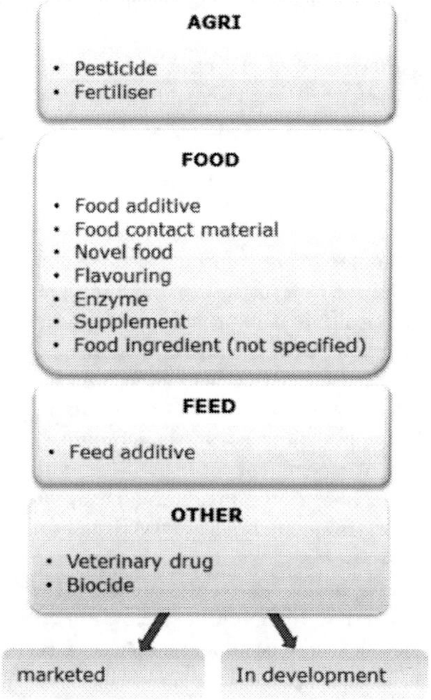


**Figure 1:** Application fields in the Nano Inventory

## 3. Current and future applications of nanotechnology in agri/feed/food

Most of the records in the Nano Inventory concern applications in food (almost 90%) with food additives and food contact materials being the most frequent types of application. A much smaller percentage of applications in the Nano Inventory is concerned with agriculture (9%) and feed (3%). 55 different types of NM, both inorganic and organic, can be found in the Nano Inventory. The most frequent types of NM are nano-encapsulates, silver and titanium dioxide (Figure 2).

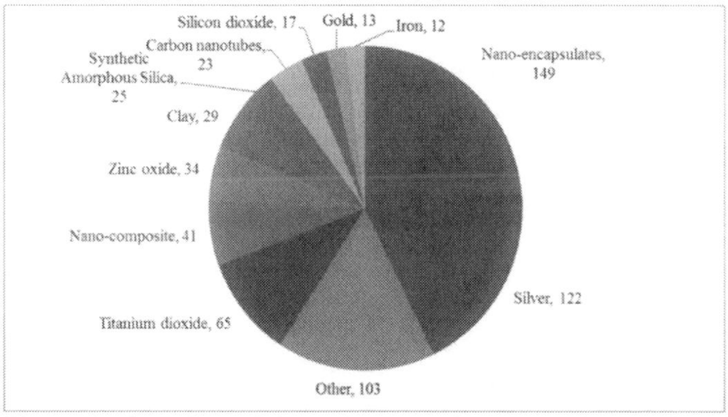


**Figure 2:** Overview of the main types of nanomaterial mentioned in the Nano Inventory (including the number of records)

About half of the identified applications are claimed to be already used in a consumer product while about 20% of the applications are still in the research and development stage. For the remaining applications the available information did now allow to determine the status.

A comparison between marketed and in development applications indicates a trend from inorganic materials (e.g. silver) towards organic materials (nano-encapsulates, nanocomposites) (Figure 3). Applications in novel food, feed additives, biocides and pesticides are currently mostly at a developmental stage.

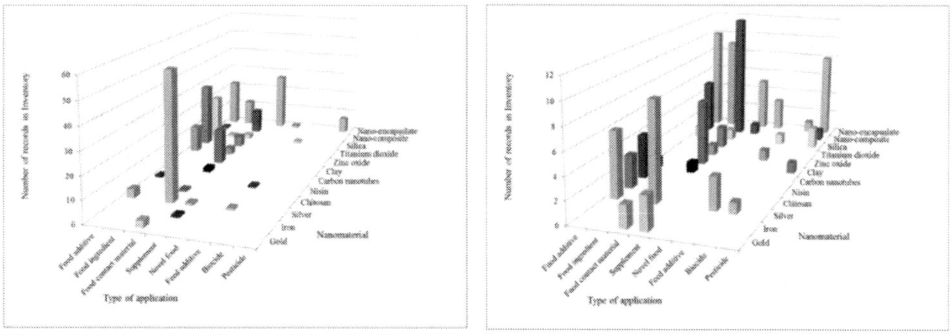

**Figure 3:** Number of records in the Nano Inventory for several combinations of nanomaterial and related application. The graph on the left shows applications that are already available on the market while the graph on the right shows applications that are still in development

Applications of NM in agriculture include improved and targeted pest management and crop protection through increased efficacy, durability, bioavailability and controlled release of pesticides and other agrochemicals [4]. This can for example be achieved by nano-encapsulation or binding active substances to solid lipid nanoparticles or porous solid particles (e.g. silica) [5, 6].
Nanotechnology can also enhance crop production and thereby potentially reduce the quantity of fertiliser that has to be applied [7]. Nanotechnology is also applied for detection of animal and plant pathogens, and for identity preservation and tracing and the slow release of nutrients or active compounds used in veterinary drugs [8, 9].

In food and feed processing, nanotechnology can contribute to new or improved tastes, textures and mouth sensations through nano-scale processing of foodstuffs or improved uptake, absorption, and bioavailability of nutrients through nanoscale-formulations [10]. Several food additives are composed of nanoparticles or may contain a fraction in the nanosized range. Synthetic amorphous silica (SAS) is used for surface coating of packaging materials, for the clearing of beverages, and mostly as a free-flow and anti-caking agent in many powdered food items (E551) [11]. It is composed of aggregates of primary particles of nanosize. Titanium dioxide which is mainly used as white pigment may contain also particles in the nanosize range.

Nanotechnology has contributed to the development of innovative packaging materials that can improve the safety and shelf life of products by providing barrier materials or detect foodborne pathogens. Some applications are already marketed while many more are at the moment in the research and development stage [1, 9].

## 4. Toxicological data and risk assessment

Data and information stored in the Nano Inventory indicate that silica, silver and titanium dioxide are the most common NM in toxicity testing and risk assessment. The most tested toxicity endpoints include genotoxicity, acute toxicity, cytotoxicity and repeated dose toxicity. Very often the physicochemical characterisation of the NM is very poor and was reported in less than 15% of the records concerning (eco)toxicity and risk assessment of NM.

## 5. Review of legislation addressing nanotechnology in agri/feed/food

To prepare the overview of existing legislation and guidance for risk assessment with regard to NM and nanotechnology in agri/feed/food in both EU and non-EU countries we collected information from websites of the European Commission, European Agencies/Authorities, the Official Journal of the European Union, national governments, and national and international organisations. In addition, we searched peer-reviewed publications as described above. Information on non-EU legislation was mainly retrieved from recently published reports, such as by WHO/FAO [12] or OECD [13]. In addition a questionnaire on "Regulation and safety assessment of nanomaterials in agri/feed/food applications" was sent to relevant stakeholders (RIKILT and JRC 2014).

The EU is the only world region where nano-specific provisions have been incorporated in legislation for agri/feed/food. Some EU legislations explicitly addressing NM include the Regulation on the Provision of Food Information to Consumers (1169/2011), the Regulation on Plastic Food Contact Materials and Articles (10/2011), the Regulation on Active and Intelligent Materials and Articles (450/2009), the Biocidal Products Regulation (528/2012) and the Cosmetic Products Regulation (1223/2009). In other sectors NM are considered implicitly covered by legislation [14]. Some pieces of legislation are currently under revision to better address NM, e.g. Novel Food Regulation (258/97) [15].

The use of nanotechnology in food production is currently covered by EC Regulation No 258/97 concerning 'novel foods' and 'novel food ingredients' [16] which addresses food not consumed to any significant degree in the EU prior to May 1997. The revised regulation will address NM and nanotechnology more explicitly by covering *'foods modified by new production processes such as nanotechnology and nanoscience and food or vitamins, minerals and other substances containing or consisting of 'engineered nanomaterials'*. A pre-market approval (safety assessment and authorization) is required for novel food as well as for food additives. Food additives on the market before 2009 are subject to re-evaluation, during which potential nanospecific data can be considered [17]. This includes also some of the common particulate food/feed additives, which have been in use for years, such as anti-caking/free-flow powders, pigments and others (e.g. silicon dioxide, titanium dioxide, iron oxide or silver).

Countries outside the EU have rather adopted a broad approach when dealing with regulation of NM in agri/feed/food. Many non-EU countries have introduced non-mandatory frameworks and consider existing regulatory frameworks able to adapt to and cover the particularities of NM (e.g. US, Australia and New Zealand, Canada). They rather build on guidance for industry.

In the EU some sectors have introduced a legally binding definition of the term 'nanomaterial' (i.e. for food, cosmetic and biocidal products). Furthermore, an EC Recommendation for a broadly applicable definition of 'nanomaterial' is available (2011/696/EU). All EU definitions of the term 'nanomaterial' intended for regulatory purposes use size (1-100 nm or below 100 nm) as main identifier. There are differences among the EU definitions with regard to inclusion criteria (e.g. only intentionally produced NM for food) or thresholds (e.g. number size distribution). An alignment of sector-specific definitions with the EC Recommendation is currently (January 2015) being discussed. No legally binding definitions could be identified outside EU and rather working definitions of NM are applied. Beside size some countries (e.g. US, Canada) consider also other properties or phenomena to define NMs or nanotechnology.

In the EU all ingredients present in food or biocidal products in the form of NM have to be clearly indicated in the list of ingredients with the names of such ingredients followed by the word 'nano' in brackets (Figure 4). A system for tracking and labelling consumer products containing NM has been introduced by some Asian countries such as Iran, Taiwan and Thailand(e.g. NanoMark system in Taiwan). This system is substantially different from the labelling requirement in the EU. While in EU the labelling of products containing nanomaterials is a mandatory requirement included in specific pieces of legislation (food, cosmetics and biocides legislations), the labelling introduced in the above mentioned non-EU countries is established within the context of "industrial standard certifications for nanoproducts" and it is aimed at eliminating fake nanoproducts on the market. Both finally aim to inform consumer and to protect their rights.

Ingredients:
..., anticaking agent E551 (nano), ...

**Figure 4:** Example of labelling of a food product containing amorphous silicon dioxide (E551) in nanoform as anticaking agent.

## 6. Conclusions

Nanotechnology applications in the agricultural, feed and food sector are growing and novel products are expected to enter the market in the near future. An up-to-date knowledge on the occurrence and type of application of NM in this sector is vital to estimate the potential human exposure to NM from food and feed. It is also becoming increasingly important that the regulatory frameworks properly address and specifically manage the potential risks of nanotechnology. Several countries over the world have been particularly active in examining the appropriateness of their regulatory frameworks for dealing with nanotechnologies but have applied different approaches to address safety issues of nano-based products in agri/feed/food: from legally binding provisions to guidance for industry.

## Acknowledgments

The information in this publication was produced as part of EFSA contract CFT/EFSA/FEED/2012/01, "Inventory of food additives and other food ingredients/food contact materials/feed additives applications in the area of nanotechnologies". Full information can be found in the EFSA report, "Inventory of Nanotechnology applications in the agricultural, feed and food sector" [3]. The authors would like to thank Teodor Ivanov and Aris Tsois, both JRC Ispra for their contribution to the inventory.

## References

1. Chaudhry, Q., et al., *Applications and implications of nanotechnologies for the food sector.* Food Additives and Contaminants - Part A Chemistry, Analysis, Control, Exposure and Risk Assessment, 2008. **25**(3): p. 241-258.
2. Kah, M., et al., *Nanopesticides: State of Knowledge, Environmental Fate, and Exposure Modeling.* Critical Reviews in Environmental Science and Technology, 2013. **43**(16): p. 1823-1867.
3. RIKILT and JRC, *Inventory of Nanotechnology applications in the agricultural, feed and food sector*, in *EFSA supporting publication* 2014, RIKILT; European Commission Joint Research Centre;: Parma. p. 125. http://www.efsa.europa.eu/en/supporting/pub/621e.htm
4. Kah, M. and T. Hofmann, *Nanopesticide research: Current trends and future priorities.* Environment International, 2014. **63**(0): p. 224-235.
5. Frederiksen, H.K., H.G. Kristensen, and M. Pedersen, *Solid lipid microparticle formulations of the pyrethroid gamma-cyhalothrin—incompatibility of the lipid and the pyrethroid and biological properties of the formulations.* Journal of Controlled Release, 2003. **86**(2–3): p. 243-252.
6. Liu, F., et al., *Porous hollow silica nanoparticles as controlled delivery system for water-soluble pesticide.* Materials Research Bulletin, 2006. **41**(12): p. 2268-2275.
7. Kole, C., et al., *Nanobiotechnology can boost crop production and quality: first evidence from increased plant biomass, fruit yield and phytomedicine content in bitter melon (Momordica charantia).* BMC Biotechnology, 2013. **13**(1): p. 37.
8. Underwood, C. and A.W. van Eps, *Nanomedicine and veterinary science: The reality and the practicality.* The Veterinary Journal, 2012. **193**(1): p. 12-23.
9. Parisi C., V.M., Rodríguez-Cerezo E., *Proceedings of a workshop on "Nanotechnology for the agricultural sector: from research to the field"* in *JRC Scientific and Policy Report* 2014, Publications Office of the European Union, 2014: Luxembourg. http://bookshop.europa.eu/en/proceedings-of-a-workshop-on-nanotechnology-for-the-agricultural-sector-from-research-to-the-field--pbLFNA26625/
10. Chaudhry, Q., L. Castle, and R. Watkins, *Nanotechnologies in Food.* RSC Nanoscience&Nanotechnology. Vol. 25. 2010, Cambridge: Royal Society of Chemistry. 229.
11. Dekkers, S., et al., *Presence and risks of nanosilica in food products.* Nanotoxicology, 2011. **5**(3): p. 393-405.
12. FAO/WHO, *State of the art on the initiatives and activities relevant to risk assessment and risk management of nanotechnologies in the food and agriculture sectors*, 2013, FAO/WHO. http://www.fao.org/docrep/018/i3281e/i3281e.pdf
13. OECD, *Regulatory frameworks for nanotechnology in foods and medical products. Summary results of a survey activity*, 2013: Paris http://www.oecd.org/officialdocuments/publicdisplaydocumentpdf/?cote=DSTI/STP/NANO(2012)22/FINAL&docLanguage=En
14. European Commission, *Communication from the Commission to the European Parliament, the Council and the European Economic and Social Committee: Second Regulatory Review on Nanomaterials*, E. Commission, Editor 2012.
15. European Commission, *Proposal for a Regulation of the European Parliament and of the Council on novel foods*, in *COM(2013) 894 final* 2013.
16. European Parliament and Council, *Regulation (EC) No. 258/97 of the European Parliament and of the Council of 27 January 1997 concerning novel foods and novel food ingredients*, in *Official Journal of the European Communities*, E. Union, Editor 1997. p. 1-6.
17. European Commission, *Commission Regulation (EC) No 257/2010 of 25 March 2010 setting up a programme for the re-evaluation of approved food additives in accordance with Regulation*

*(EC) No 1333/2008 of the European Parliament and of the Council on food additives*, in *Official Journal of the European Union* 2010. p. 19-27.

4th International Conference on Safe Production and Use of Nanomaterials (Nanosafe2014) IOP Publishing
Journal of Physics: Conference Series **617** (2015) 012033 doi:10.1088/1742-6596/617/1/012033

# nanoSTAIR: a new strategic proposal to impulse standardization in nanotechnology research

**JM López de Ipiña[1], O Salvi[2], B Hazebrouck[2], A Jovanovic[3], F Carre[4], A Saamanen[5], D Brouwer[6], M Schmitt[7], S Martin[8]**

[1] TECNALIA – Parque Tecnológico de Alava, 01510 Miñano/Spain
[2] EU-VRi - Willi-Bleicher-Straße 19, 70174 Stuttgart/Germany
[3] STEINBEIS Haus der Wirtschaft Willi-Bleicher-Strasse 19, 70174 Stuttgart/Germany
[4] INERIS - Rue Jacques Taffanel, 60550 Verneuil-en –Halatte/France
[5] FIOH - Topeliuksenkatu 41 a A, 00250 Helsinki/Finland
[6] TNO - Utrechtseweg 48, 3704 HE Zeist/The Netherlands
[7] DIN - Burggrafenstraße 6, 10787 Berlin/Germany
[8] CEN – CENELEC - Avenue Marnix 17, B-1000 Brussels/Belgium

E-mail: jesus.lopezdeipina@tecnalia.com

**Abstract.** Nanotechnology is considered one of the key technologies of the 21$^{st}$ century within Europe and a Key-Enabling Technology (KET) by Horizon 2020. Standardization has been identified in H2020 as one of the innovation-support measures by bridging the gap between research and the market, and helping the fast and easy transfer of research results to the European and international market. The development of new and improved standards requires high quality technical information, creating a fundamental interdependency between the standardization and research communities. In the frame of project nanoSTAIR (GA 319092), the present paper describes the European scenario on research and standardization in nanotechnology and presents a proposal of a European strategy (nanoSTAIR) to impulse direct "pipelines" between research and standardization. In addition, strategic actions focused on integration of standardization in the R&D projects, from the early stages of the design of a future business (Project Proposal), are also described.

## 1. The European context on standardization and research in nanotechnology

Nanotechnology has been identified as a Key-Enabling Technology (KET) in Horizon 2020 (H2020). It has rapidly promoted the development of a new generation of smart and innovative nano-enabled products and processes, creating an important growth potential for a large number of industry sectors. Nanotechnology offers substantial possibilities for improving the competitive position of the EU and for responding to key societal challenges. But ensuring nanotechnology is developed in a safe and responsible manner is a key objective.

The production of knowledge in nanotechnology is very intensive as a great amount of research projects is launched each year, both on national and European level [5]. The European Commission has identified the time between the end of a research project and the beginning of the valuable exploitation of these results from the industry — the so called "valley of death" — as one of the main obstacles for economic success. In this context, standardization has been recognized in H2020 as one of the innovation-support measures by bridging the gap between research and the market, and helping the fast and easy transfer of research results to the European and international market, providing interoperability between new and existing products, services and processes.

To support the safe development of nanotechnologies, risk management should become an integral part of the culture of the organizations involved in the supply chain, including regulatory support and risk governance. The Nanosafety Cluster published in 2013 a roadmap aims to provide an understanding of where the European nanosafety should be at the end of H2020, introducing a strategic vision for future research on the safe use and safe applications of engineered nanomaterials during 2015-2025 [9] .

The document identifies the steps and achievements needed to accomplish objectives within this time frame, describing the current status and the research needs and priorities for the coming 10 years in four main thematic areas: 1) nanomaterial identification and classification; 2) nanomaterial exposure and transformation; 3) hazard mechanisms related to effects on human health and the environment; and 4) tools for the predictive risk assessment and management including databases and ontologies. The report recognizes that standardization is highly important in promoting the success of engineered nanomaterials and technologies, and when incorporating safety as a vital issue in the standardization process.

On June 2011, the European Commission adopted a Communication entitled "A strategic vision for European standards: Moving forward to enhance and accelerate the sustainable growth of the European economy by 2020", that places in focus the important role expected from standardization in supporting the Europe 2020 Strategy for smart, sustainable and inclusive growth [3]. The document identifies European standards as an important step for bringing research results to the market and for the validation of technologies. This is particularly important for nanotechnologies, because standards in this field can facilitate the introduction of new products and contribute to the public acceptance of these innovations.

In 2010, CEN accepted 'Mandate M/461 - Standardization Mandate to CEN, CENELEC and ETSI for standardizations activities regarding nanotechnologies and nanomaterials [2]. In 2013, CEN/TC 352 'Nanotechnologies' was involved together with CEN/TC 195 'Air filters for general air cleaning' and CEN/TC 137 ' Assessment of workplace exposure to chemical and biological agents' in developing several European deliverables under this Mandate.

**Table 1.** Key players in the European context on standardization and research in nanotechnology

| Research | Standardization | Industrial | Regulatory |
|---|---|---|---|
| • Research projects and organizations: see CORDIS, NSC, OECD, databases<br>• Nanosafety cluster<br>• European Technology initiatives: ETPs, EIPs, PPP<br>• European, national and regional research funding agencies.<br>• ERA-NET on Nanosafety (SIINN)<br>• NMP NCP Network | • European Standardisation Bodies (ESB): CEN, CENELEC, ETSI<br>• National Standardization Bodies (NSB): DIN, AENOR, AFNOR, etc<br>• International Standardization Organizations: ISO, OECD, … | • Industry and industrial associations (NIA) | • European Commission<br>• EC - Advisory Committees and Agencies: SCCP, SCHER, SCENIHR, EFSA, ECHA, …<br>• Project NANoREG |

According to Nanotechnology Industries Association (NIA), in the field of nanotechnology, the development of appropriate nomenclature and standards is still in its infancy, and fast progress is urgently required so as to avoid hampering the development and acceptance of nanotechnology-based innovation. So far over one hundred key vocabularies, as well as dozens of technical specifications and reports, have been produced. Nevertheless many more are needed and active industry participation in the standardizations process is of the utmost importance in ensuring the best possible market conditions for nanotechnologies.

The regulatory uncertainties are considered as one of the main challenges in bringing nanomaterials to market in Europe. The current debate, including on the lack of regulatory clarity and on the uncertainties surrounding the potential risks of nanotechnology and nanomaterials have had a negative effect on their development, uptake and exploitation in Europe and have been clearly identified as a major barrier to innovation based on these technologies. Standards play a role for regulatory compliance, in particular in areas where regulation requires a risk assessment and risk management approach, compliance with which can be demonstrated by standards. In general, the application of standards, unlike legal texts, is voluntary. Standards can however become part of legislation, when their wording or content is taken up by legal texts (New Approach Directives and harmonized standards).

Scientific activities make a key contribution to the standardization process [5,10]. H2020 documents refer standardization as an important element to enhance the competitiveness of European industry by promoting the take-up and trade of novel technologies. Many calls for funding programs of RTD projects refer to standardization as a relevant requirement to be considered by the project. This is due to the advantages that standardization provides the project on issues such as interoperability, dissemination of results and supports the market introduction.

According to a study presented by CEN and CENELEC [12], one in eight FP7 calls made explicit reference to standardization, and that projects funded under these calls were almost three times more likely to make use of standards than projects funded through calls that did not refer to standardization. In brief, 43% of all FP6 and FP7 projects addressing standardization in some way: 39% of projects used standards as an input, 12% proposed new or revised standards as an output, and 8% contributed to the development of new or revised standards. The same study showed nearly three-quarters (73%) of FP6 and FP7 project coordinators who included standards in their previous projects said that they would be willing to address standardization again. The FP7 priority areas estimated to have the highest share of projects addressing standardization were Security (75%), Transport (including aeronautics) (66%), ICT (62%) and Energy (60%).

**Table 2.** Fifteen most relevant standardization items identified by the nanoSTAIR survey [6]

| Num | Standardization item |
|---|---|
| 1 | Protocols for the characterization of manufactured nanoparticles from aerosols and from environmental sources, including sampling, sample stabilization, agglomeration, aggregation, etc. |
| 2 | Repeatability, reproducibility and intercomparability of test methods. |
| 3 | Guide to the identification and definition of measurands required for characterizing, evaluating functional properties and performance etc., of materials and devices at the nanoscale. |
| 4 | Guidance on nano-material characterization prior to, or in association with toxicity testing. |
| 5 | Guidance on dosimetry and exposure determination in occupational settings relevant to manufactured nanomaterials. |
| 6 | Guidance on safe handling of manufactured nanoparticles and other manufactured nanoscale entities (including selection of Personal Protective Equipment). |
| 7 | Inter-Laboratory Comparisons and validated methods/techniques for measurement/control of quality, process, etc. |
| 8 | Development of Reference Materials and Certified Reference Materials dedicated to existing and new techniques, particularly for challenging and checking the functioning/calibration of nanoparticle measurement and analysis equipment. |
| 9 | Related metrology (instrumentation and techniques) for measurement and characterization of nanoparticles and other nano-objects. |
| 10 | Guidance on sample preparation for toxicity testing, toxicokinetic and ecotoxicokinetic (air, water, soil) studies on nanoparticles and other nanoscale entities. |
| 11 | Guide to the management of waste and the disposal of nanomaterials. |
| 12 | Guide to performance measurement of nanoscale materials and devices. |
| 13 | Guidance on integrated (toxicity) testing strategies (ITS) and integrated risk assessment. |
| 14 | Protocols for evaluating the effects of short and long term dermal, nasal, oral and pulmonary exposure to, elimination of, and fate determination for nanoparticles and other nanoscale entities. |
| 15 | Guidance on detection and identification of nanoparticles and other nanoscale entities (in all media types, including waste streams from manufacturing and manufacturing discharges). |

The development of new and improved standards requires high quality technical information. This creates a fundamental interdependency between the standardization and research communities. Consequently the need of direct "pipelines" to connect research to standardization has been widely recognized and promoted by European Commission, European Standardization System and other interested parties, in order to use standards to drive markets and build consumer confidence.

Integration of European research and standardization is a recurring topic, at least in the last two European research programs (FP6, FP7). In fact, agents of European Standardization System (ESS) and European Innovation System (EIS) have been involved in projects related to that topic in FP6 and FP7 programmes (e.g. FP6 -INTEREST, FP -7 nanoSTAIR, BRIDGIT).

The project nanoSTAIR carried out a survey to clearly identify the needs of European stakeholder for new standards in nanotechnology, to bridge the gap between the research objectives and the standardization needs, by facilitating the process of clustering [6]. Research and development stakeholder group was the biggest interested part, representing 76% of the answers. The table 2 summarizes the fifteen most relevant standardization items identified by the survey.

There are some important barriers that inhibit participation of research community in standardization. The list can be synthesized into three main categories [4]: 1) Resources required for participation (money, time, personnel, etc.); 2) Standardization process (length, flexibility, IPR rules, integration of input, etc.); 3) Awareness and visibility of standards and standardization (awareness of benefits, links to SSBs, helpdesks, etc). In this regard, table 3 groups all the barriers identified by literature sources consulted by project nanoSTAIR [1,4,7,8,9,11] .

**Table 3.**Barriers identified in the literature consulted by nanoSTAIR, classified in the three main categories as established by project INTEREST [7,8]

| Barriers and literature sources | Resources required for participation | Standardization Process | Awareness and recognition of standardization |
|---|---|---|---|
| Project INTEREST (FP6) (2006) | • "My expenses are not covered by my organization or by third parties." (The most important barrier by far). | • "The standardization process is not flexible and fast enough." | • "Standardization activities do not have a positive impact on the evaluation of my research."<br>• "My expertise is not relevant in the standard-setting process." |
| NMP Expert Advisory Group (2012) | • Funding for pre- and co-normative research, although relatively small, is extremely difficult to find. | • Timescale of research projects often not aligned with standardization timescale. | • Researchers and standardizers often operate in different circles<br>• Lack of knowledge on the potential contribution of the projects towards the development of new standards or to the improvement of existing standards.<br>• Little recognition and enhancement of a research career |
| NSC Strategy (2013) | • Lack of resources to invest time for standardization activities | | • Lack of awareness among researchers of what is going on in standardization<br>• Lack of incentives for researchers to invest time and efforts. |
| CEN-CENELEC (2013) | • Lack of funding to take forward the proposed standards development work<br>• Lack of resources to provide inputs to the development process | • Inherent complexity of the standardization world<br>• Non-alignment between the project and standardization 'timetables'<br>• Uncertainty surrounding the decision-making process<br>• Difficulties gaining access to Standardization Bodies and their TCs | • Competition from other competing proposals or ideas in similar or related areas<br>• Difficulty in gaining acceptance of the inputs provided<br>• Difficulties generating industrial support for new standards |
| CEN-CENELEC Project BRIDGIT (2014) | • Resource commitment required to participate in standardization activities is too high for researchers<br>• Standards bodies lack resources | • Lack of comprehension of the world of standards<br>• Operation of the standardization process | • Mutual irrelevance of standardization and research<br>• It is not clear which research is relevant<br>• Lack of recognition of researchers professional contribution |

In this context, the project FP7 nanoSTAIR - finished in 2014 – carried out research to build a sustainable process, a platform and a set of advanced tools (services) in the field of nanotechnologies, to support the transfer of knowledge gained through research to standards and standardization deliverables, in the framework of the STAIR approach promoted by CEN-CENELEC [1]. Recently a consortium consisting of 10 NSBs and CEN and CENELEC also completed the project BRIDGIT, funded by the EC and EFTA, aimed to "bridge the gap" between standardization, research and innovation.

## 2. nanoSTAIR: contributing to impulse standardization in research

Summarising of all information sources consulted by nanoSTAIR [7], it can be suggested that perception of EIS agents on the interest of standardization and its importance in research projects, closely depends on several factors: 1) Typology of research agent (e.g. university, research institute, technology centre, RDI unit of a company, other stakeholders); 2) Research area concerned (e.g. fundamental research, technology development, demonstration); 3) Existence of a previous experience or established relationship between EIS and ESS agents (e.g. agent's participation in technical standardization committees (TC), working with ESS in previous research projects, etc.); 4) Technology Readiness Level (TRL) assigned to research project; 5) Life cycle stage of project concerned (Call for proposals, preparation of the proposal, evaluation process, negotiation phase, implementation phase, stage of completion and exploitation of results); 6) IPR issues involved in the project and 7) European country, relationships between research community and NSB.

nanoSTAIR strategy [7] has been designed as a tool to impulse standardization in research, either at national or European level. The strategic approach is based on the systematic analysis of the barriers that have been identified (Table 3) and of the possible ways to overcome these barriers, involving all relevant actors and identifying the possible benefits for them.

**Table 4.** Number of strategic actions to be deployed by nanoSTAIR strategy, classified by typology of barrier and strategic objective concerned [7].

| nanoSTAIR Strategic Objectives | | Barriers | | | | |
|---|---|---|---|---|---|---|
| Main objective | 2nd Level | RA[1] | SP[2] | ARS[3] | Total | |
| SO1.- Ensure that nanotechnology standardization is visualized as a demand to be satisfied by researchers, research organizations and research projects | SO1.1.- Reinforce visibility of nanotechnology standardization in research projects | 4 | 0 | 9 | 13 | 16 |
| | SO1.2.- Foster a culture of implementation of research results in standardization deliverables | 0 | 2 | 1 | 3 | |
| SO2.- Provide support to research projects, from the European Standardization System (ESS). | SO2.1.- Facilitate contact between research projects and standardization organizations | 1 | 2 | 4 | 7 | 15 |
| | SO2.2.- Provide practical tools to solve standardization issues in research projects | 0 | 2 | 4 | 6 | |
| | SO2.3.- Develop a networking with a view of exchanging experiences and best practices among members of ESS | 0 | 0 | 2 | 2 | |
| SO3.- Provide support to ESS, from the research community | SO3.1Facilitate contact between standardization and research community | 0 | 0 | 2 | 2 | 2 |
| SO4.- Establish and deploy a sustainable platform of services - nanoSTAIR - to facilitate the interaction between research and standardization | SO4.1.- Provide research projects, advanced services to properly manage standardization issues | 0 | 0 | 19 | 19 | 20 |
| | SO4.2.- Ensure sustainability of nanoStair platform after project completion | 1 | 0 | 0 | 1 | |
| | Total | 6 | 6 | 41 | 53 | 53 |

[1] RA: Resources available
[2] SP: Standardization process
[3] ARS: Awareness and recognition of standardization

The strategy rests on five pillars: 1) Strategic focus on the research project throughout its life cycle, as a key element for bridging research (EIS) and standardization (ESS); 2) Visualization of standardization as a need to be satisfied by research projects; 3) Proximity and accessibility to standardization resources. The strategic approach pivots on NSB, due to its proximity and ease of interaction with project partners as well as their ability to provide resources to the project in the field of standardization; 4) Synergistic approach to standardization and strategic actions aimed at coordinating European and national efforts – ESS and National Standardization System (NSS) - to maximize impact and 5) Development of a private platform of services - nanoSTAIR - to facilitate integration between research and standardization.

The project identified four strategic objectives, aimed at bridging barriers and facilitate connections between research and standardization, and a set of targeted strategic activities for deployment. The table 4 and figure 1 summarize nanoSTAIR strategy and show that all types of barriers described in table 3 have been considered by the strategic approach. The total number of strategic actions considered is 53. The deployment of these actions may take place at European level only, national level or in both areas.

The largest number of actions concerns the Strategic Objective 4 (SO4: Deploying nanoSTAIR platform), followed in short order by objectives 1 and 2. These last two objectives are closely related to the satisfaction of the needs of researchers, research organizations and research projects, and both account for more than half of the strategic actions proposed by nanoSTAIR (57%).

By type of barrier concerned, the largest number of actions (78%) is directed towards the awareness and recognition of standardization, as a strategic pillar to establish direct pipelines between research and standardization. Actions to provide resources and actions to improve the standardization process represent the remaining 22%, equally distributed (11% each).

The document containing nanoSTAIR strategy will soon be available for download at the project website (www.nanostair.eu-vri.eu/).

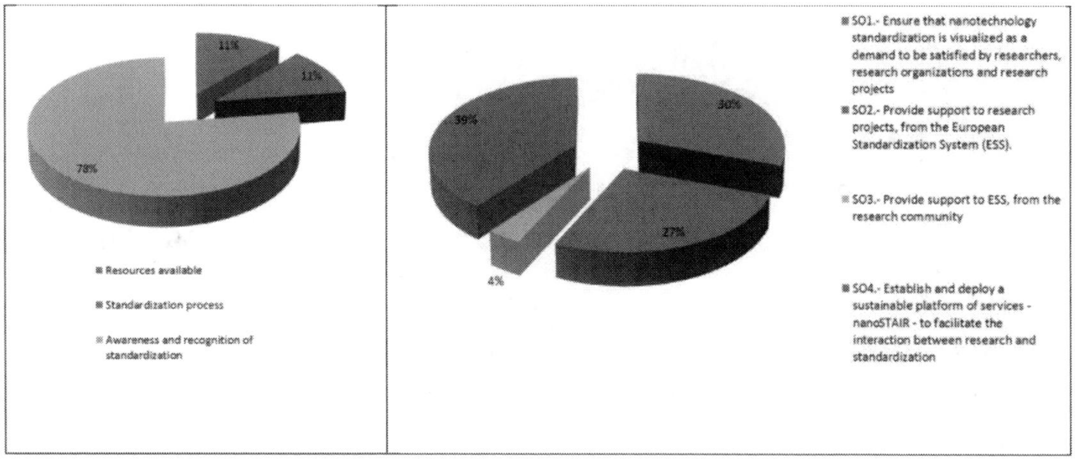


**Figure 1.** Distribution of nanoSTAIR strategic actions according to the type of barrier (Top) and the strategic objective concerned (Bottom) [7,8].

## 3. Integrating standardization in research projects from the early stages

Integrating standards and standardization in project proposals can amplify the impact of projects, facilitate the dissemination and exploitation of research results, and ensure a better evaluation. Integration of standardization in research project involves, at least, two basic actions: 1) Consider existing standards and standardization deliverables as a part of the project´s state of the art; 2) Use standards and standardization deliverables as an effective means to disseminate and exploit the project results and facilitate market access. nanoSTAIR proposes a simple procedure in five steps [8] (Figure 2).

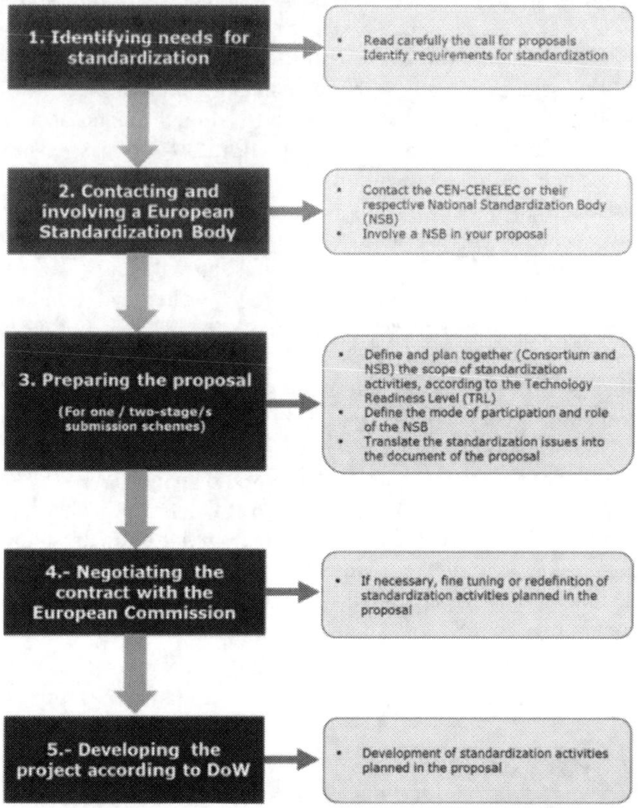


**Figure 2.** Basic procedure to integrate standardization in the research project [8]

Depending on the Technology Readiness Level (TRL) considered by the proposal, a different set of standardization activities can be defined. As an example, in Table 5, a proposal to select standardization issues according to the TRL is presented [8]. As can be seen, when the TRL level increases, the complexity of recommended standardization activities grows in parallel.

Finally, notice that CEN and CENELEC cannot become formal partner in European funded research projects, delegating this participation in their national representatives (NSB). Consequently the most effective way to support the integration of standardization in a research project is to involve a NSB.

From a higher to a lower degree of involvement, the NSB can become: 1) A partner of the consortium; 2) A subcontractor of a consortium partner, exclusively for project standardization activities; 3) A member of the project International Advisory Board or Steering Committee; 4) An organization that supports the project by providing a Letter of Intent (LOI).In this sense, to facilitate the national interaction between the research and standardization communities, most of CEN and CENELEC members have appointed a dedicated national contact point (DNCP). For members without a DNCP, you can use their generic e-mail address.

Journal of Physics: Conference Series **617** (2015) 012033 doi:10.1088/1742-6596/617/1/012033

**Table 5.** Proposal for standardization activities, according to the project's TRL [8]

| | TRL | Description | Minimum standardization activities expected or recommended | Types of proposed standardization documents |
|---|---|---|---|---|
| Basic & applied research | 1 | Basic principles observed and reported | 1.- Identification of published or under development applicable standards. | |
| | 2 | Technology concept and/or application formulated | | |
| | 3 | Analytical and experimental proof of concept established | 2.- Identification of new standardization fields. | • Terminology |
| Demonstration in lab /Simulation/real | 4 | Laboratory validation of component or process | Previous activities and: | • Testing methods |
| | 5 | Validation of component or process in simulated environment | 3.- Collaboration with existing standardization Technical Committees. | • Characterization<br>• Best practices |
| | 6 | Subsystem model or prototype demonstrated in relevant environment | 4.- Technical proposals for new or revised standardization documents. | • Specifications for products, services, processes, etc |
| Implementation, industrialization, commercializatio | 7 | Integrated pilot system demonstrated in operational environment | Previous activities and: | |
| | 8 | Actual system completed and qualified through test and demonstration | 5.- Participation and leadership in the elaboration of new or revised standardization documents. | |
| | 9 | System ready for full scale deployment, end-use operation | | |

## 4. Conclusions

Standardization is a powerful tool to transform the innovation produced from R&D projects into marketable tangible results; in future business. Especially when it comes to innovation, standardization is basic to provide confidence to consumers on new products / services and to ensure the fulfilment of applicable regulations.

The strategic approach formulated by project nanoSTAIR is based on the systematic analysis of the barriers between research and standardization, and of the possible ways to overcome them. It defines four strategic objectives and 53 strategic actions for deployment during the period 2014-2020 that may be used to promote "direct pipelines" between research and standardization at European level only, national level or in both areas.

The strategic deployment is addressed to visualize standardization as a demand to be satisfied by researchers, research organizations and research projects, and to establish effective bidirectional connections to support research and standardization communities, including the creation and deployment a sustainable platform of services - nanoSTAIR - to facilitate the interaction between them.

Within the approach nanoSTAIR a proactive strategy, considering standardization since the early stages of the design of our business (Project Proposal), is key to accelerating the adoption and entry into the market of new products / services and avoid potential trade barriers.

## References

[1] CEN-CENELEC 2014 *CEN and CENELEC response to the public consultation on the Europe 2020 Strategy*, 11 pp.

[2] European Commission 2010 *Mandate M/461 - Standardization Mandate to CEN, CENELEC and ETSI for standardization activities regarding nanotechnologies and nanomaterials*.

[3] European Commission 2011 *A strategic vision for European standards: Moving forward to enhance and accelerate the sustainable growth of the European economy by 2020* COM (2011) 311 final.
[4] INTEREST 2007 *Integrating Research and Standardization. Guide to standardization for R&D organisations and researchers*, 21 pp.
[5] NANOFUTURES 2012 *Integrated Research and Industrial Roadmap for European Nanotechnology*, 119 pp.
[6] nanoSTAIR 2013 *Snapshot of the needs of the stakeholders and drivers for standardization.* Project deliverable 2.1, 28 pp.
[7] nanoSTAIR 2014 *Innovative strategies and procedures to translate EU nano-research into standards*. Project deliverable D4.1, 48 pp.
[8] nanoSTAIR 2014 *nanoSTAIR practical guideline (NPG)*. Project deliverable D4.3, 73 pp.
[9] Nanosafety Cluster 2013 Savolainen, K et al. *Nanosafety in Europe 2015 - 2025: Towards Safe and Sustainable Nanomaterials and Nanotechnology Innovations, 212 pp.*
[10] Nanosafety Cluster 2014 *Compendium of Projects in the European NanoSafety Cluster 2014*, 250 pp
[11] Optimat 2014 *Research Study on the Benefits of Linking Innovation and Standardization*. Final report. 88 pp.
[12] Technopolis Group 2013 *Study on the contribution of standardization to innovation in European-funded research projects* Final Report, 108 pp.

**Acknowledgment**
Research carried out by project nanoSTAIR was made possible thanks to funding from the European Commission, through the Seventh Framework Programme (GA 319092).

Journal of Physics: Conference Series **617** (2015) 012034 doi:10.1088/1742-6596/617/1/012034

# Miniature nanoparticle sensors for exposure measurement and TEM sampling

**Martin Fierz[1,2], Dominik Meier[1], Peter Steigmeier[2] and Heinz Burtscher[2]**

[1] naneos particle solutions llc, Dorfstr. 69, CH-5210 Windisch
[2] University of applied sciences northwestern Switzerland, Klosterzelgstr. 2, CH-5210 Windisch

martin.fierz@fhnw.ch

**Abstract.** Nanoparticles in workplaces may pose a threat to the health of the workers involved. With the general boom in nanotechnology, an increasing number of workers is potentially exposed, and therefore a comprehensive risk management with respect to nanoparticles appears necessary. One (of many) components of such a risk management is the measurement of personal exposure. Traditional nanoparticle detectors are often cumbersome to use, large, heavy and expensive. We have developed small, reliable and easy to use devices that can be used for routine personal exposure measurement in workplaces.

## 1. Introduction

Exposure measurements should be a part of a general risk-management strategy for workplaces where nanoparticles are produced or processed. Since the prime exposure route is over inhalation, measuring airborne nanoparticles in such workplaces seems like a sensible idea, and it can be done in different ways: (I) measurements by an industrial hygienist (II) with stationary equipment that is always present at a sensible location (similar to a fire detector), or (III) as personal exposure measurement with small instruments carried by the workers, similar to dosimeters in the nuclear industry. Ideally, for personal monitoring, the instruments should be small, lightweight, affordable, simple to use and reliable, and provide an online measurement so that an alarm can be raised if a threshold value is exceeded. Unfortunately, most current particle detectors have severe drawbacks for this type of application: Optical detectors are generally simple to use, and low-maintenance, but are not sensitive to nanoparticles – sub-100nm particles cannot be detected. Condensation particle counters that first grow the particles to larger sizes before an optical detection are very precise devices, but need a working fluid that needs regular replenishment, and they need to be held level to prevent internal spilling of the working fluid. They are thus reasonable for occupational hygiene professionals, but cannot be used for measurements of types II and III. Filter samples can be taken, but are not real-time, and need a lot of effort after the sampling. Fortunately, there is one class of particle instruments that is nano-sensitive, reliable, and real-time – electrical particle detectors. This class of instrument has been known for a long time, and already exists in the market. Our new instruments are similar in performance to existing electrical particle detectors, but much smaller and easier to use.

Journal of Physics: Conference Series **617** (2015) 012034 doi:10.1088/1742-6596/617/1/012034

## 2. Standard electrical detection by unipolar charging of aerosols

Unipolar charging of nanoparticles is at the basis of many current aerosol instruments (e.g. TSI FMPS, Cambustion DMS500, Dekati ELPI, Pegasor PPS etc). The simplest possible instrument that can be built based on unipolar charging is known as a diffusion charger (DC). A scheme of such an instrument is shown in figure 1.

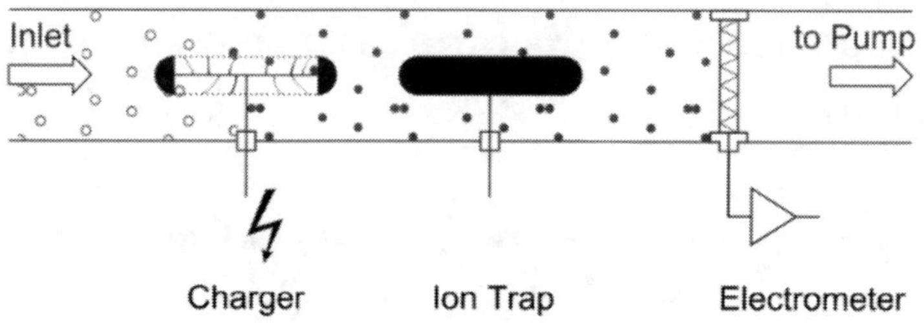


Figure 1: Scheme of a standard diffusion charger (DC)

The diffusion charger consists of 3 basic elements: the unipolar charger (usually a corona charger), where ions are generated and mixed with the particles to charge them, the ion trap, where excess ions are removed, and finally a filter, where all particles are captured and the current deposited by the particles onto the filter is detected with a sensitive electrometer, which must be able to measure currents in the order of fA. This type of instrument has been commercially available for some time (Matter LQ1-DC, Ecochem 2000-DC, TSI EAD, TSI NSAM, TSI Aerosotrak 9000). It measures the final charge state of the aerosol after the unipolar diffusion charger. The particles take up a charge that can well be described by a power law:

$$q(d) = c \cdot d^x$$

where x is usually in the range of 1.1 – 1.2. The signal of the DC is thus nearly proportional to an "aerosol length concentration" (with units e.g. mm/cm$^3$), and this is true over a very wide size range, from about 10 nm to 10 µm. Diffusion charging is unspecific, i.e. it hardly depends on the particle material, so all particles are detected with this instrument. An interesting interpretation of the DC signal was given by Wilson et al. [1]: for particles in a more limited size range (about 20nm – 350nm), the DC signal is approximately proportional to the lung-deposited surface area (LDSA) concentration of nanoparticles in the human lung, expressed in units of µm$^2$/cm$^3$, indicating how much particle surface area in µm$^2$ is deposited in the lung for every cm$^3$ of inhaled air. Since many toxicological studies have shown that particle surface area usually correlates better than particle mass or particle number with health effects, the LDSA concentration is a potentially highly interesting particle metric.

In summary, the DC is thus an interesting instrument, because it is simple (and thus robust and potentially low-cost), measures a probably relevant quantity, operates over a very large size range and needs no consumables.

## 3. Electrical detection by induced currents

We have developed a novel technique to measure LDSA by induced currents, rather than particle-deposited currents on a filter, and implemented it in an instrument called the partector. It works as follows:

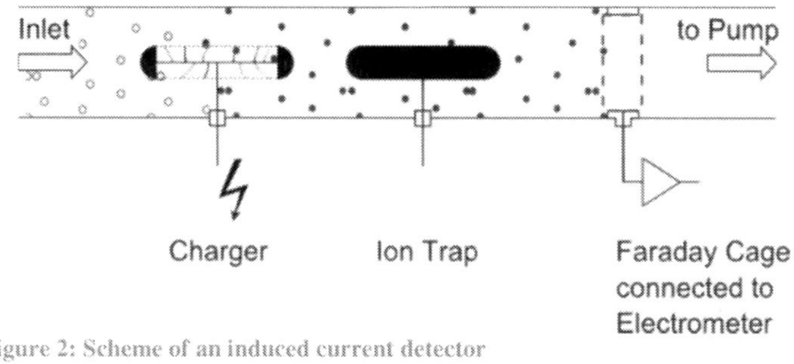


Figure 2: Scheme of an induced current detector

The only visible difference to the DC is the lack of the particle filter. It is replaced by an empty Faraday cage, which is connected to an electrometer, and thus virtually grounded. Since a grounded Faraday cage will not allow any electric field lines to escape, the total charge inside the cage and on the cage must be zero by the law of Gauss. Therefore, by measuring the charge on the cage, an indirect measurement of the charge in the cage is made. However, in our setup we are only measuring the current flowing to/from the cage, i.e. the time derivative of the charge, dQ/dt. Thus, a constant charge in the cage is not detectable. In order to always have a measureable signal, a temporal variation of the charge must be introduced. We do this be pulsing the unipolar charger, which turns on and off at a rate of 0.5 Hz. In this manner, we can detect an oscillating signal at the electrometer, whose amplitude is proportional to the charge transfer to the aerosol (see Fig. 3).

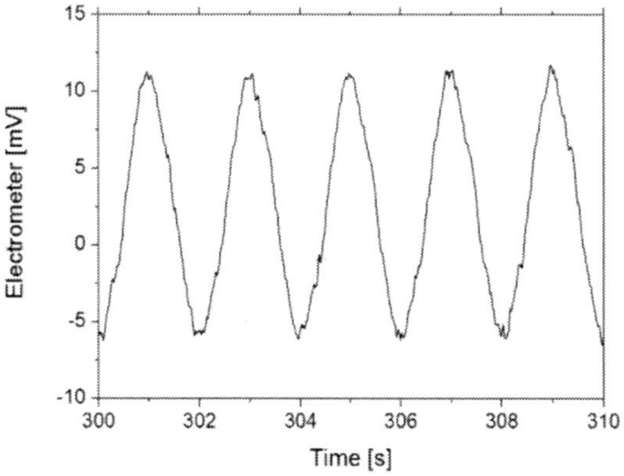


Figure 3: example electrometer signal from the Faraday cage

This approach has some important differences to the standard DC approach:

- It measures the charge transferred to the aerosol in the unipolar charger, and not the final charge state. In practice, this is mostly identical, since aerosols are usually neutral. However, if a highly charged aerosol is measured, this difference will be clearly visible. In particular, negatively pre-charged aerosols will give a larger signal than neutral aerosols, while positively pre-charged aerosols will give a smaller signal (since the unipolar charger uses positive charging).
- An amplitude of an AC signal is measured rather than a DC signal. This is much more robust, since all electrometers exhibit some zero offset that may drift with time and/or ambient conditions (such as temperature and humidity). In traditional electrometer-based aerosol instruments, warmup periods and regular zero checks are necessary, which are not needed here.
- A better monitoring of the instrument health status is possible in this version: (1) the electrometer zero offset is constantly monitored, because it is just the period-average of the

electrometer signal. Although a drift or high offset in itself does not influence the measurement, it is an indicator that something bad is happening to the instrument. The firmware can thus display a service warning while the instrument is still operational. (2) In the unipolar charger, the charging current reaching the ion trap is measured, and the high voltage is regulated for constant current. It is possible that a leak current flows to this counterelectrode via insulator due to water condensation or, over long times, contamination with particles. In traditional instruments, such leak currents are indistinguishable from real currents, whereas here, the charger is in an off-state half of the time, where we can monitor if the charging current is really 0 as it should be.

- Service intervals are longer. Since no particles are collected nominally, there is no need for cleaning or filter exchanges. Of course, particles are always deposited by diffusion and impaction in any instrument, so some service is required at some point. Nevertheless, service intervals are much longer than in previous instruments we have built.
- Finally, because this is a non-contact method of aerosol detection, the particles are still available after the measurement in a charged state, and can be further manipulated. We have built a version of the instrument where the particles can be deposited on a TEM grid for further analysis. This is described in section 4.

Overall, the Faraday cage variant of a DC seems to have more advantages to us than drawbacks. We used this principle to build an instrument designed for occupational hygiene / workplace safety, i.e. our goal was to build a small instrument that would be simple to use. The partector, has the following specifications:

- Small (13.4 x 7.8 x 2.9 cm)
- Light-weight (400 g)
- No consumables, such as working fluids in handheld CPCs
- Battery life of 14 hours
- Alarm capability when a threshold concentration is exceeded
- Only 16 s warmup time, and a time resolution of 1 s
- Concentration range: 1 – 20'000 $\mu m^2/cm^3$ (This corresponds to about 500 – $10^7$ particles/$cm^3$)
- Size range: 10nm – 10μm; however, the interpretation as LDSA is only valid in a more limited size range of about 20-350nm. Nevertheless the instrument can be used to measure larger particles.

Summing up, the partector has a very wide detection range both in terms of particle concentration and particle diameter, making it suitable even for workplaces with very high concentrations of nanoparticles (e.g. welding). The electrical charging principle is very sensitive even to nanoparticles, which cannot be detected with optical means.

## 4. TEM sampling

The partector can serve as a simple instrument in occupational hygiene to easily make a quick survey of nanoparticle concentrations. However, like all instruments measuring physical parameters (mass, number, size distribution etc), it cannot distinguish between different types of particles – whether they are harmless (e.g. salt particles) or potentially dangerous (e.g. metal particles, fibers), they are all seen by the instrument. Furthermore, as always in occupational hygiene, there is the general difficulty of distinguishing between the always present background aerosol and nanoparticles released by a process in the workplace. In such cases, a deeper analysis of the particles is necessary. The transmission electron microscope (TEM) is the most powerful analytical technique for analysing nanoparticles; in particular for morphology and single-particle chemical analysis. We thus designed a second instrument (partector TEM) which adds a sampling stage for TEM probes. Since the particles passing through the partector are already charged, it is natural to use electrostatic deposition to collect the particles on the TEM grid. The partector TEM contains a small insert that holds the TEM grid, and a strong electric field can be applied to deposit the particles on the grid. Since TEM grids are notoriously difficult to handle due to their small size, the entire insert is exchanged at once when a new sample needs to be taken – the instrument comes with 6 inserts which can be prepared with TEM grids before the actual sampling.

One of the biggest challenges in TEM sampling is to produce samples with optimal coverage. If too many particles are sampled, they will form agglomerates on the grid, and it remains unclear whether they were agglomerated in the gas phase. If only very few particles are sampled, then the sample is not very representative. For traditional methods to collect TEM samples (on holey grids by flow-through filtering, by thermophoretic or electrostatic deposition), it is therefore necessary to either take multiple samples with different sampling time, or to estimate the necessary sampling time from a concentration measurement made with a second instrument. The partector TEM is unique, because it can directly estimate the area of the grid covered by nanoparticles by integrating the electrometer signal. This works because the time-integrated signal is approximately proportional to the area coverage of the grid: for the area A covered with N nanoparticles of diameter d we can write:

$$A = N \cdot d^2 = \eta \cdot nt \cdot d^2 = d^{-1} \cdot nt \cdot d^2 = nd \cdot t$$

Where n is the nanoparticle concentration in the air, $\eta$ is the deposition efficiency, and t is the sampling time. Because smaller particles have a higher electric mobility after diffusion charging than larger particles, the deposition efficiency is approximately proportional to $d^{-1}$, and thus the area covered is proportional to the instrument signal integrated over time. This calculation once again breaks down for large particles (d >~ 300nm) where the mobility becomes more or less constant, so large particles will lead to a higher coverage of the grid. Nevertheless, especially for particles in the nano size range, integrating the LDSA signal over time gives an excellent estimate for when to stop sampling.

The partector TEM is thus a very interesting instrument, as it allows an occupational hygienist to first make a quick survey of nanoparticle concentrations in a workplace, and then, if there is some concern about the levels or the types of particle that might be produced, to take a sample with a single push of a button. TEM analysis is expensive due to the high cost of the instrument and the skilled operator involved, so, although the partector TEM is much more expensive than simple filter samplers, it can easily pay off through decreased operating costs in the TEM analysis.

## 5. Applications and Conclusions

We have developed a new method of aerosol detection, by induced currents. This method is sensitive to nanoparticles, and non-collecting, thus leading to long service intervals. As demonstrated, it can be used to create miniature nanoparticle detectors that are very easy to use. These instruments can for example be used for personal exposure monitoring in the workplace, but also, due to the long service intervals, for ambient monitoring where it seems that 1 year 24/7 operation is possible. Another application is to combine it with a CPC measurement; the ratio of the two instrument signals is then proportional to the average particle diameter over a very wide size range, allowing (crude) particle sizing in real time (1s) from 10nm to 10µm [2].

## References

[1] Wilson W.E., Stanek J., Han H.-S., Johnson T., Pui D.Y.H., Turner J., Chen D.-R. and Duthie, S. (2007). Use of the Electrical Aerosol Detector as an Indicator of the Surface Area of Fine Particles Deposited in the Lung. J. Air Waste Manage. Assoc, 57:211-220.

[2] Bukowiecki N., Kittelson, D.B., Watts, W.F., Burtscher, H., Weingartner, E. and Baltensperger U. (2002). Real-time Characterization of Ultrafine and Accumulation Mode Particles in Ambient Combustion Aerosols, J. Aerosol Sci. 33: 1139-1154

Journal of Physics: Conference Series **617** (2015) 012035 doi:10.1088/1742-6596/617/1/012035

# Strategies, methods and tools for managing nanorisks in construction

**J M López de Ipiña[1], C Vaquero[1], D Boutry[2], J F Damlencourt [2], P Neofytou[3], M Pilou[3], E Jankowska[4], I Larraza[5], R Pina[5], S Fernández[6], S Contreras[6], A Romero[6], M Calderon[6], P Swiezewsk[7], K Otkallo[7], A Pintea[8], C Salazar[9], T Oroz[9], B Hargreaves[10], R Ciobanu[11], A Tabrea[11], B Hazebrouck[12], O Salvi[12], H Stockmann-Juvala[13], V Vaananen[13], D H Y Pui[14] Drew Thompson[14]**

[1] TECNALIA – Parque Tecnológico de Alava, 01510 Miñano/Spain
[2] CEA – Grenoble, 17, rue des Martyrs 38000 Grenoble/France
[3] DEMOKRITOS - Patriarchou Gregoriou Str., 15310 Aghia Paraskevi/Greece
[4] CIOP-PIB - Czerniakowska 16, 00701 Warsaw/Poland
[5] ACCIONA - Avenida de Europa 18, 28108 Alcobendas/Spain
[6] AENOR - Genova 6, 28004 Madrid/Spain
[7] MOSTOSTAL - Ul Konstruktorska 11a, 02673 Warszawa/Poland
[8] ROSSAL - Bogdan Dragos 119, 61001 Roman/Romania
[9] TECNAN - Area industrial Perguita C/A, 31210 Los Arcos/Spain
[10] NETCOMPOSITES - 4A Broom Business Park, Bridge, S41 9QG Chesterfield/United Kingdom
[11] ICECON - Sos Pantelimon 266 Sector 2, 21652 Bucuresti/Romania
[12] EU-VRi - Willi-Bleicher-Straße 19, 70174 Stuttgart/Germany
[13] FIOH - Topeliuksenkatu 41 a A, 00250 Helsinki/Finland
[14] UMN-PTL - Oak Street SE 200, Minneapolis 55455 2070/United States

E-mail: jesus.lopezdeipina@tecnalia.com

**Abstract**. This paper presents a general overview of the work carried out by European project SCAFFOLD (GA 280535) during its 30 months of life, with special emphasis on risk management component. The research conducted by SCAFFOLD is focused on the European construction sector and considers 5 types of nanomaterials ($TiO_2$, $SiO_2$, carbon nanofibres, cellulose nanofibers and nanoclays), 6 construction applications (Depollutant mortars, self-compacting concretes, coatings, self-cleaning coatings, fire resistant panels and insulation materials) and 26 exposure scenarios, including lab, pilot and industrial scales. The document focuses on the structure, content and operation modes of the Risk Management Toolkit developed by the project to facilitate the implementation of "nano-management" in construction companies. The tool deploys and integrated approach OHSAS 18001 – ISO 31000 and is currently being validated on 5 industrial case studies.

## 1. Introduction

The European construction industry (2013) contributes 8,8 % to the EU-28 GDP and employs 13,9 million people - 6,4 % of total employment and 29 % of industrial employment - in 2,9 million enterprises, most of which – 95 % - are SMEs with less than 20 operatives [7].

The use of manufactured nanomaterials (MNM) and nano-enabled products (NEP) in construction and the related infrastructure industries is an increasing reality, mostly in cement or concrete products, coatings or insulation materials and to a lesser extent in road-pavement products, flame retardant materials or textiles [1,12,13].

A majority of workers and their employers in the construction sector (~75%) are not aware that they actually work with MNMs and NEP. Detailed information about the product composition and their possible nano-specific health and safety issues is generally lacking and the information available for the raw material manufacturer is often lost while stepping down the user chain. As a consequence, it is very difficult for average construction companies to conduct a proper risk assessment and organize a safe workplace for its employees [27].

In this context, project SCAFFOLD is a three years industrial oriented idea (2012-2015), funded by EC under FP7, specifically addressed to provide practical, robust, easy-to-use and cost effective solutions for the European construction industry, regarding current uncertainties about the occupational exposure to nanomaterials. The aim of project is to develop, test, validate in real conditions and disseminate a new holistic, consistent and cost effective Risk Management Model (RMM), by integration of a set of innovative strategies, methods and tools developed by the project into consistent state-of-the-art OHS management systems (OHSAS 18001 and ISO 31000) (Figure 2).

The research conducted by SCAFFOLD is focused on construction and considers five types of nanomaterials ($TiO_2$, $SiO_2$, carbon nanofibres, cellulose nanofibers and nanoclays), six construction applications (Depollutant mortars, self-compacting concretes, coatings, self-cleaning coatings, fire resistant panels and insulation materials) and twenty-six exposure scenarios, including lab, pilot and industrial scales (Figure 1).

## 2. Overview of project results in the areas of risk prevention, risk assessment and risk protection

Within SCAFFOLD project, different safe-by-design strategies has been designed in order to increase the safety of nanomaterials and nano-enable products for construction. Concretely, the first strategy involves the preparation of highly concentrated dispersions ($TiO_2$, $SiO_2$) instead of using powders directly for safer transport and handling. Thus, apart from achieving excellent results regarding particle size and stability with the mentioned new dispersions, the effect and performance of the nanomaterials was perfectly maintained. Additionally, specific safe-by-design strategy was used in order to modify some nanoclays so that their incorporation in fire resistant panels results in lower toxic emissions.

In addition, we studied the effects of mixing conditions and clay loading level on mechanical properties and fire performance of glass fibre composites. Clays with different organomodifier content and type were also investigated. For the materials studies, there were no differences in fire performance and mechanical properties with different mixing conditions. The positive conclusion from this is that workers could use lower energy processes to incorporate clays into polymer resins, helping to limit the amount of dust produced. With regard to the different clays produced, there was found to be very little difference between in the fire performance of the resulting composites. Anecdotally, one of the novel clay samples was observed to be less dusty than the others, which could have potential "safe-by-design" implications; however further qualitative analysis is required in order to confirm this finding.

Specimens containing either $TiO_2$ or $TiO_2$-on-sepiolite in bulk and sol-gel coatings have been prepared and tested – a slight increase in mechanical properties was observed in bulk additions, but there was no significant difference between the $TiO_2$ and $TiO_2$-on-sepiolite. $TiO_2$-sepiolite is a safer additive than $TiO_2$ free nanoparticles due to the fact that this special clay is used to improve the dispersion of $TiO_2$ nanoparticles and to attach the nanoparticles on the surface. Samples of self-compacting concrete with nano-$SiO_2$ (1 and 5 %wt.) have been prepared and tested. In this case it exist also a slight increase in mechanical properties. Nano-$SiO_2$ has been supplied in a dispersion that is safer than same material in powder.

SCAFFOLD project measured particle release and occupational exposure (inhalation, dermal) at pilot scale, lab scale and in real scenarios, considering: five NOAAs ($TiO_2$, $SiO_2$, CNF, CeNF and nanoclays), six applications (depollutant mortars, self-cleaning coatings, self-compacting concretes, fire-retardant panels, coating laminates and insulations) and five main scenarios (Manufacture of NMs, manufacture of products containing MNMs and application, machining (drilling, sawing), demolition and accidental fires). It is important to highlight that in the construction sector the exposure to mixed types of dust is extremely common and results in much higher levels than the exposure to MNMs, which are generally included in a very low concentration in the products (typically not exceeding 1,5 %). In addition, attention is more focused on traditional contaminants as e.g crystalline silica (carcinogenic to humans, IARC).

The highest release of particles was measured during machining (drilling and sawing) of samples of hard materials like concrete and laminate coating, with no clear difference between conventional and nano-enabled materials, and with no observation of free nano-objects (SEM analysis). Occupational exposure measured (particle concentration of NOAA) was below proposed NRV limits. The highest mass concentration has been measured during the cleaning task in the manufacturing process of nano-$TiO_2$. Mass concentration of the specific nano-object at the personal breathing zone has been measured for the scenarios related to nano-$TiO_2$ and CNF and was below proposed OELs by NIOSH and SCAFFOLD. New measurement campaigns are scheduled during the development of industrial case studies (Figure 3).

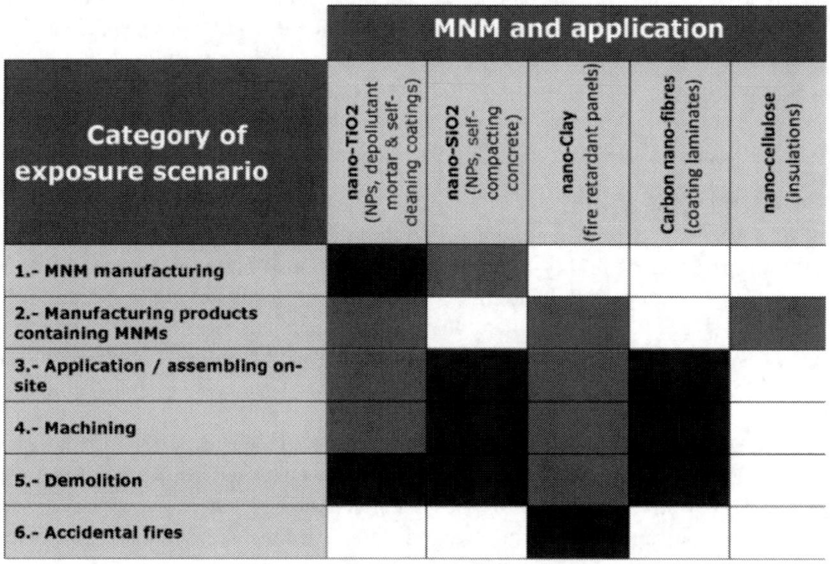


**Figure 1** Exposure scenarios selected by project SCAFFOLD (Dark blue, lab and pilot scales; light blue, industrial scenarios).

Reaction to fire tests performed with conventional and nano-enabled materials showed differences on fire behaviour linked to the presence of the nano-objects. However, none of the nano-objects introduced in the materials were identified in the effluents from the combustion, with the possible exception of nano-clays from the fire retardant panels.

Based on the available data on the toxicological properties of the nanomaterials, recommendations for occupational exposure limit values have been proposed for $TiO_2$, $SiO_2$, Carbon Nanofibers (CNF), Cellulose Nanofibers (CeNF) and Nanoclays.

Concerning the collective protection, investigations were carried out during experimental campaigns, in nine rooms with different ventilation systems. It was confirmed that only in the room with positive pressure ventilation and when works were conducted in the glove box, particles from the processes were not transferring to the room air.

The efficiency of different types of current respiratory protective devices intended for use in construction as well as three types of clothes generally used by workers on construction sites, were tested with different nanoaerosols in order to classify them and enable a proper selection for different workplaces and hazards. The Total Inward Leakage (TIL) of nanoparticles was realized with a breathing manikin simulating natural human movements and speech. The highest TIL levels were found for speech simulation and simulation of up and down head movement. The highest effectiveness of protection was recorded for the full-face mask used with P3 filters and the TH2 powered filtering device incorporating a hood.

The observation of PPEs involved in real scenarios at industrial partners' workplace showed that the actual gloves, masks and Tyvek clothes are efficient towards nanoparticles incorporated at realistic concentrations (between 0,4 and 1,7%) in a material. All the mortars studied contained just few percentage of nanoparticles, just enough to bring beneficial effects to the material. Whether in powder form (synthesis of nanoparticles, manufacturing of the mortar) or in solid state (mortar with water, applying on a wall) or in sol-gel state (liquid mortar), we never observed $SiO_2$, $TiO_2$, nanoclay or nanocellulose inside PPEs. Regarding the nature of the clothes, the rain coating was the most efficient material, with no diffusion observed at all (aerosol and liquid). The polyester 65%/cotton 35% material was efficient only for $SiO_2$ aerosol and the fleece jacket for $SiO_2$ aerosol and $TiO_2$ aerosol.

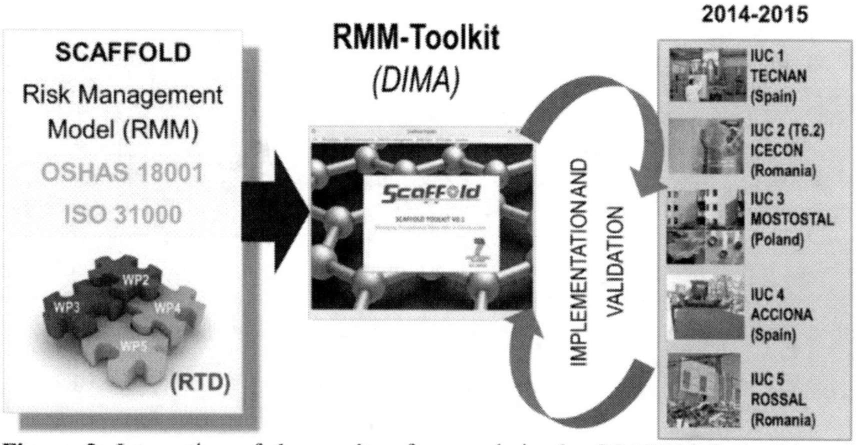


**Figure 2.** Integration of the results of research in the SCAFFOLD-Toolkit and validation.

## 3. SCAFFOLD approach for managing nano-risks in construction

### *3.1. Background*

The most common approach and usually the first choice to tackle the uncertainly related to risk assessment and management of MNM is Control Banding (CB) [2,5,6,10,14,25]. This approach is particularly useful in work situations in which information on hazards, exposure levels and risks are limited. It is also valuable for SMEs for its simplicity.

There are some CB tools already available [2], but they have not been developed for the construction sector, which is a rather different kind of work environment compared to e.g. the chemical industry. EU-OSHA [6] developed a specific document introducing a number of such control banding risk management tools, to aid in the selection of appropriate workplace prevention measures. An international standard has been also recently produced by ISO (ISO 12901-2: 2014) [10].

From the viewpoint of management systems, OSHAS 18001 [3,4] is the internationally recognized and adopted standard for Occupational Health and Safety, designed to help organizations to implement a framework that identifies and controls health and safety risks, reduce potential accidents, aids legislative compliance and improves overall performance.

OSHAS 18001 implements an approach based in the PDCA model, fully compatible and easy integrable with other management standards like ISO 9001 and ISO 14001. One of the key aspects of the management system is the continuous improvement, which allows companies to improve their performance in a continuous way, taking into account the technical and economical limitations, but also the possibilities brought by the new technology, new knowledge and new risks. This is a key issue, as nanotechnology knowledge is subject to continuous changes due to technical and scientific advances, and because of that is it necessary to provide companies with a tool capable not only to manage current risks buts also to identify and manage the new ones (coming, for instance, from new nanomaterials or just from the improvements made in detection capabilities).

Near 55.000 companies have been already certified according with OHSAS 18001, and the increase in the number of certificates is exceptional: between 2007 and 2009 the increase was as high as 73%. OSHAS has had a great impact in SMEs and some studies show that more than 80% of certified companies are SMEs, and that particularly the 50 to 250 workers interval represents more than 50% of companies. Regarding construction, this sector is leading the implementation of OHSAS (37%), showing the commitment of the construction companies, especially SMEs, to the improvement of occupational health in Europe. In fact, according with the above-mentioned study, 97% of the companies certified according to OSHAS 18001 where already certified according to ISO 9001 and 91% according to 14001. These figures show the importance of the compatibility in the integration of the environmental, quality and labour safety issues in the organizations [16].

However, the capability of OSHAS 18001 to deal with new risks, and particularly those related to nanomaterials manufacturing, use and disposal has yet to be tested. In this respect, the European project SCAFFOLD will answer those two questions: the capability of OHSAS 18001 to deal with "nano-risks" and eventually its gift for being implemented by SMEs.

ISO 31000 is the international standard for risk management. It provides principles, framework and a process for managing risk [9]. However, ISO 31000 cannot be used for certification purposes, but does provide guidance for internal or external audit programmes. This international standard also helps to boost health and safety performance, establish a strong foundation for decision making and encourage proactive management in all areas. Organizations using it can compare their risk management practices with an internationally recognized benchmark, providing sound principles for effective management and corporate governance.

With over 127 countries currently using OHSMS standards [15], there's a worldwide need to harmonize health and safety management systems using an international standard and share best practices. Inspired by the well-known OHSAS 18001, ISO is currently developing ISO 45001 [11]. There are 50 countries and international organizations, including the International Labour Organization, involved in this work. The future standard will also be aligned with ISO 9001 and ISO 14001. The new standard is expected to be published in late 2016.

### *3.2. Control banding approach*

The Finish Institute of Occupational Health tested the applicability/usefulness of the CB approach [18], more specifically the freely available control banding tool *Stoffenmanager Nano 1.0* [5,25], in assessing and managing risks of nanomaterials in the construction sector. The results indicate that the tool is applicable in some of the studied industrial workplaces. In other cases, the evidence that the raw material contained engineered nanomaterials was not available. Thus, the hazard assessment was the most difficult part, because it was difficult to identify the nanomaterial from the products.

Furthermore the ISO 12901-2 approach has been implemented in the Toolkit [20] and one of the partners of the project consortium is currently finishing a customized tool for the sector.

### *3.3. The Risk Management Model (RMM)*

Although there are some private schemas dealing with nanomaterials H&S management, particularly the CENARIOS model [26], SCAFFOLD will avoid creating new systems, which are annoying to implement and probably not the result of the consensus process, and will try to show the opportunity of using OSHAS 18001, with the necessary specific elements and recommendations, specially for its implementation in SMEs, in the identification and management of nano-related risks. Moreover, SCAFFOLD project will follow a new approach to incorporate the principles for risk management contained in the recent global standard reference ISO 31000 [9].

The RMM presents itself and innovative conceptual work taking into account derived from ISO 31000 requirements and OSHAS 18001 [17]. It is based on a conformity assessment extended scheme and the PDCA cycle like OSHAS 18001. The outline is very similar to the most implemented management systems ISO 9001 and ISO 14001. This design for the model allows the integration with the other management systems especially for the common requirements (systems requirements as document control, audits, training, etc). The model has been designed using requirements of OHSAS 18001 (structure, elements, etc.) with additional requirements derived from the guidelines established in ISO 31000.

The RMM includes specific considerations on initial review, monitoring and audit and can be certificated. The implementation of this RMM will allow the organization to consider MNMs risk into OHSAS system [19]. On the other hand, if the organization has no experience on a systematic approach for managing its occupational health and safety risks, the MNMs RMM implementation could be the first step for a more complete and organized perspective of OH&S risks. One organization that proves the successful implementation of this model should ensure all interested parties that has an appropriate MNMs risk management.

The main difficulties encountered in the model design were related with · the fact that ISO 31000 is not a requirement standard but a guideline, then does not include specific rules it is made on recommendations; the conversion of recommendations from ISO 31000 into specific requirements; · the analysis of this requirements that should be added to complete OSHAS 18001; · the decisions on the risk assessment approach in the model taking into account that both references have different definitions; ·the design to be applied to all type, size and context organization with particular emphasis in SMEs.

The level of detail and complexity of the RMM, the extent of documentation and the resources devoted to it depend on a number of factors, such as the scope, the nature of its activities, the organizational culture and also the size of an organization. This is the case in particular for SMEs that could need a specific and adapted Toolkit for RMM implementation. In this respect, a specific guide to establish criteria for interpreting the RMM and facilitate implementation in the SME construction companies in to the entire life cycle of the construction sector, has been drafted and implemented in the Toolkit (Customized approach for SMEs) [22,23] .

### *3.4. The Toolkit*

The RMM-Toolkit represents the integration of all the solutions developed for risk management during the project in a software tool, friendly, easy to use and customizable for SMEs. It consists in a standalone desktop application for the Windows platform [20, 21, 22]. The Toolkit is structured into: 1) Five operational modules (Library, Customization, Risk management, Tools and Help) (Table 1), 2) two setups, a general setup for large or advanced companies in risk management, and a customized setup for SMEs and 3) Two operational modes (Table 2), learning and risk management. In tables 1 and 2 the main characteristics and content of the modules and operating modes are displayed.

The main tools for risk management in the Toolkit are [20,21]:

1) Risk management: opens the checklist for diagnostic, implementation or audit. The checklist enables the user to enter comments and generate bar charts. The Toolkit will include two check list depending on the set up decided by the company, the general check list (275 questions) is for general set up and the reduced check list (150 questions) is principally for SMEs set up.
2) Risk assessment: opens the risk evaluation tool. This window displays a list of processes, tasks and scenarios. Each scenario can be characterized with both quantitative and qualitative methods. The quantitative method allows the user to enter the exposition and reference values.

The tool will then calculate the exposure value and, according to the configured thresholds, the risk level. The qualitative method implements the control banding approach ISO 12901-2. The user can navigate through these charts to get a hazard band (A – E) and an exposure band (1 – 4). This characterization leads to another risk level. For each scenario, a set of control measures can be selected, allowing the user to specify whether they are already implemented or not.

3) Planning: opens a tool to schedule the implementation of the control measures specified in the risk evaluation tool. For each control measure, the user can select the expected implementation date, the actual implementation date, the progress, the responsible and the associated cost. This planning can be exported to Excel.
4) KPIs: allows the definition, customization, calculation and visualization of Key Performance Indicators.
5) Documents and templates: gives access to a list of Word templates with procedures, instructions, registers and OHS manuals.

**Table 1.** SCAFFOLD Toolkit: Software modules [20, 21, 22]

| Module | Description |
|---|---|
| 1. Library | It provides a library with documentation to help the companies of the construction sector to deal with the risks arising from MNM. |
| 2. Customization | It allows companies to customize the application to their processes, tasks, scenarios and size. It uses the Module 1 to facilitate data input and generate the company profile. |
| 3. Risk Management | It enables the initial assessment, implementation and audit of RMM guided by a step-by-step dialog. This module deploys two different setups, depending on the company profile (Large company or SME). |
| 4. Tools | It contains the toolbox for nanosafety management. |
| 5. Help | It gives access to miscellaneous options: file management, configuration, and help (User manuals). |

**Table 2.** SCAFFOLD Toolkit: Operation modes [20, 21]

| Operation mode | Description |
|---|---|
| Learning | The toolkit is used for training (e.g. toolbox), general information and communication (e.g. NOAA, hazards, control measures, good practices, etc). Only modules 1 and 4 are operating. |
| Risk Management | Customized mode. The toolkit is used for diagnosis, implementation, monitoring, auditing and improving the management of nanorisks in a specific construction company. All modules are operating. |

### *3.5. The library of solutions for risk management*

Project SCAFFOLD will produce by April 2015 a Library of Solutions to facilitate the diagnosis, implementation and audit of nano-risks management in construction companies (large companies and SMEs). The library will include: 1) a handbook for managing nanoriks in construction, 2) four quick guides (risk prevention, risk assessment, risk protection and risk management) and 3) the Toolkit (software).

Recently the CEN/TC 352 "Nanotechnologies" decided to accept a New Work Item (NWI) Proposal relating to CEN/TS "Manufactured nanomaterials (MNMs) in the construction industry. Guidelines for occupational risk management", promoted by project SCAFFOLD.

### *3.6. Testing and validation stage*

The SCAFFOLD approach is currently under validation in five Industrial Use Cases (IUC). The IUC have been selected to cover different applications, size of companies, exposure scenarios, countries with different safety cultures and different OHS needs (Table 3).

Since the IUCs show different characteristics and needs, every IUC deploys its own customized plan [24]. All these activities will be developed by using the RMM-Toolkit. In all cases, the plan consists of five basic steps: 1) Information and training first; 2) Diagnosis of the OHSMS, 3) Implementation, monitoring and internal audit, 4) External audit and finally 5) Reporting of the company and auditor (Figure 3).

**Table 3.** The five Industrial Use Cases (IUC) of project SCAFFOLD [24]

| IUC | Company | Country | Size | Exposure scenario | MNM |
|---|---|---|---|---|---|
| 1 | TECNAN | Spain | SME | Manufacturing nanomaterials : nano $SiO_2$, powder | $SiO_2$ |
| 2 | ICECON | Romania | Large | Manufacturing nano-enabled products: Fire resistant panels | Nanoclay |
| 3 | MOSTOSTAL | Poland | Large | Use of nano-enable products in building construction: Application of coatings with three methods: brush, roller or spray gun) | $TiO_2$ |
| 4 | ACCIONA | Spain | Large | Use of products containing nanomaterials in civil construction: Construction of a concrete slab | $TiO_2$ |
| 5 | ROSSAL | Romania | SME | End of life of nano-enabled products: Demolition of fire resistant panels | Nanoclay |

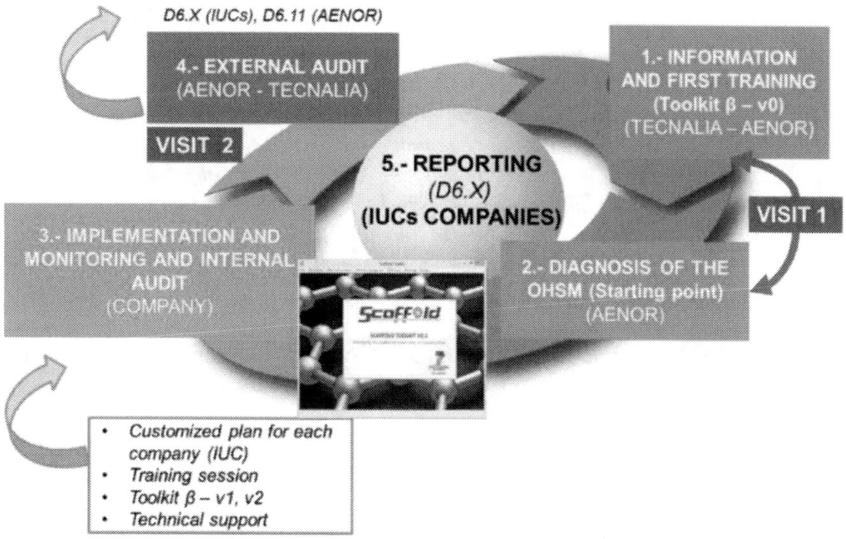


**Figure 3.** Stages for the deployment of the Industrial Use Cases (IUC) in construction companies, during the demonstration stage of project SCAFFOLD [24]

## 4. Conclusions

Nanotechnology is making its advance faster than the safety management related to it. Development of new methods, strategies and tools for risk management based on solid scientific knowledge may take a long time. But the European construction industry is already manufacturing and handling nanomaterials and nano-enabled products and workers are exposed to nano-risks.

Consequently, until having the total knowledge available, efforts should be made to provide the construction industry with intermediate management solutions, based on the state of the art, to make decisions with minimal uncertainties. It means the need to translate and encapsulate the results of current research in a battery of practical methods, strategies and tools for the management of nano-risks, directly usable by industry and companies that provide services to industry. All these tools should be updated with the evolution of the state of the art.

The project SCAFFOLD follows this approach and is ready to provide the construction industry with a package of practical solutions, to enable the management of nanorisks in large companies and SMEs.

## References


[1] Broekhuizen P, Broekhuizen F, Cornelissen R, Reijnders L 2011 Use of nanomaterials in the European construction industry and some occupational health aspects thereof. *J Nanopart Res* (2011):13 447-462.

[2] Brouwer D. 2012 Control Banding Approaches for Nanomaterials. *Ann. Occup. Hyg.* Vol. 56, No. 5, pp. 506-514.

[3] BSI 2007 *OHSAS 18001:2007, Occupational health and safety management systems — Requirements.*

[4] BSI 2008 *OHSAS 18002:2008, Occupational health and safety management systems — Guidelines for the implantation of OHSAS 18001.*

[5] Duuren-Stuurman B et al 2011 Stoffenmanager Nano Version 1.0: A Web-Based Tool for Risk Prioritization of Airborne Manufactured Nano Objects. *Ann. Occup. Hyg.* Vol. 56, pp. 525-541.

[6] EU-OSHA 2012 *Tools for the management of nanomaterials in the workplace and prevention measures.* E-Facts 72.

[7] FIEC 2014 *Key figures. Activity 2013. Construction in europe.*

[8] ILO 2011 OSH *Guidelines on occupational safety and health management systems*, 28 pp.

[9] ISO 2009 *ISO 31000:2009, Risk Management — Principles and guidelines.*

[10] ISO 2014 *ISO 12901-2 Nanotechnologies -- Occupational risk management applied to engineered nanomaterials -- Part 2: Use of the control banding approach.*

[11] ISO/CD 45001 2014 *Occupational health and safety management systems – Requirements.*

[12] Lee J, Mahendra S, Alvarez PJ Nanomaterials in the construction industry: a review of their applications and environmental health and safety considerations. *ACSNANO* vol 4, nº 7, 3580-3590.

[13] Pacheco-Torgal F, Labrincha, JA The futiure of construction materials research and the seventh UN Millennium Developmnet Gola; a few insights. *Construction and building materials* 40 /2013), 729-737.

[14] Paik S Y, Zal M, Swuste P 2008 Application of a pilot control banding tool for risk level assessment and control of nanoparticle exposures. *Annals of Occupational Hygiene*, 52(6), 419-428.

[15] Robson LS, Clarke JA, Cullen K, Bielecky A, Severin C, Bigelow P L, Mahood Q 2007. The effectiveness of occupational health and safety management system interventions: a systematic review. *Safety Science*, 45(3), 329-353.

[16] Sanchez-Toledo A, Fernandez B, Montes JM, Vazquez CJ 2009 Spanish survey reveals motivations, obstacles and benefits of OHSAS 18001 certification. *ISO Management Systems* July-August 2009, pp. 36-40.

[17] SCAFFOLD Project 2013 *RMM architecture and specifications*, project deliverable D5.1

[18] SCAFFOLD Project 2014 *Results of application of the Stoffenmanager Nanotool in the construction work area*, project deliverable D4.14.

[19] SCAFFOLD project 2014 *Guide for RMM implementation in construction,* project deliverable D5.2
[20] SCAFFOLD project 2013 *RMM toolkit: architecture and specifications*, project deliverable D5.6
[21] SCAFFOLD project 2014 *Software v0 and technical report: RMM tolkit and operating manuals*, project deliverable D5.7.
[22] SCAFFOLD project 2014 *RMM Toolkit architecture–specifications for SMEs*, project deliverable D5.8.
[23] SCAFFOLD project 2014 Guide for implementation and audit in SMEs, project deliverable D5.10
[24] SCAFFOLD project 2015 Specifications and implementation and audit plans for IUCs, project deliverable D5.5
[25] TNO (2011) *Stoffenmanager nano: Description of the conceptual control banding model. TNO Report V9216.*
[26] TUV 2008 *Certification Standard CENARIOS®* (Parts A to E), 60 pp.
[27] Van Broekhuizen FA, Van Broekhuizen JC 2019 *Nanotechnology in the European Construction Industry -State of the art 2009*, FIEC-EFBWW final report, 109 pp.

**Acknowledgment**
Research carried out by project SCAFFOLD was made possible thanks to funding from the European Commission, through the Seventh Framework Programme (GA 280535).

# A standardized non-instrumental tool for characterizing workstations concerned with exposure to engineered nanomaterials

**Guseva Canu I[1], Ducros C[2], Ducamp S[1,3], Delabre L[1], Audignon-Durand S[3], Durand C[3], Iwatsubo Y[1], Jezewski-Serra D[1], Le Bihan O[4], Malard S[5], Radauceanu A[5], Reynier M[5], Ricaud M[5], and Witschger O[5]**

(1) French Institute for Health Surveillance (InVS), Occupational health department (DST), 12, rue du Val d'Osne 94415 Saint-Maurice, France (2) CEA – Grenoble, 17, rue des martyrs 38000 Grenoble, France (3) University of Bordeaux, Inserm Center U897, Occupational Health and Environment Team, (Essat-DST/LSTE), 146, rue Léo Saignat, 33076 Bordeaux, France (4) INERIS, rue Jacques Taffanel, 60550 Verneuil-en-Halatte, France (5) INRS, 65 boulevard Richard Lenoir, 75011 Paris, France

i.guseva-canu@invs.sante.fr ; irinacanu@hotmail.com

**Abstract**. The French national epidemiological surveillance program EpiNano aims at surveying mid- and long-term health effects possibly related with occupational exposure to either carbon nanotubes or titanium dioxide nanoparticles ($TiO_2$). EpiNano is limited to workers potentially exposed to these nanomaterials including their aggregates and agglomerates. In order to identify those workers during the in-field industrial hygiene visits, a standardized non-instrumental method is necessary especially for epidemiologists and occupational physicians unfamiliar with nanoparticle and nanomaterial exposure metrology. A working group, Quintet ExpoNano, including national experts in nanomaterial metrology and occupational hygiene reviewed available methods, resources and their practice in order to develop a standardized tool for conducting company industrial hygiene visits and collecting necessary information. This tool, entitled "Onsite technical logbook", includes 3 parts: company, workplace, and workstation allowing a detailed description of each task, process and exposure surrounding conditions. This logbook is intended to be completed during the company industrial hygiene visit. Each visit is conducted jointly by an industrial hygienist and an epidemiologist of the program and lasts one or two days depending on the company size. When all collected information is computerized using friendly-using software, it is possible to classify workstations with respect to their potential direct and/or indirect exposure. Workers appointed to workstations classified as concerned with exposure are considered as eligible for EpiNano program and invited to participate. Since January 2014, the Onsite technical logbook has been used in ten company visits. The companies visited were mostly involved in research and development. A total of 53 workstations with potential exposure to nanomaterials were pre-selected and observed: 5 with $TiO_2$, 16 with single-walled carbon nanotubes, 27 multi-walled carbon nanotubes. Among the tasks observed there were: nanomaterial characterisation analysis (8), weighing (7), synthesis (6), functionalization (5), and transfer (5). The manipulated quantities were usually very small. After analysis of the data gathered in logbooks, 30 workstations have been classified as concerned with exposure to carbon nanotubes or $TiO_2$. Additional tool validity as well as inter-and intra-evaluator reproducibility studies are ongoing. The first results are promising.

## 1. Introduction

### *1.1. EpiNano, the French surveillance program for workers potentially exposed to engineered nanomaterials*

The development of EpiNano surveillance program is conducted by the French Institute for Public Health Surveillance (*Institut de Veille Sanitaire, InVS*) at a joint request of the French Ministries of

Health and of Labour [1,2]. EpiNano aims at surveying mid- and long-term health effects possibly related with occupational exposure to either carbon nanotubes or titanium dioxide ($TiO_2$) nanoparticles, aggregates and agglomerates in workers employed in the nanotechnology-related industrial or research and development facilities in France. EpiNano consists of a registry of workers likely to be exposed to engineered nanomaterials and a prospective epidemiological cohort study [3]. The protocol of the EpiNano program received an approval from the French authority of privacy and individual rights protection (*Commission nationale de l'informatique et des libertés, CNIL*) for next 20 years of follow-up.

Carbon nanotubes and $TiO_2$ nanoparticles, aggregates and agglomerates were chosen as priority engineered nanomaterials based on the considerations as follow [2,3]:

- available toxicological data,
- quantities manufactured in France and projected for production development,
- the choice of France in the framework of the sponsorship program for the testing of engineered nanomaterials sponsored by the Organisation for Economic Co-operation and Development (OECD),
- social perception factors.

### *1.2. Identification of workers eligible for EpiNano program*
Workers potentially exposed to carbon nanotubes or $TiO_2$ are identified using a 3-level approach [3]:

1. identification and selection of companies dealing with corresponding engineered nanomaterials (based on compulsory declaration and questionnaires [3,4]),
2. company in field visit and identification of the workstations concerned with exposure to engineered nanomaterials,
3. identification of workers involved in jobs and tasks performed in workstations identified as concerned with exposure to engineered nanomaterials and invited into the program.

## 2. Tool for characterizing workstations concerned with exposure to engineered nanomaterials

### *2.1. Aim and methodological development*
To cope with high variability of companies in terms of size, activity, industrial process, work size as well as with operation and exposure conditions, a standardization of a second step of the worker identification method was extremely important. Moreover, additional criteria the EpiNano method should meet with were:

1. non-instrumental assessment based on the state-of-the-art methodology [5],
2. easiness of usage by EpiNano team members (i.e. epidemiologists and industrial hygienists non-specialized in engineered nanomaterial exposure),
3. inexpensiveness and friendly-using format,
4. extensiveness to all necessary information for workers' individual exposure assessment in upcoming epidemiological studies, and
5. potential usefulness for company occupational safety and health staff (managers and/or occupational physicians).

A working group *Quintet ExpoNano* was created including national experts from the leading French institutes (French institute for public health surveillance (InVS), the French institute for occupational health and safety (INRS), the Atomic energy commission (CEA), the French institute for industrial safety and environmental protection (INERIS), and the University of Bordeaux Segalen) specialized in nanoparticle metrology, industrial hygiene, occupational medicine, and epidemiology. The working group reviewed available methods and tools for in-filed observations and inspections, measurement technics and exposure measurement data, and compared their respective practices of in-field studies. The recommendations for characterizing potential emissions and exposure to aerosols released from nanomaterials in workplace operations published by INRS, INERIS and CEA [5] were respected. The

method integrated the first three stages of the general five-stage procedure, the fourth and fifth stages being dedicated to a measurement campaign [5].
At term of 6-months collaboration, a first version of the tool was proposed and tested in field. After a series of additional format improvements and rewordings, an agreement of the working group has been reach on a final version of the tool. This version was used on 10 workplaces during four months, before its final validation [6].

*2.2. Tool description*
The method consists of identifying within each company the workrooms and activities that work with engineered nanomaterials in order to identify the workstations possibly causing exposure to them and to assess this potential exposure semi-quantitatively. It is based on a technical inspection of the plant (in-field visit), interviews with workroom supervisors, and observation of the activity at each workstation. This inspection is based on the ***Onsite technical logbook*** [4,6]. This tool enables evaluators to standardize the in-field observation and data collection. Two versions of the Onsite technical logbook, in French and in English, are available [4,6].

The tool is structured in 3 parts:

1. Company: activity and process description;
2. Workrooms: type and dimensions, air flow, efficacy of the ventilation system, local maintenance, staff and workstations, potential sources of non-manufactured ultrafine particles emissions (background aerosols);
3. Workstations: instruments, techniques, equipment, process enclosure, details about incoming and outgoing products; presence of collective protection, personal protective equipment (PPE), tasks and operation performed in the workstation, quantity of product handled per operation, frequency and duration of operation.

A short questionnaire [3,4] sent to a company occupational safety and health manager prior to the onsite company visit allows to prepare the in-field visit, and to gather all potentially useful documents (the plant's blueprints, certificates of control and maintenance of the collective protective equipment, annual declaration reports and supplementary materials such as nanomaterial characterization data and results of the exposure measurement campaigns) to be consulted onsite during the in-field visit.

An in-field visit is generally organized over one or two days. It begins with an exchange of information with representatives of the company in a conference room, about the EpiNano project (objectives, procedures) and about the company (its activities and work processes). The discussion makes it possible to fill in the first part of the Onsite technical logbook on company's activities and processes implemented. The discussion is followed by a study of the plant's blueprints to locate the circulation of materials in the premises and thus identify the workrooms where nanomaterials are present.

The technical inspection, in the strict sense of the term, makes it possible to visit workrooms and to observe the workstations and real activity. This step enables to describe the use of nanomaterials in detail. During the inspection, the EpiNano team members (2 or 3 people, including at least one industrial hygienist and one epidemiologist) must be accompanied by the plant's director of hygiene and safety, the laboratory or department director, and the occupational physician. During the inspection, the items of the second and third parts of the Onsite technical logbook are completed, in order, so that the workstations possibly causing exposure can be identified and the potential exposure further assessed [6].

After the inspection, verification and data entry of the information in the logbook, a report of the inspection is sent to the company. This report includes the conclusions of the workstation evaluations and a list of the workstations that potentially cause exposure to nanomaterials, aggregates and agglomerates. A copy of the computerized data from the logbook is attached to the report.

### *2.3. Implementation of the method and first results*

The method is designed for non-instrumental exposure assessment by non-specialized users. It was tested and further used for tracking workstations concerned with exposure to engineered nanomaterials and recruiting potentially exposed workers. Ten first companies which accepted to participate in EpiNano program were visited from January through May 2014 [6]. The visited workplaces had in average six workrooms (Min=1, Max=13) and 2 workstations per workroom (Min=1, Max=4). The mean number of workstations where carbon nanotubes or $TiO_2$ nanoparticles, their aggregates or agglomerates could be handled is around eight depending on company activity, with up to 27 workstations in a largest industrial workplace. In total, fifty three workstations were observed and resulted in completed Onsite technical logbooks. Among these workstations, there were twenty-five (47%) workstations in private companies and 28 (53%) in public workplaces, mostly academic research and development laboratories. Carbon nanotubes were most frequently handled material encountered in 43 of the observed workstations (single-wall carbon nanotubes in 16 (30%) and multi-wall carbon nanotubes in 27 (51%) workstations respectively), while $TiO_2$ was handled in 5 (9.4%) workstations. In 18 workstations (34%) multiple types of engineered nanomaterial were handled.

*2.3.1. In epidemiology and qualitative exposure assessment* In EpiNano system the identification of workstations with exposure concern is preformed regardless the use of personal protective equipment [2,6]. Workstations where a worker could experience a direct contact with engineered nanomaterial (including aggregates and agglomerates) that gives potential for inhalation or cutaneous contamination are classified as workstations concerned with exposure. The information about personal protective equipment, amount of engineered nanomaterial handled during an operation as well as frequency and duration of handling is gathered from workers' individual EpiNano inclusion questionnaire. This information will be accounted for in workers' individual exposure score for workers involved in workstations identified as concerned with exposure to engineered nanomaterials [2,3,6].

Overall, in ten workplaces visited till May 2014, 30 workstations (57%) were classified as concerned with exposure to either carbon nanotubes or $TiO_2$. Figure 1 presents the types of operations and tasks performed in the observed workstations and in workstations classified as concerned with exposure to engineered nanomaterial. Among the parameters assessed during the in-field visits, dustiness and humidity of the engineered nanomaterial seem to be the most important determinants of the possible exposure in a workstation [6].

*2.3.2. In industrial hygiene and risk management.* The data collected through the Onsite technical logbook are computerized and sent to the company. This data might be directly used by companies for risk management proposes, for instance by implementing control banding approach to assess and control exposure to engineered nanomaterials in different workstations. Several tools of control banding have been proposed specifically for engineered nanomaterials [7-9]. The Onsite technical logbook contains all essential parameters for implementing any of these tools for assessing exposure bands in workplaces. While there is no consensus on an appropriate exposure metric to be measured for assessing individual exposure to engineered nanomaterials in workers, International organization for standardization (ISO) recommends using control banding approach in workplaces dealing with engineered nanomaterials [9]. Consequently, our method may be straightforward and helpful for both exposure characterization and risk management which might be further improved with more accurate and quantitative exposure measurement data.

### *2.4. Method validation*

A validation study is required in order to address the reliability of the proposed method and reproducibility of the exposure assessment results based on it in order to prevent bias in risk estimator in the epidemiological study [10]. The inter-method [11] and intra-method [12] comparisons of the exposure assessment were performed; the methods and results will be presented in the upcoming publication (manuscript in preparation). According to the results of these comparisons, the proposed method presents a substantial agreement with a more precise expert exposure assessment (Cohen's Kappa=0.69) and a good agreement based on intra-method repeatability test.

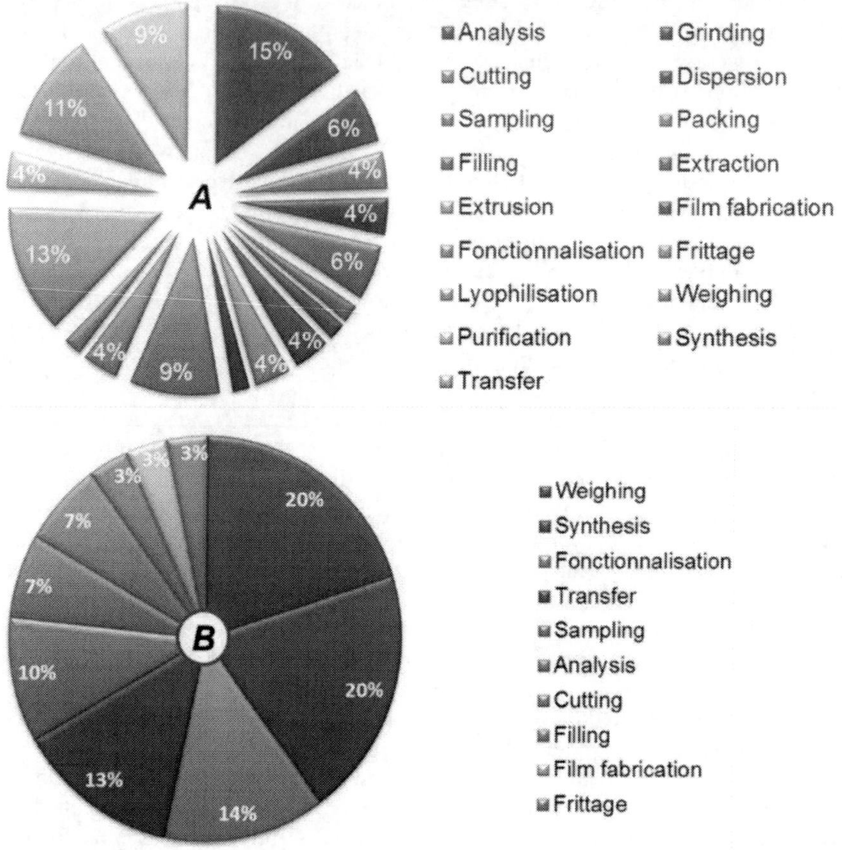


**Figure 1.** Distribution of workstations according to the type of operation performed:
***A***-all workstations observed between January and May 2014 (n=53);
***B***-workstations classified as concerned with exposure to carbon nanotubes or $TiO_2$ nanoparticles, aggregates and agglomerates (n=30).

In conclusion, the method and the tool (the Onsite technical logbook) presented in this paper were developed by the French institute for public health surveillance (InVS), the French institute for occupational health and safety (INRS), the atomic energy commission (CEA), the French institute for industrial safety and environmental protection (INERIS), and the University of Bordeaux Segalen, as part of the partnership entitled ExpoNano Quintet. This tool makes it possible to collect all of the information necessary to identify and characterize workstations that might cause occupational exposure to carbon nanotubes or $TiO_2$ nanoparticles, aggregates, and agglomerates. It is part of a semi-quantitative method to characterize the potential for exposure to intentionally produced nanomaterials in different workstations [2,3]. This practical method makes it possible to follow the recommended procedure for assessing potential emissions and characterizing occupational exposure during operations involving nanomaterials [5]. The results of validation studies are promising and will be provided in a forthcoming publication. This method, which is simple and does not require an instrument (no sampling, no aerosol measurements), is designed to be usable as part of the EpiNano program of epidemiologic surveillance of workers potentially exposed to nanomaterials in France [3,6]. Moreover, it can be useful for risk management purposes in companies, for instance in frame of implementation of the control banding approach to assign exposure bands to the different workstations concerned with exposure to carbon nanotubes or $TiO_2$ nanoparticles, aggregates and agglomerates.

**References**

[1] Boutou-Kempf O. Feasibility elements for epidemiological surveillance of workers exposed to nanomaterials intentionally produced. Saint-Maurice: French Institute for Public Health Surveillance; 2011. 69 p.

[2] Boutou-Kempf O, Marchand JL, Radauceanu A, Witschger O, Imbernon E, André JC, et al. Development of a French epidemiological surveillance system of workers producing or handling engineered nanomaterials in the workplace. J Occup Environ Med 2011;53:S103-S107.

[3] Guseva Canu I, Boutou-Kempf O, Delabre L, Ducamp S, Iwatsubo Y, Marchand JL, et al. French registry of workers handling engineered nanomaterials as an instrument of integrated system for surveillance and research. Journal of Physics: Conference Series 2013;429(1).

[4] InVS. EpiNano project web-site. http://www.invs.sante.fr/Dossiers-thematiques/Travail-et-sante/Epinano-Dispositif-de-surveillance-epidemiologique-des-travailleurs-potentiellement-exposes-aux-nanomateriaux

[5] Witschger O, Le Bihan O, Reynier M, Durand C, Marchetto A, Zimmermann E, et al. Recommendations for characterizing potential emissions and exposure to aerosols released from nanomaterials in workplace operations. Hygiène et sécurité du travail 2012;(2255):1-15.

[6] Guseva Canu I, Ducamp S, Delabre L, Audignon-Durand S, Ducros C, Durand C, et al. Method for indentification of workstations with exposure to nano-objects and their aggregates and agglomerates in companies dealing with engineered nanomatrials [in French]. Archives des Maladies Professionnelles et de l'Environnement; 2015 (submitted)

[7] Brouwer DH. Control banding approaches for nanomaterials. Annals of Occupational Hygiene 2012;56(5):506-14.

[8] Fleury D, Fayet G, Vignes A, Henry F, Frejafon E. Nanomaterials risk assessment in the process industries: Evaluation and application of current control banding methods. Chemical Engineering Transactions 2013;31:949-54.

[9] International organization for standardization. International technical specification. Nanotechnologies -- Occupational risk management applied to engineered nanomaterials. Part 2: Use of the control banding approach. ISO/TS 12901-2:2014.

[10] Rajaraman P, Samet J. Quality control and good epidemiological practice. In: Ahrens W, Pigeot I. eds Handbook of epidemiology. Berlin: Springer, 2005:503–56.

[11] Cohen J. Weighted kappa: nominal scale agreement with provision for scaled disagreement or partial credit. Psychol Bull 1968;70:213–20.

[12] Landis JR, Koch GG. The measurement of observer agreement for categorical data. Biometrics 1977;33:159–74.

Journal of Physics: Conference Series **617** (2015) 012037 doi:10.1088/1742-6596/617/1/012037

# QUALITATIVE RISK ASSESSMENT DURING POLYMER MORTAR TEST SPECIMENS PREPARATION – METHODS COMPARISON

**F Silva[1,2], S P B Sousa[3], P Arezes[2], P Swuste[4], M C S Ribeiro[3,5], J S Baptista[5]**

(1) Technological Centre for Ceramic and Glass, Coimbra, Portugal, (2) School of Engineering, University of Minho, Guimarães, Portugal, (3) Institute of Mechanical Engineering and Industrial Management, Campus da FEUP, Porto, Portugal, (4) Delft University of Technology, Delft, The Nederlands, (5) Faculty of Engineering, University of Porto, Porto, Portugal

Email: fsilva@ctcv.pt

**Abstract**. Polymer binder modification with inorganic nanomaterials (NM) could be a potential and efficient solution to control matrix flammability of polymer concrete (PC) materials without sacrificing other important properties. Occupational exposures can occur all along the life cycle of a NM and "nanoproducts" from research through scale-up, product development, manufacturing, and end of life. The main objective of the present study is to analyse and compare different qualitative risk assessment methods during the production of polymer mortars (PM) with NM. The laboratory scale production process was divided in 3 main phases (pre-production, production and post-production), which allow testing the assessment methods in different situations. The risk assessment involved in the manufacturing process of PM was made by using the qualitative analyses based on: French Agency for Food, Environmental and Occupational Health & Safety method (ANSES); Control Banding Nanotool (CB Nanotool); Ecole Polytechnique Fédérale de Lausanne method (EPFL); Guidance working safely with nanomaterials and nanoproducts (GWSNN); Istituto Superiore per la Prevenzione e la Sicurezza del Lavoro, Italy method (ISPESL); Precautionary Matrix for Synthetic Nanomaterials (PMSN); and Stoffenmanager Nano. It was verified that the different methods applied also produce different final results. In phases 1 and 3 the risk assessment tends to be classified as medium-high risk, while for phase 2 the more common result is medium level. It is necessary to improve the use of qualitative methods by defining narrow criteria for the methods selection for each assessed situation, bearing in mind that the uncertainties are also a relevant factor when dealing with the risk related to nanotechnologies field.

## 1. Introduction

The need for a building material with high strength and durability afforded the development of a relatively "new" composite material: polymer concrete (PC) [1]. Although the PC advantages, its usage is only competitive in applications in which durability, strength or other particular requirements render conventional materials unusable, where the initial higher price of PC is mitigated by their superior short and long-term performances. PC present, however, some drawbacks brought by resin matrix like high sensitivity to high temperatures and creep phenomena, and mainly, the deficient behaviour under fire [2-4]. Polymer binder modification with inorganic nanomaterials (NM), leading to nanocomposites (NC), could be a potential and efficient solution to control matrix flammability without sacrificing other important properties. Although the NM use as flame retardant (FR) is promising, this technique is still giving the first steps [5, 6]. The majority of research work on the subject of polymer flame retardancy by means of NM has been carried out on polymer-clay NC. Other emerging nanoparticles that have also shown likely effects on polymer thermal degradation are metal oxide nanoparticles such as aluminium oxides ($Al_2O_3$). The few studies focusing on this issue show promising results attesting that alumina nanoparticles incorporation can improve thermal stability and other relevant properties of final composite [5-9]

There is a lack of knowledge on the effects of engineered nano-$Al_2O_3$ in the human body, as existing studies are divergent. In a comparative inhalation study, it was clearly showed that $Al_2O_3$ nanoparticles (20 nm) induced an inflammatory reaction in rat lungs; however, others similar studies showed that there were no or only slight acute effects on animals due to nanoalumina inhalation, even after high doses [10]. For a final assessment additional research will be needed [10].

The global market for NC will be worth 3.000 billion dollars in 2015, since they have an increasingly decisive role in various industries [11]. However, this new technologies rise is related with new human and environmental risk factors [12]. Occupational exposures can occur all along the life cycle of NM and "nanoproducts" from research through scale-up, product development, manufacturing, and end of life. Quantitative toxicity studies on engineered NM are still relatively sparse and until now there are not known direct implications to the humans health [13].

Risk assessment is an important part of the process to achieve safer workplaces in the nanotechnology field. Because of the limited amount of data on NM, many of the assumptions and estimations are based on the traditional chemical risk assessments. Recently, several exposure assessment approaches have been developed, since the traditional risk and exposure assessment methods seem to be not fully adequate to assess risk related to nanotechnologies, mainly due to existing doubts on the most adequate dose metrics and to the lack of data on the chronic health effects [14]. In this scenario, qualitative risk assessment helps on supporting decisions in the risk management process [15]. The validation of the different proposed qualitative risk assessment methods is an on-going work in the scientific community, and it is expected that modifications could occur [16].

Meanwhile, it is necessary to raise the awareness of the nanotechnology potential risks and one of the ways is through qualitative risk assessment. Under this framework, the present study is an effort to contribute to an enhanced performance and use of qualitative risk assessment methods during the production of polymer mortars (PM) materials with NM. The research question underlying this study is: Do different qualitative risk assessment methods identify comparable risk levels for the same tasks?

This research work was developed at INEGI laboratories, and the tasks were performed in typical work conditions.

## 2. Materials and Methods

### *2.1. Tasks under evaluation*

In the production process of PM, a commercially available unsaturated polyester resin (Aropol® FS3992, Quimidroga Portugal-Produtos Químicos Unipessoal Lda) was applied as polymer binder. Alumina NM (NanoDur®, $Al_2O_3$, 99.5% purity), for polymer binder modification, was purchased from Cymit Quimica S.L. (Spain), with 45 nm average size and 36.0 $m^2.g^{-1}$ specific surface. Siliceous foundry sand (SP55, Fundipor), with an average diameter of 245 μm, was applied as mineral aggregate. Manual stirring and ultrasound sonication techniques were used to disperse the NM into the resin system (2.5% by weight of resin), and afterwards, the mixture was added to the sand aggregates (with a binder to sand ratio of 1:4) and thoroughly mixed in a mechanical mix device. The final mixture was casted into standard prismatic moulds.

The laboratory scale production process was divided in 3 main phases: 1 - Pre-production: handling, weighing, adding nanoalumina to the resin and cleaning; 2 - Production: stirring, pouring the mixture into the mould and cleaning; and 3 - Post-production: demoulding, cutting and cleaning. During the production process, the nanoalumina is present in powder (phase 1), suspension (phase 2), or inserted in a cured polymer matrix (phase 3), which allowed testing the assessment methods in three different situations.

The production process only involved a single operator who used collective protection measures existing in the lab (mechanical ventilation/general exhaust), and personal protective equipment (two pairs of nitrile gloves, three latex gloves, a mask with ABEK1 P3 filters, protection goggles, type 5 category III disposable coverall and disposable polypropylene shoe covers).

### 2.2. *Risk assessment methods*

The risk assessment involved in the manufacturing process of PM was made by using the qualitative analyses based on: French Agency for Food, Environmental and Occupational Health & Safety method (ANSES), from France; Control Banding Nanotool (CB Nanotool), from United States of America; Ecole Polytechnique Fédérale de Lausanne method (EPFL), from Swiss; Guidance working safely with nanomaterials and nanoproducts (GWSNN), from. The Nederlands; Istituto Superiore per la Prevenzione e la Sicurezza del Lavoro (ISPESL), from Italy; Precautionary Matrix for Synthetic Nanomaterials (PMSN), from Swiss; and Stoffenmanager Nano, from The Nederlands.

#### 2.2.1. *French Agency for Food, Environmental and Occupational Health & Safety method (ANSES)*

ANSES is a risk assessment method by control bands. The risk values are obtained by overlapping the hazard bands and emission potential bands. The hazard bands are defined according to the severity level of the hazard, resulting from the analysis of the available information of similar chemicals. The hazard levels may assume five classifications from HB1 - very low (no significant risk to health) to HB5 - very high (severe hazard requiring a full hazard assessment by an expert). The exposure bands are defined according to the nanomaterial emission potential, whether raw or included in a matrix. The exposure bands can assume four levels: EP1 – solid, EP2 – liquid, EP3 – powder, and EP4 – aerosol. From the resultant matrix it can be defined the control level that corresponds to technical solutions for collective prevention to be implemented at the workplace (CL1 - natural or mechanical general ventilation to CL5 - full containment and review by a specialist required) [17].

#### 2.2.2. *Control Banding Nanotool (CB Nanotool)*

CB Nanotool is a four by four factors matrix that relates severity parameters on one-axis and probability parameters on the other. The severity parameters consider that the physicochemical and general properties of NM are often unknown. Adding information about the parent material solves partially this problem. The overall severity score is determined based on the sum of all the points from the severity factors. The probability axis fits with traditional information. The probability scores are based on factors determining the extent to which employees may be potentially exposed to NM. The obtained control bands by risk level can be classified in RL1 – general ventilation to RL4 – seek specialist advice [18].

#### 2.2.3. *Ecole Polytechnique Fédérale de Lausanne method (EPFL)*

EPFL method consists in a decision tree for "nano-laboratories" with three risk classes, which correspond to similar approaches applied to other hazards types (biological, chemical or radiation). This decision tree analyses the established collective protection measures, NM form/state handling typology, NM quantity use, possibility to release dust or aerosol and NM agglomeration ability. The risk classification can be Nano1 (low) to Nano3 (high). With the risk classification it can be defined several safety measures [19].

#### 2.2.4. *Guidance working safely with nanomaterials and nanoproducts (Guidance)*

GWSNN risk assessment method analyses different scenarios through a three by three decision matrix, informing the policy options and procedures to guarantee safe working conditions with NM. The hazard category can be classified as: 1 - soluble NP, 2 - synthetic, persistent NM, and 3 - fibrous, nonsoluble NM. The exposure classification is made based on the NM potential exposure in the different activities related with the polymeric NC production: I - no emission of free NP due to working in full containment, II - emission of NP embedded in a matrix is possible, and III - emission of free NP is possible. The recommended control measures range from level A (applying sufficient ventilation, if needed local exhaust ventilation and/or containment of the emission source and use appropriate personal protective equipment) to level C –(the hierarchic Occupational Hygienic Strategy will be strictly applied and all protective measures that are both technically and organizationally feasible will be implemented) [20].

*2.2.5. Istituto Superiore per la Prevenzione e la Sicurezza del Lavoro, Italy method (ISPESL)*
The ISPESL risk assessment method is based on ten different factors (A to J). The aforesaid factors are denominated "factors level risk" and each one of them may assume three increasing values, 1 (low) to 3 (high), referred to as "risk levels". Since the use of NM presents uncertainty about danger level, the risk assessment takes into consideration these aspects through the index denominated "corrective factor" (within the range 0.5 and 2.0) [15]. The risk is estimated through the "factor level risk" (flr) sum (from A to J) and then multiplied by the "corrective factor" (cf). The evaluation result consists in three risk levels (risk level "low" 5-15, "medium" 16-35, and "high" 36- 60) [21].

*2.2.6. Precautionary Matrix for Synthetic Nanomaterials (Precautionary Matrix)*
The PMSN estimates the precautionary need that represents the relation between the parameters: "Nano-relevance according to the precautionary matrix" (N), "Potential effect" (W), "Potential exposure" (E), and "Specific framework conditions" (S). The precautionary matrices are logically completed and evaluated in two iterative steps: 1$^{st}$ - a rapid evaluation to demonstrate knowledge gaps and uncertainties, which leads to a preliminary precautionary matrix; 2$^{nd}$ - exact clarifications on the fundamentals of the results from 1$^{st}$ step and from the specific answers to knowledge gaps that afford a finished and definitive evaluation of precautionary matrix. The potential risk can be classified into class A (the nanospecific need for action can be rated as low, even without further clarification) or class B (nanospecific action is needed; existing measures should be reviewed, further clarification undertaken and, if necessary, measures to reduce the risk associated with development, manufacturing, use and disposal implemented in the interest of precaution) [22].

*2.2.7. Stoffenmanager Nano*
The Stoffenmanager Nano is a risk-banding tool to prioritize health risks occurring as a result of exposure to NP for a broad range of occupational scenarios and to assist implementation of control measures to reduce exposure levels. In order to prioritize the health risks, it is made the combination between the available hazard information of a substance with a qualitative estimate of potential for inhalation exposure. Input parameters for the hazard assessment of NP are selected based on the available information (e.g.: safety data sheets, product information sheets). The method was converted into an online tool that offers a practical approach for risk prioritization in exposure situations where quantitative risk assessment is currently not possible. The obtained priority bands by risk assessment can be classified in 1 (highest priority) to 3 (lowest priority) [23].

## 3. Results and discussion

### *3.1. Results*

*3.1.1. Inputs for the assessment – risk determinants*
From the previous sections it is possible to verify that the different methods consider different characteristics from the nano-objects (hazard band) and take in consideration different aspects from the tasks performed (exposure likelihood band). Are also relevant the scale considered to each

parameter, and the weight that each parameter assumes in the final result of the assessment. These differences will influence the obtained results with each method.

In short it is possible to have a wide view of the considered hazard factors of each method in Table 1.The toxicity of the nanomaterial, or its bulk form, is directly considered by five of the methods. Solubility of the nanoparticles and fibrous form are also considered by the majority of the methods, although the toxicity itself could result from these factors, and also from reactivity and size, considered as relevant factors in three of the seven methods. The ISPESL method includes the fire and explosion hazard factor which is relevant for safety risk assessment, biasing the result when only the health effects are considered.

**Table 1** – Hazard band factors

| Hazard factor | ANSES | CB Nanotool | EPFL | Guidance | ISPESL | Precautionary Matrix | Stoffen-manager |
|---|---|---|---|---|---|---|---|
| Toxicity (nano and/or bulk material) | • | • | | • | • | | • |
| Solubility | • | • | | • | | • | • |
| Fibre form (particle shape) | | • | • | • | | | • |
| Reactivity | • | • | | | | • | |
| Size | | • | | | • | | • |
| Fire and explosion | | | | | • | | |

In Table 2 the factors that could determine the worker's exposure are presented.

**Table 2** – Exposure (or likelihood) band

| Exposure factor | ANSES | CB Nanotool | EPFL | Guidance | ISPESL | Precautionary Matrix | Stoffen-manager |
|---|---|---|---|---|---|---|---|
| Quantity | | • | • | | | • | • |
| Duration/ frequency (time) factor | | • | | | • | • | • |
| State of material (e.g. solid, liquid) | • | | • | | | | • |
| Release of nano-objects (e.g. dustiness) | • | • | | • | | • | • |
| Aggregation/ agglomeration | | | • | | • | | |
| Embedded in a matrix | | | | • | | • | • |
| Number of workers | | • | | | • | | |
| Risk control / organization | | | | | • | | • |
| Containment | | | • | • | | | |
| Type of process/task | • | | • | | | | • |

The ISPESL and Precautionary Matrix methods include also general factors that influence the final result of the assessment, taking in account the uncertainty on the information gathered. These factors are related to the knowledge on the materials and, in the Precautionary Matrix also on the size and state of aggregation/agglomeration of the nanomaterials present. CB Nanotool leads with the uncertainty, introducing a score for "unknown" in several parameters corresponding to 75% of the scale.

### *3.1.2. Inputs for the assessment – information gather*

The information on nanoalumina and the tasks performed is of major importance for the risk assessment result. The aluminium oxide is a relatively inert material. For the risk assessment it was considered that nanoalumina is a non-soluble, non-fibrous material and its carcinogenicity, mutagenicity and toxicity for reproduction are unknown. In Phase 1, nanoalumina is a solid (powder), in Phase 2 it is in liquid suspension and finally, in Phase 3 it is embedded in a matrix. Based on the assumption that the materials in nano-form are more reactive than its bulk material, nano-$Al_2O_3$ is considered more reactive than the parent material. Some doubts arise due to information lacking about dustiness, styrene evaporation rate influence on NP release and agglomeration/aggregation of the airborne nanoparticles.

The total amount of nanoalumina used in each phase is 30 g and the tasks durations are around 50 min for Phase 1, 1h50 min for phase 2 and around 30 min for Phase 3. For the frequency factor it was considered that the tasks were performed on a monthly basis.

Figure 1 shows the obtained results of the application of the 7 different risk assessment methods for all the considered stages.

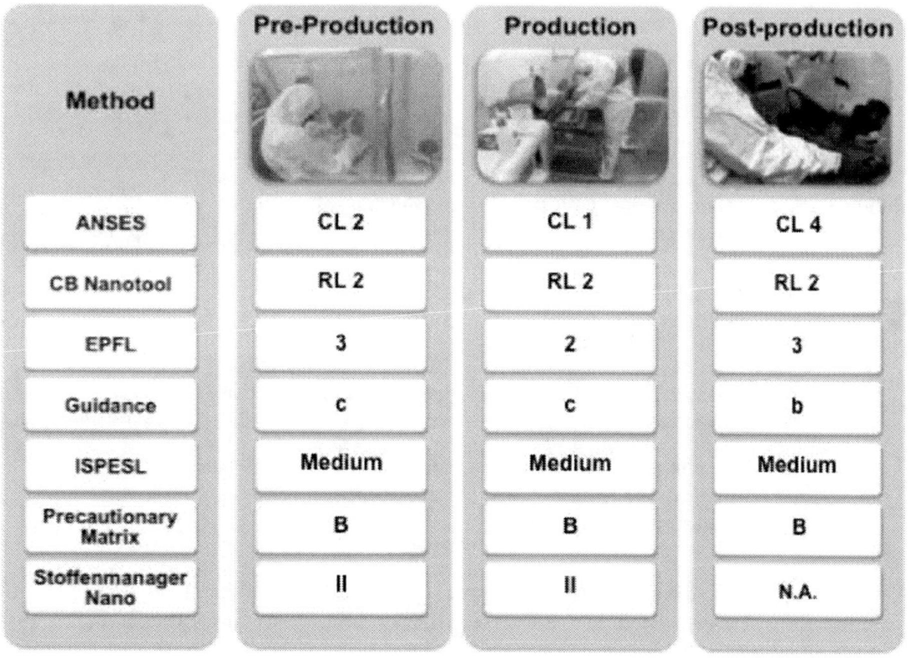

| Method | Pre-Production | Production | Post-production |
|---|---|---|---|
| ANSES | CL 2 | CL 1 | CL 4 |
| CB Nanotool | RL 2 | RL 2 | RL 2 |
| EPFL | 3 | 2 | 3 |
| Guidance | c | c | b |
| ISPESL | Medium | Medium | Medium |
| Precautionary Matrix | B | B | B |
| Stoffenmanager Nano | II | II | N.A. |

**Figure 1** - Comparing the results of the different risk assessment methods (N.A. – Non Applicable)

### *3.2. Discussion*

Analysing the results, it was found that the most critical operation in the PM processing is the pre-production phase, because it deals with NMs in powder state and may lead to typical scenarios of

Journal of Physics: Conference Series **617** (2015) 012037 doi:10.1088/1742-6596/617/1/012037

exposure. However, it was verified that the used qualitative risk assessment methods had different final results (Figure 1).

In phases 1 and 3 the risk assessment tends to be classified as medium-high risk, while for phase 2 the more common result is medium level. The consideration of different assumptions as risk determinants could explain some of the found differences. It is also relevant to consider the sensibility of the methods regarding the different exposure scenarios, as some of them give the same risk level for the three different phases.

In respect to phase 3, divergent results are obtained: while ANSES and EPFL methods point to an increasing risk level, when comparing with the other phases, with the Guidance and Stoffenmanager Nano lower risk levels are obtained (CB Nanotool, ISPESL, and Precautionary Matrix present the same risk level, despite some variations on the exposure scores).This divergence arises from the consideration that NP embedded in a matrix are released together with other material in large particles, thus not resulting in exposure to free NP, assumed by the Guidance and Stoffenmanager Nano methods.

Bearing in mind that these assessment methods are risk management tools it is also relevant to compare their outputs considering the risk control measures. In Table 3 the control measures recommended or necessary to low the risk to an acceptable level, according to each different method.

**Table 3 –** Risk control measures

| Method | Phase 1 | Phase 2 | Phase 3 |
|---|---|---|---|
| **ANSES** | Local ventilation | Natural or mechanical general ventilation | Full containment |
| **CB Nanotool** | Fume hood or local exhaust ventilation | Fume hood or local exhaust ventilation | Fume hood or local exhaust ventilation |
| **EPFL** | High level technical, organizational and individual control measures | Medium level technical, organizational and individual control measures | High level technical, organizational and individual control measures |
| **Guidance** | Mandatory technical and organizational protective measures | Mandatory technical and organizational protective measures | Technical and organizational protective measures considering economical feasibility |
| **ISPESL** | Existing controls are sufficient | Existing controls are sufficient | Existing controls are sufficient |
| **Precautionary Matrix** | Review existing control measures and if necessary improved | Review existing control measures and if necessary improved | Review existing control measures and if necessary improved |
| **Stoffenmanager Nano** | Medium priority, enclosure necessary to reduce exposure level | Medium priority, enclosure necessary to reduce exposure level | N. A. |

As observed on Table 3, the risk control measures recommended (when there are specific recommendations) by each method are different. Depending on the method considered, the user could be led to implement technical, organizational or individual control measures with different degrees of efficacy and/or complexity.

The use of various parameters and/or differences in their interpretation can lead to differences in risk level results. These methods should be reviewed in order to give more convergent results avoiding unequal final outcomes, and their selective and/or complementary application could also help to improve risk management in the workplaces.

## 4. Conclusions

The occupational risks are a key issue to be considered especially in the early stages of any new material production. By studying proactively emerging risks, one can prevent future problems. The risks are inherent to any technology and nanotechnology is no exception.

Within this scope, in this study the different qualitative risk assessment methods were used to evaluate the risk during the production stages of PM with nanoalumina. The experimental study was conducted in a laboratory environment but provides an overview of the measures that could be applied during the PM production at an industrial scale.

The results obtained with the qualitative methods were divergent; however, generally the methods consider that the pre-production stage is the one with the higher risk.

The results of this study highlight the need to improve the qualitative methods, together with the definition of narrow criteria for the methods selection for each assessed situation, bearing in mind that the uncertainties are also a relevant factor when dealing with the risk related to nanotechnologies field. Thus, more research work is needed to fill existing gaps.

## 5. Acknowledgments

The authors wish to acknowledge FCT, COMPETE and FEDER (under PTDC/ECM/110162/2009), for funding the research and Quimidroga Portugal-Produtos Químicos Unipessoal Lda., for kindly providing the unsaturated polyester resin Aropol® FS3992.

## 6. References

[1] Ribeiro, M.C.S., New polymer mortar formulations: development, characterization and application forms, in PhD thesis - Mechanical Engineering. 2006, Faculty of Engineering of the University of Porto (FEUP): Porto.
[2] Ribeiro, M.C.S., et al., Fire Behaviour Enhancement of Epoxy Polymer Mortars Modified with Flame Retardant Systems. Advanced Materials Forum, 2008. **587-588**: p. 903-908.
[3] Fowler, D.W., Polymers in concrete: a vision for the 21st century. Cement and Concrete Composites, 1999. vol. **21**(no. 5-6): p. pp. 449-452.
[4] Tavares, C.M.L., et al., Creep behaviour of FRP-reinforced polymer concrete. Composite Structures, 2002. **57**(1–4): p. 47-51.
[5] Ribeiro, M.C.S., et al., Effects of micro and nano-sized Al2O3 particles on mechanical behaviour and fire reaction properties of epoxy polymer mortars, in 13th International Congress on Polymers in Concrete. 2010: CS Madeira Atlantic Resort, Funchal, Madeira.
[6] Ribeiro, M.C.S., et al., Fire Reaction and Mechanical Performance Analyses of Polymer Concrete Materials with Micro and Nano Alumina Particles. Restoration of Buildings and Monuments, 2013. **19**(2/3): p. 195-202.
[7] A. Laachachi, M. Ferriol, M. Cochez, J-ML Cuesta, D. Ruch, A comparison of the role of boehmite (AlOOH) and alumina (Al2O3) in the thermal stability and flammability of poly(methyl methacrylate), Polym. Degrad. Stabil. **94** (2009)1373.
[8] R. Baskaran, M. Sarojadevi, C. Vijayakumar, Unsaturated polyester nanocomposites filled with nano alumina, J. Mater. Sci. **46** (2001) 4864.
[9] D.C. Moreira, L.A. Sphaier, J.M.L. Reis, L.C.S. Nunes, Experimental investigation of heat conduction in polyester-Al2O3 and polyester-CuO nanocomposites, Exp. Therm. Fluid Sci. **35** (2011)1458.
[10] Som, C., et al., Environmental and health effects of nanomaterials in nanotextiles and façade coatings. Environment International, 2011. **37**(6): p. 1131-1142.
[11] Kiliaris, P. and C.D. Papaspyrides, Polymer/layered silicate (clay) nanocomposites: An overview of flame retardancy. Progress in Polymer Science, 2010. **35**(7): p. 902-958.

[12] Beaulieu, R.A., Engineered Nanomaterials, Sexy New Technology and Potential Hazards in 2009 Safety Analysis Workshop. 2009, Lawrence Livermore National Laboratory (LLNL), Livermore, CA Las Vegas, NV, United States. p. 14.

[13] Soto, K., K.M. Garza, and L.E. Murr, Cytotoxic effects of aggregated nanomaterials. Acta Biomaterialia, 2007. **3**(3): p. 351-358.

[14] Kuempel, E.D., Geraci, C.L. and Schulte, P., 2012. Risk assessment and risk management of nanomaterials in the workplace: translating research to practice. *The Annals of occupational hygiene*, **56**(5), pp.491–505

[15] Schulte, P.A. et al., 2010. Occupational exposure limits for nanomaterials: state of the art. *Journal of Nanoparticle Research*, **12**(6), pp.1971-1987

[16] Brouwer, D., 2012. Control Banding Approaches for Nanomaterials. *Annals of occupational hygiene*, **56**(5), pp.506–514.

[17] ANSES, Development of a specific Control Banding Tool for Nanomaterials. 2010, French Agency for Food, Environmental and Occupational Health & Safety: France.

[18] Zalk D., Paik S., Swuste P. (2009). Evaluating Control Banding Nanotool. *Journal of Nanoparticles Research* **11**:1685-1704.

[19] Groso, A., et al., Management of nanomaterials safety in research environment. Particle and Fibre Toxicology, 2010. **7**(1): p. 40.

[20] Cornelissen, R., et al., Guidance working safely with nanomaterials and – products, the guide for employers and employees. 2011, FNV, VNO-NCV and CNV. p. 1-17.

[21] Giacobbe, F., L. Monica, and D. Geraci, Nanotechnologies: Risk assessment model. Journal of Physics: Conference Series, 2009. **170**(1): p. 012035.

[22] Höck, J., et al., Guidelines on the Precautionary Matrix for Synthetic Nanomaterials. 2011, Federal Office of Public Health and Federal Office for the Environment: Berne.

[23] Van Duuren-Stuurman, B., et al., Stoffenmanager Nano Version 1.0: A Web-Based Tool for Risk Prioritization of Airborne Manufactured Nano Objects. Annals of Occupational Hygiene, 2012. **56**(5): p. 525-541.

# Regulation and innovation dynamics for nanoresponsible development: The case of the French code de l'environnement L 523-1 to L 523-5

**C Auplat [1] and S Ben Slimane[2]**

[1] Novancia Business School Paris, 3 rue Armand Moisant, 75015 Paris, France and PSL, Université Paris-Dauphine, DRM - MLab, UMR CNRS 7088, F-75016 Paris, France
[2] Novancia Business School Paris, 3 rue Armand Moisant, 75015 Paris, France.

E-mail: cauplat@novancia.fr

**Abstract.** This paper examines one aspect of innovation dynamics for nanoresponsible development: the links between regulation and innovation dynamics. It focuses on the case of the French Code de l'environnement, Articles L. 523-1 to L. 523-3. Articles L. 523-1 to L. 523-3 of the French environment code provide for the obligation to declare the quantities and uses of substances at nanoscale produced, distributed or imported in France. This procedure is intended to improve knowledge of these substances and their uses as well as of their markets and volumes sold, to ensure traceability and to collect available information on their toxicological and ecotoxicological properties. The paper builds on recent work on the emergence of a regulatory framework for nanotechnologies to take stock of the current situation in France, in the EU and globally and to explore how this specific law package may influence innovation and the shaping of new markets for nanobased materials. The study shows that nano-regulation does have an impact on innovation. However, the impact is not the same with EU regulation and with French regulation, and while EU regulation seems to create a favourable context for innovation, French regulation seems to do the opposite. With this study we hope to bring new perspectives to the field of the strategic management of innovation, and also to shed some light on the roles and challenges of institutions to facilitate nanoresponsible development.

## 1. Introduction

The notion that innovation should go hand in hand with resource efficiency and a greener economy as part of responsible development is at the heart of much debate that can be characterized by two opposing visions: Those who contend that this approach is detrimental to competition because it is altogether more costly to aim for resource efficiency and a greener economy, and those who support the idea that the players aiming for resource efficiency and a greener economy will end up being more competitive than their contenders because their approach will have allowed them to innovate in more radical ways.

Porter's research [1] has had a strong focus on competitiveness, and much of his work analyses the links between innovation and competitiveness. Porter and van der Linde [2] examined more specifically the connections between regulation and competitiveness, and this publication led to the well-known 'Porter hypothesis' that strict environmental regulation can help improve commercial competitiveness. Although opponents of strict environmental regulation contend that it stifles innovation because of the added costs it entails, Porter's hypothesis suggests that it does the opposite. This happens through the 'innovation effect process': by triggering the introduction of cleaner technologies and environmental improvements, environmental regulation leads to more efficient production processes and in the end the cost savings achieved by better efficiency compensate the costs induced by compliance and innovation.

Environmental regulation is defined by UNEP as a system of treaties, conventions, statutes, common law, and policies that concern the protection of the natural environment as it is impacted by human activity. Some environmental laws regulate the quantity and nature of impacts of human activities while others seek to assess possible impacts before the human activities can occur. The precautionary principle underlies much of environmental regulation. It is a moral and political principle that states that if an action or policy might cause severe or irreversible harm to the public, in the absence of a scientific consensus that harm would not ensue, the burden of proof falls on those who would advocate taking the action.

We consider that the French 'Code de l'Environnement' is part of environmental regulation. It is a package of laws that started being enacted in 2005 and that defines in seven separate books the French regulatory framework relating to the environment. Our decision to focus on its Articles L. 523-1 to L. 523-3 and on the French case to study links between regulation and innovation dynamics for nanoresponsible development comes from the fact that the scope of this mandatory nanolegislation was a first, not only in the EU, but in the entire world.

As a policy maker, the European Commission considers that innovation is at the heart of competitiveness, and that it should be supported by strong financial commitments, as well as rooted in the concept of sustainable development. Indeed, in its landmark document *Europe 2020: A strategy for smart, sustainable and inclusive growth* it proposed as a major target that 3% of the EU's GDP should be invested in R&D and it put forward two mutually reinforcing priorities:

1. The development of an economy based on knowledge and innovation for smart growth.
2. The promotion of a more resource efficient, greener and more competitive economy for sustainable growth.

Why should we focus on nanotechnologies? OECD defines nanotechnology as a collective term for a set of interlinked sciences and technologies which contribute together to the understanding and control of matter and processes at very small scales (typically in the range of 1-100 nanometers, one nanometer being one billionth of a meter). Nanotechnology is therefore essentially the engineering of matter down to the level of individual atoms. For reasons that remain partially unexplained today, when particles of a given chemical substance are of nanometric size, they present properties that are not observed at a larger scale, when they are in bulk. The range of changes in the properties of elements is extremely broad and scientists are only beginning to explore a new world of possibilities where they can design new products which possess the specific characteristics they want them to have. Alongside a wide range of likely societal benefits like providing renewable energy, clean water, and improving the environment or human longevity and health, nanotechnologies are extremely interesting because they are very diffuse technologies that develop globally in diverse industrial sectors. Whether they constitute an industry per se is a subject of debate: more than being the source of entirely new sectors of business activity, they tend to diffuse in existing products and services. Indeed, nanoproducts are mostly used to enhance the qualities or specific characteristics of existing products and services. A comparison with plastic is useful: just as there is a plastic industry that produces and markets different forms of plastic, there is an industry that creates nanomaterials that are 'raw materials', that is the powders and substances that will be integrated in other components. But then, just as plastic is omnipresent in an extremely large number of products in various industrial sectors, nanomaterials are found in a growing number of products.

Taking into consideration the patenting activity of the largest R&D industry players Delemarle et al. [3] showed that a majority of the world largest R&D actors (as accounted for by the UK DTI scoreboard) were involved in nanotechnologies, covering all fields from electronics to food. The field of nanotechnologies therefore appears as particularly relevant to study innovation dynamics. The very differences between materials in their nano- and bulk- formats that make them of interest in new applications also suggest that these materials may interact differently with ecosystems and living cells. This raises questions linked to the management of the unknown and of potential harms to the environment or to human beings. Nanotechnologies pose new challenges to policy makers because they possess entirely new properties, and because it is therefore difficult to evaluate the way they react during their life cycle. There are two major reasons for this. Firstly, scientists are only beginning to develop nanostructures in a stable and measurable way and toxicity assessment standards and protocols are still largely non-existent. Secondly, nanoparticles are so small that it is very complex to trace them and to assess their inherent risks. One speaks here of monitors as well as of modelling capacities. Methods to analyze their impact on humans and the environment are still in the making and regulation is a key issue, mostly over the necessity – or not – to devise new laws that apply specifically to nanotechnologies [4].

## 2. Methods

In our paper, we seek to answer the question: 'what are the links between French nanoregulation, EU nanoregulation and innovation dynamics'? Our methodology rests mostly on archival work, completed by some focused interviews. Archival work includes the study of a comparative database of existing regulatory frameworks, both inside and outside the EU, a synthesis of public statements, reports and public inquiries intended to measure the potential impact of regulation on innovation, as well as a study of the corporate and annual reports of CAC 40 firms. The CAC 40 (Cotation Assistée en Continu) is a benchmark French stock market index, which represents a capitalization-weighted measure of the 40 most significant values among the 100 highest market capitalizations on Euronext Paris. Euronext Paris is itself part of the Euronext Group, one of the major stock exchanges globally. Traditionally, the CAC 40 index represented the most powerful French companies and it is essentially composed of French-domiciled firms. CAC 40 firms are all operating under EU law and are illustrative of how markets could shape around the development of nanobased materials. In order to explore if and how EU regulation that bears on nanotechnologies affects innovation, we study the evolution of where CAC 40 firms locate their R&D centres globally. The link between innovation and R&D, as well as the importance of the geographical location of R&D in a globalized world has been documented in the literature [5], [6], [7]. Besides, nanotechnologies have been described as diffuse technologies that require integration to fulfil their economic promises [8] and most key industrial players carry out nano-related activities in their R&D centres [9]. Our hypothesis is that if CAC 40 firms feel that EU regulation has a negative impact on their freedom to innovate, either by restraining some fields of research, or by enforcing costly and restrictive compliance measures, they will opt to relocate R&D centres outside Europe. Taking our information from institutional and company websites, annual reports and analysts' reports, we listed all the various locations of R&D centres, in the reference dates of 2007 and 2012. We chose year 2007 because it preceded the 2008 financial crisis and was therefore not affected by it, as shown by figure 1.

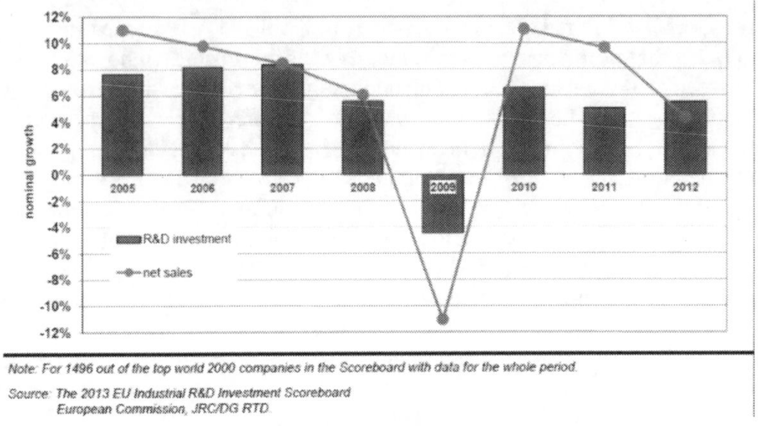


Figure 1: One-year R&D investment and net sales growth
Source: (Hernandez Guevara et al. 2013), p 6.

Year 2007 also marked the beginning of the marketing of nanobased products in a measurable way. Production before that year was too small to be analysed in a significant way. Year 2012 was chosen as the closest to the time of writing. Because there is always a gap of at least one year to obtain corporate information of the type needed, data gathered in 2013 and 2014 was too incomplete to be useful. We surveyed the 40 firms, and excluded 8 firms for which the study was not applicable because their activities did not involve any technology R&D (see appendix).

To study the evolution of regulation and analyse what is specific about the regulatory context of the EU and of France we drew from literature on the regulatory environment of nanotechnologies [10]. We updated the information with the most recent developments concerning nanoregulation, in the EU and worldwide.

## 3. Nanoregulation

At the date of writing, there was no comprehensive international mandatory regulation for nanotechnologies. The United Nations Environment Program (UNEP) introduced the Strategic Approach to International Chemicals Management (SAICM) at the 2002 World Summit on sustainable development. SAICM was conceived as a global policy framework aiming to promote chemical safety around the world and its “2020 goal” was adopted as part of the Johannesburg Plan of Implementation. SAICM’s overall objective was to achieve the sound management of chemicals throughout their life cycle so that, by 2020, chemicals were produced and used in ways that minimized significant adverse impacts on human health and the environment. In the early 2010s SAICM meetings started producing non-binding resolutions on nanotechnology, but in 2014 SAICM had not produced any binding regulation.

The EU was the most advanced organization to enforce a binding law of international impact: EU Regulation 1223/2009 on cosmetic products came into force in 2013. This law – the first international law with some parts specifically designed for nanotechnologies – concerned the EU market. It included a review of the safety of nanomaterials. All cosmetic products were subject to a safety assessment and to a premarket notification and approval procedure. A new and important feature of the regulation was that in order to establish clear responsibilities, every product must be linked to a "responsible person" who was responsible for ensuring compliance with the regulation, and who must keep a product information file for a period of 10 years after the date when the last batch of the cosmetic product was placed on the market. The “responsible person” must have an address in the EU.

It could be the company that manufactured the cosmetic, or had it manufactured, or the local office or agent for a manufacturer based outside the EU. If there was no local agent for a non-EU manufacturer, then the first company importing the product into the EU was the "responsible person".
The regulation also required traceability of a cosmetic product throughout the whole supply chain, as well as clear labelling including the name and address of the responsible person, and the presence of all ingredients containing nanomaterials, with their names followed by (nano).

The other law that originated in the EU was REACH, Registration, Evaluation, Authorization and Restriction of Chemical Substances - EC 1907/2006 – 2006. REACH is a comprehensive regulatory framework which was introduced by the European Commission in 2006 to replace some 40 existing legal acts and to create a single system for all chemical substances in Europe. Its main objective was to improve the protection of human health and the environment through the better and earlier identification of the intrinsic properties of chemical substances, and this included the plan to find substitutes for the most dangerous chemicals in use. The law entered into force on 1 June 2007. One of its characteristics was that it transferred the burden of proof for demonstrating the safe use of chemicals from EU Member States to industry.
REACH has two major components. Starting in June 2008 and before the deadline of June 2018, all firms that manufacture or import more than one ton of a chemical substance per year are required to register it in a central database, the European Chemical Agency (ECHA). Besides, all new chemical substances that were produced or imported in the EU after the cut-off date of 1981 have to undergo a strict risk assessment process. This means that all chemical substances produced before 1981 do not need to undergo the safety assessment process again, and that firms that produce or import less than one ton of a given chemical per year do not need to register these substances. The REACH provisions have been planned to be phased-in gradually over a period of 11 years because of the complexity of the process and because of the need to find replacements to some chemicals that were considered too dangerous to be retained.
REACH is a general framework and it does not apply specifically to nanosubstances. Critics of the law say that because most nano substances are so small, they are produced in quantities that are below one ton per year, which means that they go unregulated. They also criticize the fact that there are no specific provisions for them considering their novel nature. A general review of the scope of REACH including a review of the specific information requirements took place in 2012, and a new public consultation was organized in 2014 to identify and develop the most adequate means to increase transparency of nanomaterials on the market.
REACH can in fact apply to substances produced or imported in volumes below 1 ton per year if they are considered to be of very high concern. This means in effect that risks from certain nanoscale substances could be addressed through REACH if they were identified as being "substances of very high concern" as defined in Article 57, for example as being persistent, bio-accumulative and toxic (PBT) substances. Although REACH had still in 2014 no specific provision for nanotechnologies, the modification introduced in 2008 by Regulation EC 987/2008 opened the way to nano-specific regulation. Indeed, Regulation EC 987/2008 established that some chemicals which used to be exempt from regulation because they were considered well-known and safe (Regulation (EC) No 1907/2006, Article 2(7) (b)) had to be removed from the lists of exemption (annexes IV and V). This concerned, among other substances, carbon and graphite. The chemical structure of carbon and graphite was well known. Yet it was judged that there was insufficient data on their forms at the nano-scale to consider that "their use caused minimum risk because of their intrinsic properties". So the substances registered as forms of carbon or graphite at the nano-scale on the European Inventory of Existing Commercial chemical Substances (Einecs) or on the Chemical Abstracts Service (CAS) had to be removed from the list of exemptions and be treated as new substances.
France took EU regulation one step further, and introduced the strictest law in the world concerning nanomaterials. This came as part of the Grenelle 2 law package, in the French 'Code de l'environnement, L 523-1 to L 523-5. The full title of this specific regulation translates as Book V: prevention of pollution, risk and nuisances. Title II: chemicals, biocides and nanoparticulate

substances; Chapter III: Prevention of risks for health and the environment resulting from exposure to nanoparticulates". As of July 2013, the law required manufacturers, importers or distributers of nanoparticulates in France to inform relevant authorities, and to provide information about the substances involved on a specific website called R-Nano (https://www.r-nano.fr/). The requested information included intended use of substance, quantities involved, identity of the professional users, and danger relative to exposure in terms of health or of environmental risks. The data provided could be made available to the public.

## 4. Discussion and conclusions

Our hypothesis is that if environmental regulation impacts innovation negatively by encouraging firms to locate R&D centres in countries that are not affected by this regulation, then French firms will displace their R&D centres away from France. To test this hypothesis, we surveyed where the 32 CAC 40 firms that had R&D centres had set up their research centres in 2007 and in 2012. We surveyed the 40 firms of the CAC 40 index, and excluded 8 firms for which the study was not applicable. In 2007, the 32 firms that we retained had a total of 164 R&D centres worldwide. This number rose to 229 in 2012. This increase of nearly 40% shows the importance of the perceived role of R&D centres in corporate strategy.

The introduction of the French Code de l'Environnement L 523-1 to L 523-5 was accompanied by a change in the geographical location of the R&D centres of the CAC40 firms. In 2007, 45% of the research centres were in France, but in 2012, there were only 41% of them. The drop was even larger in relative terms, considering the overall increase in the number of R&D centres (Figure 2).

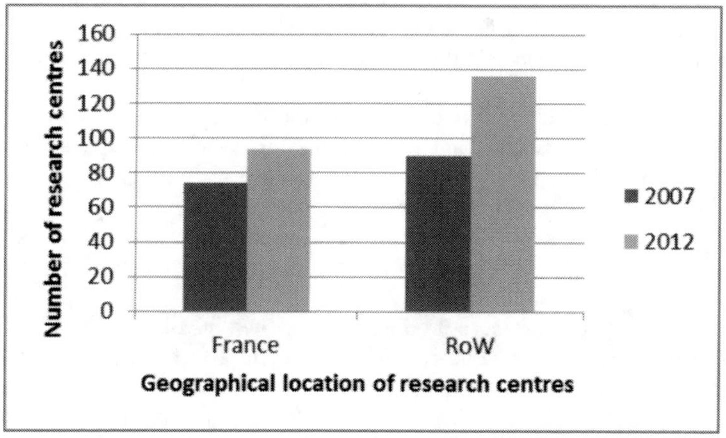


Figure 2: Evolution of the geographical location of R&D centres of the CAC 40 firms between 2007 and 2012: France vs Rest of the World. Source: authors.

This change seemed to indicate a trend toward moving out of the too strict regulation. However, things are slightly different if we look at a more global perspective (Figure 3).

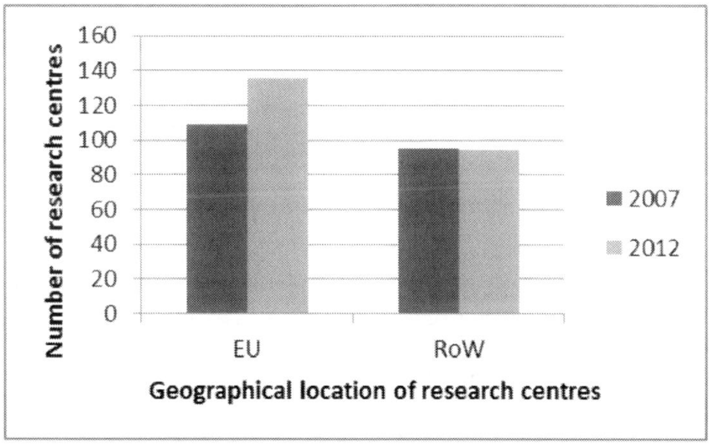


Figure 3: Evolution of the geographical location of R&D centres of the CAC 40 firms between 2007 and 2012: EU vs Rest of the World. Source: authors.

Indeed, the CAC firms did establish R&D centres outside France, but they did not leave massively the EU zone. On the contrary, in 2012, 59% of the R&D centres of the CAC 40 firms were located in the EU as opposed to 51% in 2007. When they offshored their R&D, CAC 40 firms clearly looked for diversification, as they spread between the Americas, Asia, and Africa.

The conclusion of this study is that nano-regulation does have an impact on innovation. However, the impact is not the same for EU regulation as for French regulation, and EU regulation seems in a way to boost innovation while French regulation stifles it. This can be explained by one major factor: the size factor. When regulation happens at different levels, from the local one to the international one, the terms "international", "regional", "national" and "local" usually refer to the bodies that pass the said regulations, not to the areas covered by them. For example, "Regional regulation" refers to regulation passed by a regional institution, like the EU, which includes regulatory bodies of different nations united by a specific agreement. On the whole the geographic scope of the regulation corresponds to the level at which it is passed: national regulation will apply to a specific country, and regional regulation to a region (like the EU). However, the geographic scope of some regulations goes beyond that of the body that passed them. To continue with the example of EU regulation, when a specific law targets the products or substances manufactured or imported in the EU, its scope may in effect be much larger since it may impact producers globally. Indeed, it is often too costly to run separate supply chains, and manufacturers who wish to do business with the EU may have to adopt EU regulation for all their production processes. This is what is happening with cosmetics. The stricter EU nanoregulation for cosmetics is in fact affecting all cosmetics producers globally, and once producers start testing their products to comply with EU regulation, they tend to adopt the same protocols for their global production. However, with France, the situation is different: the geographic scope of the French nano decree relative to the Code de l'environnement is wider than France since it requires manufacturers, importers or distributers of nanoparticulates in France to fill the R-Nano file wherever they operate from. The problem is that while the size of the EU market is large enough to jeopardize a firm's business if it cannot sell on this market, the size of the French market is small, and so producers of nanoparticulates will probably prefer to do without it rather than risk having their data exposed to public scrutiny, which might endanger competitive advantage based on original processes. Indeed, one of the most controversial aspects of the French decree is that the data provided by firms relative to quantities and intended use can be made available to the public.

A good illustration that EU regulation may serve as a facilitator of innovation in spite of the constraints it involves is the decision of the Organization for Economic Cooperation and Development (OECD) to terminate one of its programs, called Cooperative Chemical Assessment Program

(CoCAP). This program has long been working to assess the hazards of chemicals across international lines. OECD has now decided that CoCAP has served its purpose and must be replaced by a different organization at the end of 2014. One of the stated reasons for the change is that the activities of the EU's REACH program have rendered CoCAP's current activities unnecessary. By introducing a large-scale framework adopted de facto by the European market, EU regulation sets rules that govern innovation and EU countries have a competitive advantage by being ready for the new rules. Besides, a number of countries outside of the European Union have started to implement REACH regulations or are in the process of adopting the framework (Croatia, Serbia, Switzerland, Turkey and even China are considering such a move).

This paper indicates that when the EU decides to work as a whole entity, it can act as an environmental regulator while facilitating innovation at the same time. The lesson to draw is to move as a large block and not as a small entity. The French initiative in this respect looks out of touch with reality, because the additional constraints it imposes apply to a market that is very small. French CAC 40 firms are becoming more and more global, and they have well understood this. This is why they decide to reduce their R&D activities in France and to strengthen their presence in other EU locations.

**Acknowledgements**

This work has been carried out in the framework of the Labex SERENADE (ANR-11-LABX-0064) and of the A*MIDEX project (ANR-11-IDEX-0001-02), funded by the «Investissements d'Avenir» French Government program managed by the French National Research Agency (ANR).
The views expressed in this paper do not necessarily reflect the views or policy of the Labex SERENADE or of its individual members. Comments are welcome, directed to cauplat@novancia.fr

**Appendix**

List of Cac40 firms: (N/A: no R&D center).

Accor (N/A); Air France – KLM; Air Liquide; Alcatel-Lucent; Alstom; Arcelor Mittal; AXA (N/A); BNP Paribas (N/A); BOUYGUES ; Capgemini (N/A). Carrefour (N/A); Credit Agricole (N/A); Dexia ; EADS ; EDF ; Essilor ; France Telecom ; Gaz de France ; Groupe Danone ; L'Oreal; Lafarge; Lagardere; LVMH; Michelin; Pernod Ricard; PPR ; PSA Peugeot Citroen ; Renault ; Saint-Gobain ; Sanofi-Aventis ; Schneider Electric; Societe Generale (N/A); STMicroelectronics ; Suez ; Total ; Unibail-Rodamco (N/A); Vallourec ; Veolia Environnement ; VINCI ; Vivendi

**References**

[1] Porter M 1980 *Competitive Strategy: Techniques for Analyzing Industries and Competitors*, Free Press
Porter M 2008 *On competition, updated edition*, Harvard Business School Press
[2] Porter M and van der Linde C 1995 Toward a New Conception of the Environment-Competitiveness Relationship. *Journal of Economic Perspectives* **9** 97-118
[3] Delemarle A, Kahane B and Larédo P 2010 Dynamics of Nano Sciences and Technologies: Policy Implications. *STI-Science, Technology Industry Review* **1** 43-62
[4] van Teunenbroek T et al. 2013 NANoREG – A common European approach to the regulatory testing of nanomaterials. EU Framework 7 Programme, Grant agreement number 310584.
[5] Audretsch D and Feldman M 1996 R&D spillovers and the geography of innovation and production *The American economic review* 630-640
[6] Gerybadze A and Reger G 1999 Globalization of R&D: recent changes in the management of innovation in transnational corporations *Research Policy* **28** 251-274
[7] Doz Y and Keeley W 2012 *Managing global innovation : frameworks for integrating capabilities around the world* Harvard Business Review Press
[8] Avenel E, Favier A, Ma S, Mangematin V and Rieu C 2007 Diversification and hybridization in

firm knowledge bases in nanotechnologies *Research Policy* **36** 864-870
[9] Hernandez Guevara H, Tuebke A, Hervas Soriano F, Vezzani A, Amoroso S and Dosso M 2013 *The 2013 EU Industrial R&D Investment Scoreboard* Institute for Prospective and Technological Studies, Joint Research Centre (Luxembourg)
[10] Auplat C 2012 The Challenges of Nanotechnology Policy Making PART 1. Discussing Mandatory Frameworks *Global Policy* **3** 492-500

# AUTHOR INDEX

# AUTHOR INDEX

# AUTHOR INDEX

**Institute of Physics**
Dirac House, Temple Back
Bristol BS1 6BE UK